EXISTENCE:
WHAT IT IS AND
WHAT WE THINK IT IS

VICTOR SENCHENKO

Copyright © 2017 Victor Senchenko.
All rights reserved.

No part of this publication may be altered nor may it be reproduced in any form or by any means without the prior permission of the author and his estate, even if initially obtained as a free ebook.

Senchenko, Victor
    Existence: what it is and what we think it is

ISBN 978-0-646-96758-5

Printed by Lulu

To all who want to know
their place in all the physical existence

CONTENTS

PART 1. WHY HUMANS ARE HUMAN? ... 1
  1. The meaning of life forms ... 2
  2. What makes humans human ... 8
      The early childhood ... 15
      Conscio personality development ... 17
      Senses of brain communication ... 20
      Us, human conscios ... 23
  3. Understanding fear ... 26
  4. Sex, conscio and human brain ... 45
  5. Living with conscio ... 53
  6. Conscio and elements of sleep ... 60
  7. Conscio and brain conflicts and addictions ... 68
  8. Conscio and the supernatural ... 77
  9. Human dimensions of thinking ... 84
  10. Conscio fear of living with change ... 92
  11. Conscio making choices ... 101
  12. Conscio pursuit of happiness ... 112
  13. Talking to your brain ... 119

PART 2. DISCLOSING THE SHAM OF
         HUMAN CONSCIO PERCEPTIONS ... 123
  14. Imagination: the path to deception ... 124
  15. Greed with no boundaries ... 134
  16. Monuments built to lies ... 146
  17. The ugly truth ... 157
  18. Converting a Jewish political concept into a new religion ... 160
  19. The Muhammad deception ... 188
  20. The Eastern way ... 202
      India: acceptance of all gods ... 203
      The Buddha: giving it all away ... 205
      China: gods for the masses; philosophy for intellectuals
        and Emperors ... 209
      The Way of Tao ... 209
      Confucius: humanity of all for all ... 215
      Chinese Buddhism: the Way of the Zen Buddha ... 218
  21. The way of 'dominant male' ... 221
  22. The God business ... 227
  23. Thinking aloud ... 234

PART 3. DISCLOSURE OF EVERYTHING	237
24. A brief look at all physical existence	239
The past with no beginning	239
How and why inunens look as they do	241
The four physical phases in an inunen cycle	255
Phase 1: being a 'uniting energy' inunen	255
Phase 2: being a 'united energy' inunen and	
Phase 3: being a 'dispersing energy' inunen	259
Light and heat	268
Phase 4: being a 'dispersed energy' Inunen	275
The two systems of the physical process of change	277
Summary of the four inunen phases	281
Gravitation: The attractions and repulsions of combined mass	283
Physical behavior of 'energy bodies' (electrons)	293
Visualizing atoms ('united bodies')	301
From intelligence to life	326
The absolute of all active physical existence	330
Directions of human space travel	331
25. When science deceives as much as religion	336
Deceptions of 'Time' and 'time travel'	336
Theories of nonsense and Deception of 'black holes'	366
Deception of space voyaging	380
26. Deception in the name of gods	387
27. The ultimate truth of all physical existence	400
Index	408

PART 1

WHY HUMANS ARE HUMAN?

We, humans, have always been trying to work out as to who we are and why we behave as we do. Thousands of books had been written on this subject in attempts to do explain our thinking and behaviour. There are institutes, university departments and laboratories dedicated to such research. There are hundreds of thousands of practicing professionals in fields of psychiatry and psychology attempting to apply what they think they know. However, to this date none of these endeavors have come even close to discovering the reason for the paradox that we, humans, are.

Part 1 of this book is dedicated to doing just that: giving full explanation of why we are the way we are. In providing this information, there is no fabrication or presumption involved. Unlike the usual human method of using imagination to invent an interpretation to anything that is unknown, everything revealed in this book is based on physical reality and physical facts as they apply to human existence.

In order to provide an explanation of the physical processes by which we, humans, come to be us, an intentional effort had been made to present this knowledge as simply as possible, so that it may be understood by anyone, irrespective of their levels of education and interests in life.

The one important factor relating to this book, which should be understood from the beginning, is that it is not a "survival manual" to any individual human life. It merely explains exactly who we actually are, as humans, and why we behave as we do, as well as disclosing what all the physical existence – of which we are part of – actually is, and how and why it all functions as it does. So whatever each and every single reader does with this knowledge is up to that individual.

1

The meaning of life forms

In order to understand us, humans, it requires an understanding of all life forms, be they flora or fauna. While humans have been studying botany of flora and biology of fauna for hundreds of years, they still have not recognized the very basics of these life forms.

For now it had been presumed that fauna, such as animals, have brains and flora, trees, grass and the like are brainless. This perception is wrong. The fact is, all life forms, be they flora or fauna have a brain. The difference between these two categories is in the structure of the brain.

The best way to explain that difference is to classify flora: seaweed, flowers and trees, as being body-brains and fauna as brain-bodies. Flora had been classified as such because while they have no singular brain, their whole structure functions as unique intelligence, like that of a brain. Therefore, they are bodies developed as brains: body-brains.

If this is hard to accept, consider this:

– Plants were early life forms, flourishing in the oceans before relocating onto land hundreds of millions of years prior to any appearance of fauna life forms having made their appearance in water, and much later on land. Incidentally, it was plants (flora) that provided fauna with sustenance, both in water and on land. All these diverse body-brain formations and evolutionary developments have one element and one element only to thank: their own intelligence.

– Flora shares many DNA codes with animals, even humans. DNA is not something that had emerged out nothing but as an intended, and therefore, an intelligent method of transferring body instructions from a parent plant to its offshoot.

– Body-brains, as flora life forms, have a basic function shared by all life forms: survival by adaptation to changing environmental circumstances. One important method of adaptation and evolution is by transfer of their location. For that reason, while incapable of personal movement, flora had devised means by which they could influence the spread of their species. This applies to underwater plants and those established on land.

The techniques that flora (body-brains) employs on land includes sending runners from the root systems, which then begin to grow as a new offshoot of the original plant or a tree. Other means of propagation include the use of wind, water, or the invited assistance of animals, birds, and insects; all these methods being an intelligent process of spreading (dispersing) their pollen and seeds to new regions.

Other plants, such as lavender for example, contain aromatic oils, which humans value so much that it, as a body-brain species, has assured its existence thanks to its clever ability to be wanted.

– Flora is not the only form of body-brains. Unicellular microorganisms, which have cell walls but lack an organized nucleus, such as bacteria and germs are also body-brains. While many of these body-brain microorganisms are important to the wellbeing of global ecology, some, such as viruses, can form harmful-to-life infectious diseases. The difficulty of controlling many infectious diseases by means of medicinal chemicals, such as antibiotics, is that they adapt to drugs, developing a resistance and mutating into different strains, all due to their body-brain intelligence.

– Anyone studying botany would be aware of the ingenious systems and structures by which all plants, be they small grasses or huge trees, attain nourishment essential to their life, using all the elements available to them, including sunlight, water, and minerals in the soil. None of these internal systems had occurred out of the blue. They were devised and developed by the very plants themselves, using their very own individual body-brain intelligence.

Were all the body-brains to lack intelligence then there would be no bacteria, no organisms and no flora life forms.

Of course, the intelligence of body-brains is not to the high level of the mental intelligence of current human beings, but though it is different it remains intelligence nonetheless.

Unlike the body-brains developing their shapes and structures that simultaneously maintain intelligence, the brain-bodies are life forms that comprise of singular intelligent brains, which have devised, implemented and evolved their own bodies that cover them, the brains.

It was very powerful intelligence that made it possible for brains to build precise bodies around themselves. That is why fauna animal life forms have been classified as brain-bodies: brains who devise and implement bodies for themselves.

This, of course, applies directly to us, humans. Our ancient human ancestor brains had decided on what kind of bodies they would have, and here we are, human brains, encased in particular sculls in particular heads, with particular skeletons and organs forming our bodies to this very day!

PART 1. WHY HUMANS ARE HUMAN?

Brain-body fauna did not emerge out of ether or nothingness. At some point of early life development on Earth brain-bodies began to evolve from body-brain mono cells. At first these brain-body primitive animals developed into multitude of species in the water, including fish, before more advanced aquatic species ventured onto land. In doing so, the brains of these land occupiers became responsible for all the dinosaurs, reptiles, birds and mammals.

The fundamental reason why this happened is because brain-bodies of fauna had devised an ability that the body-brains of flora did not: the ability to move as independent bodies.

Brain-bodies chose to live in motion, needing vigorous forms of movement for relocation, and navigation. Brain-bodies, or simply brains, required, and require larger quantities of nutrition, the source of which they need to reach quickly. Brains also required mobility to evade threats of physical danger, arising from acts of nature and from other brain-body animals that would want to feed on them.

Such stringent, never-ending demands on the brains to sustain and adapt their bodies to varying environments had caused them to develop differing bodies, where the intention of every brain of any species was, and is, to have a physical body best suited to its individual need for survival (of feeding and reproducing) into future.

Any long-duration evolutionary process is not easily perceived in its entirety. It may slow down, but never ceases for all species. As far as the brains are concerned it is always ongoing and never stops. The brains continue to explore developments to their bodies, in efforts to have bodies that would best sustain them and physically protect them.

For us, for instance, one of the more noticeable body developments had been in the enlargements of our brain and body sizes over the past centuries, which still continues. Our brains are making our bodies larger. This can be assessed by the actual army and navy uniforms, and ladies' dresses, displayed in multitudes of museums, which we worn by humans of past ages. If compared to current body sizes those old clothes would only fit early teens of today.

Irrespective of the species, animal brains – as the drivers of their bodies – became located at the front and the top of their bodies; that location having the best and the closest vantage point to manage observation with eyes and feeding with mouths.

A forward position, however, has a shortcoming of being exposed to possible approaching danger. For that reason animal brains had taken precautions of protecting themselves by devising various protrusions for the scull structures of their bodies. These defensive and offensive devices include various antlers, horns, spikes, scales, and any other bone re-enforcements for foreheads. Similarly, many brains devised sharp teeth, fangs, and tusks for their jaws, both as a protection in defence and assistance in offence for biting and tearing.

As mentioned earlier, a process of evolution may slow down, lasting for millions of year, as is the case with crocodiles. While their environment remains constant – that is, while they can successfully maintain their nourishment and reproduction – their brains will not contemplate any major changes to their bodies. However, once they notice a gradual change in their environment, they then proceed to consider how they can overcome (navigate) change by applying change to their bodies. If the environmental change is perceived to be drastic, then the affected brains are prepared to be radical with the new development of their bodies.

One important characteristic of all brains is that they retain in their memory any physical changes they had devised, and devise, for their bodies; never discarding previous body designs when developing a new body for themselves. That is because the amount of calculation required to produce an absolutely new body design within a few generation would be beyond any intelligence in existence. Therefore, when evolving new bodies the brains had always preserved their previous calculations of body parts retained in genes.

By retaining the essential elements and forms of their bodies the brains could, can, and do gradually reapply these components more calculation-efficiently in order to produce altered bodies from formerly-designed components that previously may have provided different functions. It is by such means that the evolving-from-recycling early life forms developed more complex organs from what previously were whole bodies. By gene retention and refinement, animals developed eyesight from molluscs that initially attempted to convert light into food. Such shrewd use of body-building had allowed the brains to alter fins to limbs or wings, and from limbs back to fins, in order to take advantage of all available living space existing on the planet.

When, for instance, the original aquatic animal brains saw a potential of life on land some of them gradually moved onto land, by converting gills to lungs and fins to legs, and adapting their bodies accordingly – in the process becoming land animals. Very much later, some land-living mammal brains did the opposite, when the brains of animals such as Tetrapod (ancestors of the whales, dolphins, and multitude of other marine mammals, such as the sea lions and fur seals), predicted a better potential for their life back in the sea, and duly chose to return to water.

PART 1. WHY HUMANS ARE HUMAN?

Those brains accomplished this by navigating themselves through millions of years of body redevelopment of streamlining their bodies and providing them with flippers and fins, so as to be perfect specimens for life in the aquatic environments of sea or fresh water.

Other mammal brains chose a different path of navigating space. Those mammal brains had retained some of their mammal features for their heads and upper torsos, while enhancing their hearing at the expense of their eyesight, and converting their front legs into wings: evolving into flying bats.

Once the brains of a species decide to change their bodies, so as to adapt in the best way possible to their changing environment, they proceed to do so irrespective of their present size and behaviour. Many brains of various animals with large bodies, like the Castoroides (bear sized beavers), chose to become small beavers. Alternatively, the brains of small-sized animals, such as those of Eohippus and Hyrachius, enlarged their bodies to evolve respectively into horses and rhinoceroses.

The problem with major body changes is that they require generational changes lasting hundreds of thousands, if not millions of Earth years before this is achieved. If given an opportunity to make gradual, long-term body changes brains can evolve into new species whose bodies can excel in their new environment for very long durations. However, when the speed of environmental change is very rapid, this prevents the brains from matching that rate of change to that of their bodies, and so they can fail to sustain their life and breeding, eventually becoming extinct.

When environmental changes are not extreme the external physical evolutionary changes to bodies may either appear very gradually in any species, with only minute physical differences from one generation to the next, or they may occur in generational leaps: appearing at birth as an unexpectedly pronounced physical alternatives.

But irrespective of how quickly and effectively brains can achieve generational changes to their bodies, it is essential for us, humans, to recognise that the task of performing evolutionary changes are conducted by no other entity, or entities, but the brain-bodies themselves.

Any major or minor undertakings by the brains to alter their bodies are made by means of mental calculations, requiring many years to accomplish. Apart from the efforts to assess the direction of the environmental change, the brains also need to calculate the altered body structures, including everything from skeleton frame changes to a nerve system refinement. Gradually their calculations for body changes are incorporated into their bodies by the instructions encoded as DNA and RNA on nucleic acids, for the genes to begin the process of altering bodies of the future young.

Such body transformations are possible because every brain is capable of operating on separate levels simultaneously. By doing so, brains are able to successfully disguise their intentions of enacting any long-term ongoing body changes from the rest of the functions they perform on behalf of their individual bodies. Without exceptions, the body-building development work conducted by the brains is always private. This applies to all brains, including those of humans.

From the earliest efforts of the early ape brains (which led to human brains) the fundamental objective of all brains was, and continues into the present, to remain hidden – unknown and out of sight – while they devise and calculate any body transformations for environmental adaptations. In doing this preparatory work for change, all brains – including those of humans – receive no outside assistance in how they calculate body change. No species that have ever existed on this planet, or those that still exist – including humans – had ever obtained help in this process from any supernatural beings, spirits, deities, or gods. All the responsibility for physical body changes belongs to the brains, and no one else.

Despite that human brains developed their human bodies with some difference of skin colors and other racial physical characteristics (according to their environmental adaptations), they are all similar in their ability to function. And that is what all human brains do: function. Always fulfilling their daily duties for their bodies and for themselves (the brains), without ever disclosing themselves and their full abilities.

Human brains have been responsible for everything that had been invented, influenced, devised, developed, produced, and built on this planet. Indeed, all the civilizations with their agricultures, commerce and buildings; all methods of transport; all methods of communication; all manner of human laws and technologies – all of these achievements were made possible only by the work of human brains, and no-one else.

But just as human brains have been responsible for kindergartens, schools and universities in efforts of elevating human knowledge, similar human brains were responsible for oppressive concentration camps and gas chamber murders.

So how then could human brains be responsible for all the very best and the very worst of humanity, simultaneously? The answer to that is because human brains do what they are instructed to do.

2

What makes humans human

From the earliest days we, humans, have been seeking answers as to who we are and why we behave as we do. In most instances, the reason for this quest had been based on a desire to control other humans by understanding their behaviour, something that is still of great importance to politicians and business enterprises.

For the past century and a half, psychoanalysts, psychiatrists and doctors had been attempting to learn all about the 'unconscious mind', so as to assist people with mental issues. In the last century, or so, vast sums of money and billions of hours have been spent on researching the human brain.

Despite all the research and study that has been undertaken, everyone had failed to grasp the real mental construct of human consciousness.

From the perspective of science, a human brain is considered to be an organ, like the heart, essential to overall human life but still just an organ.

This also happens to be the perception of the brain by most humans, who continue to accept the notions of a soul, from god, that supposed to control human life and behaviour. Or else it is some spirit, or the heart, or fate that are responsible for individual human existence.

As it happens, none of these views are even close to understanding what makes humans human, and why. Despite that all the answers had always been plainly before us, because of we are the way we are, we have always chosen to reason with use of imagination rather than reality.

So here then is the reality of us, humans, which shall provide all the reasons for us being the way we are.

For all the intelligence of their brains, all animals predating early humans had (and continue to have) a limited understanding of their surrounding space. Apart from breeding (for which, for example, some migrating birds fly half way around the planet), they looked at their environment and divided it only into what could and could not be eaten. Other elements did not interest them. They lived in the wild, exposed to all elements, surviving by the prowess of their bodies' attributes to evade being eaten and their ability to nourish themselves and their young. In an overall standing, their day-in-and-day-out existence held (and holds) no security, no guarantees of survival.

The early human brains were far more ambitious than that. They wanted to reduce the levels of risk that applied to other animals. In their search for a better way to sustain and guarantee the safety of their existence, the early human brains had chosen to take a radical approach to their own, and their bodies' development.

What the early human brains did was so radical and amazing it should be recognised for being the most extraordinary feat ever, in all existence. For what they had decided to do, on a permanent basis, was to dedicate more of their mental abilities to a mental system of higher awareness. The intent of this higher awareness was meant to expand their conscious awareness of their surrounding space not just in the present moment but beyond that present moment.

The pure genius was in how they were prepared to implement this higher awareness consciousness.

What they did was to initiate a **secondary** awareness, or consciousness, which would function **outside** a brain's private and personal self-awareness, or consciousness, with which a brain controls its body and works on its body's development.

(In order to achieve this ability, human brains had to enlarge themselves, as brains. This need of enlarging themselves required other physical changes: a proportional enlarging and streamlining of their bodies in order to become carnivore meat eaters of other animals.)

By having an external secondary consciousness the brains could cross-reference their own consciousness with that of their external secondary consciousness. A bit like having two eyes instead of one to maintain a three-dimensionality of visual perception.

From the very beginning, human brains intended to keep their external secondary consciousnesses totally separate from their own internal consciousness. In fact, as far as human brains were concerned, their external secondary consciousnesses would have absolutely no idea of the existence of the brains' own consciousnesses.

There is a very good reason why human brains chose to keep their external secondary consciousnesses in ignorance of them, the brains.

Human brains retain their existence to themselves, so that the external secondary consciousness functions without having any notion, or knowledge, of their brains' internal mental work. Had human brains allowed their external secondary consciousnesses share the internal consciousnesses of the brains, then that would have meant revealing their existence, and then having to let

PART 1. WHY HUMANS ARE HUMAN?

their external secondary consciousness know exactly how they (the brains) perform their internal, secret and private calculations for their own and their body's physical functions and changes.

Were they to have done this then the external secondary consciousness could, on a whim, make their own efforts to implement body changes by by-passing the brains' control and their agreed basic physical structures for an ongoing species. This could have resulted in humanity comprising of a vast multitude of different bodies. For instance, there could have been human bodies covered with feathers, or with a protection of scales, or with other deformities that are depicted in science fiction.

And had the brains simply increased their own consciousness and self-awareness, then that would have meant a larger mental ability with no three-dimensionality of perception. More brain: no gain.

Therefore, by using multi-consciousness, comprising of their own internal consciousness and the external secondary consciousness human brains intended to maintain their own internal functions, while the external secondary consciousness would be used to consciously observe and assess the environments outside their bodies and send this assessment back to the brains (without being aware of this), so that together these two consciousnesses could direct human brains towards future projection awareness, by which to have an ability of conducting pre-emptive actions by "thinking ahead". It is this ability that separates humans from all other animals in existence.

Such mental system must have seemed perfect to human brains: they were hidden from their external secondary consciousness, which could presume that it was 'itself' who was responsible for its and the body's existence as a being. Sadly for human brains their implementation of external secondary consciousness turned out to be not quite what they anticipated.

So what went wrong for the human brains? Well, they miscalculated the level of fundamental prerogative to all life: survival. And survival comprises of selfish wants.

It is selfishness that the early human brains failed to take into account. They failed to foresee that the external secondary consciousness would become more dominant through its constant selfish demands of wants. By spending most of their existence in efforts to fulfill those demands of wants, every brain and its internal consciousness became a servant to its own external secondary consciousness. Or, to put it another way: the external secondary consciousness came to rule its brain.

In order to explain how the external secondary consciousness does this, the mental system of external secondary consciousness has been given a description of: 'commanding override negotiator, selectively controlling input-output'; abbreviated to an acronym of 'conscio'.

This 'conscio' mental system is THE entity responsible for making every one of us humans human. It is the external secondary consciousness of 'conscio' – maintained by every human brain – that "makes us behave as we do".

(Those who speak Italian will recognise the term 'conscio' as an Italian word for conscious. This coincidence makes no difference to my meaning of 'conscio', because as an acronym 'conscio' describes exactly its functions in its relationship with its brain, without any direct knowledge of this, as that is how the early human brains chose, developed, applied and continue to maintain this mental system.)

Here then is a brief explanation of conscio (commanding override negotiator, selectively controlling input-output) functions:

* A conscio can 'Command' its brain with 'wants' and 'desires', even if these commands are irrational and impossible for the brain to achieve. Still, every human brain assesses each 'want' and 'desire' demanded or commanded by their conscios in an effort to fulfill them.

* If the 'want' or 'desire' issued by conscio to its brain is of illegal nature or highly dangerous to the brain and the body, the brain can relate a warning to conscio that this 'want' or 'desire' should not be attempted. In such instance a conscio can choose to 'Override' its brain's wisdom and insist that its 'want' or 'desire' must be attempted, no matter what.

* When this occurs, and the brain still refuses to participate in such venture, a conscio can endeavor to have its brain devise strategies by which it (conscio) can 'Negotiate' with its brain by means of twisted logic.

* Every human brain spends every second of its life observing and recording to its memory all the events surrounding it, provided by conscio. This function can be classifies as 'Input'. This input – of which the brains are aware – is not just a direct recoding to memory, because the brains have allocated the responsibility of input management to their conscios. This means that conscios act as 'gatekeepers' to all the information that the brains can gather, using their discriminative authority to judge what is of importance to it (conscio) and what is not. The information deemed vital (e.g., meeting with old friends)

is stored as a priority; that which is considered not important (e.g., a sound of a train whistle) is stored in periphery.

Having the 'Selective Control' of its brains input and output (that is: what information the brain accumulates and what information it provides to its conscio) that conscio is capable of cutting into the stream of incoming information, or leaving it. This is something that a conscio can do whenever it chooses.

For instance, while concentrating on some tasks at work, a conscio may ignore information coming in from the senses, such as the background noise of a group of colleagues talking amongst themselves, until a word in their conversation stimulates its interest. It will then have its brain lock onto that point of interest, that is, the conversation. Once a conscio recognizes that it has nothing to gain from overhearing that conversation, it will have its brain stop paying attention to the conversation and return to conscio's previous activity.

As another example: when taking dictation a conscio will have its brain shift its task of concentration back and forth, between that of listening to the words in a dictation and the writing down of those words.

* Along with a selective control of input, conscios have a selective control over the 'Output' of information that a brain mentally sends to it (conscio). In such way a conscio receives information that it had requested of its brain, such as a solution to a problem.

But a brain is not responsible for the actions of its conscio where the output is concerned. A brain presents its output and it is up to the conscio as to what it does with it.

On occasions, due to some influence, oversight, or just a lack of attentiveness, a conscio may exercise insufficient diligence in supervising its brain's output, resulting in an inadvertent output of thoughts unintended for disclosure. Such indiscreet incidents are usually termed as: "slip-ups", "slip of the tongue" and "blunders", which conscios dislike and for which they usually, and unjustly, reprimand themselves (that is, their brains) for having allowed such an error to occur.

Conscio abilities to obtain 'wants' and 'desires' from its brain is the entity responsible for making us, humans, selfish beyond the selfish wants of other animals. While the basics of conscio requirements are relatively simple and straightforward, that is: "I want," it is in the overall relationship with its brain that complexities and intricacies arise, because every brain also has expectations and requirements of its conscio. These resulting tug-of-war activities between the brains and their conscios have continued to the present day, and no doubt shall go on. Not that the brains really welcome this.

It is possible that by understanding just how both the brains and their conscios behave may provide the means for humans (brains and conscios) to come to a better resolve for the benefit of all humanity.

A conscio (commanding override negotiator, selectively controlling input-output), an external secondary consciousness – constantly maintained by a brain's mental abilities – has the means to experience a consciousness of its own consciousness. This consciousness of consciousness allows a conscio (commanding override negotiator, selectively controlling input-output) consciousness to have a perception of being aware of being aware. For example, humans can observe their surroundings while being aware of doing just that. Or they can mentally examine a thought, while being aware of doing this. This occurs because of the brains' ability to use both consciousnesses to produce a three-dimensional perception of self-awareness, be that of surrounding space or mental thoughts.

This consciousness of consciousness is the reason for human ability to possess a conscious self-awareness that allows conscio mental system (by efforts of its brain) to 'think' of 'itself' as being a physical entity in physical existence; referring to 'self' or 'I' in first person. This mental perception remains in existence of 'self' or 'I' only as long as its human brain actively maintains its system of conscio (commanding override negotiator, selectively controlling input-output).

In order to have its mental entity of 'self' or 'I' be real and relevant, the system of conscio is given by its brain full control over itself (the brain), with that brain being prepared to follow the wants, requests or orders from its 'self' or 'I' conscio. If a conscio, for instance, wants to dance, its brain will control the body to make dancing poses and steps possible. If a conscio wants to sit down, its brain makes this happen by having the body sit down. If a conscio wants solutions to problems, its brain attempts to work them out and presents them as mental solutions, to which the conscio may have its brain instruct its body to shout, "Eureka!" without realizing that it is its brain that had actually solved the problem (and no one else and nothing else but the brain).

Human brains do everything necessary to make their conscios presume that it is they (the mental entities of 'self' or 'I') who are in control of their bodies. Human brains do this by seamlessly blending the functions from their bodies with the 'feelings' felt by their conscios, so that the mental entity of 'self' or 'I' is consciously felt to be itself; conscious of its physical body, and conscious of itself being in a physical environment beyond restrictions of its physical body.

PART 1. WHY HUMANS ARE HUMAN?

An important fact to remember is that the brains do the mental blending so well that conscios, as an external secondary consciousness (maintained and supported by the brains) do not actually recognise themselves being merely a mental form of a consciousness. Instead they, with a presumption of 'me' or 'I', consider themselves to be the principle entity in control of the body.

The requirement of human brains to maintain such illusion means that they have to provide responses to conscio 'wants' so quickly that conscios do not even consciously notice their 'wants'. For example a human conscio feeding the body with a cup of beverage while simultaneously receiving information from its brain (as thoughts) may absentmindedly (without conscious awareness) fail to notice how the body's arm reaches for the cup of the beverage sitting on a table, bring it to the lips, have the lips take a sip, then return the cup back to the table. Once the conscio returns its attention to the cup of beverage, it will receive its brain's confirmation of having drunk some of the beverage without being aware of this but absolute certainty of having done so. This represents a brain's instant ability to fulfil its conscios 'want' in a form of a mental 'auto-drive', by knowing what that 'want' is, even though its conscio remains unaware of it (the brain) doing this.

There is a very good reason why human brains allow their conscios to consider themselves as the entity in control of the body: they actually do want their conscio to take care of the body in the external environment. For example, when a human body acquires a common cold the body's immune systems, under supervision of the brain, begins to develop and implement chemicals necessary to attack the intruded bodies of the viral disease. The disease, which is a living body-brain, propagates itself by causing the body to expel its expansion, within the body, by means of runny nose, coughing, and sneezing: its intention being to infect any other nearby human bodies. So what the brain also does in such situations is to pass onto its conscio's consciousness all the 'feelings' that the body and the brain are undergoing in their conflict against a viral intruder, so that the conscio experiences aches, fever, and the discomfort of sore throat, runny nose and sneezing. The brain does this for a specific purpose: to have conscio seek support for the body from outside the body – in a form of human nursing and medicine – so as to obtain external assistance for the body's and the brain's recovery.

By directing the transfer of body's nerve pulses to conscio's awareness allows a brain to prime its conscio to be concerned for the body's welfare, because when a body is damaged or hurt in any way, for any reason, that brain makes sure that its conscio feels the pain, and learns from such experience to protect the body from harm.

But just because human brains give so much control over themselves to their mental system of conscio (commanding override negotiator, selectively controlling input-output – a mental system of their own making and maintenance) this does not mean that human brains give conscios absolute control over themselves. They retain the right to be a conscience to their conscios. As consciences, they operate a secret and private controlling mechanisms, by which they try to maintain a counter-control of their conscios. The reason why human brains need to do this is because of a fundamental difference between themselves and their conscios. Human brains can easily find contentment and fulfillment, while their conscios do not.

Every human brain is very much like any other animal brain, which are content once their requirements for nourishment and reproduction are met in a location they temporarily occupy. This is not the case with conscios. These 'self' or 'I' mental entities – that presume themselves to be physical personalities with egos – are selfish, wanting all that they can possibly obtain for themselves. And while human brains do benefit from such selfish actions (being supplied with daily nourishment and shelter for the body) for many of the 'self' or 'I' egos of conscios no amount of possessions are ever enough. Well past the point at which their brains are more than satisfied, conscios continue to want more.

This conscio inability of being satisfied is what human brains try to counter-control by 'conscience' and 'fear'. 'Conscience' is a brain's attempt to have its conscio become aware of its obligations to its body (and with that to the brain as well) by protecting them from harm, and avoiding any exposure to unnecessary risk.

Should a conscio – in its self-interest – reject or override this advice, it is punished by 'feelings' of guilt and regret. And when a conscio does reduce the levels of selfishness by becoming less demanding and less self-centred, and by offering consideration, courtesy, and charity to others so as to protect the body (and the brain) from the displeasure of others, the brain rewards it with a sense of righteousness and possibly even joy.

The early childhood

Fear and joy are not easily or instantly understood by conscios, and for that reason human brains are required to teach conscios the mental language of 'fear' and 'joy'. In order to do this effectively, human brains begin to teach their conscios at an early age exactly what fear and joy feel like, through mental conversion of physical experiences. Human brains do this intentionally, from the necessity of needing fear and joy to be implanted into conscio conscious awareness, so as to be a mental navigating and control device by which they (the brains) can counter-control and managed their conscios.

The following explains how human brains go about this:
At birth, a human brain of a newborn has no active conscio, though the inactive conscio system is already in place. Even before birth, a growing infant brain begins to teach its conscio the basic mental language of fear and joy. This mental language, used only between a brain and its inactive conscio, comprises of agreed communication signals relayed and retained in memory as an 'infant fear code'.

An 'infant fear code' language comprises of a set of mental scales based upon expectant comfort zone level, by which a newly born brain can judge from physical experience what it likes and dislikes. By experiencing for itself the difference between comfort and discomfort, the infant brain learns mentally exactly what these polarities represent, retaining them as 'infant fear code', so as to employ it later as mental tool of management and control. Without having learnt these physical 'feelings' for itself, a brain would not be able to define them when teaching its conscio.

As the inactive conscio gradually becomes activated by the growing and developing infant brain, the learnt mental physical comfort zones obtained from physical experience by the infant brain (and retained as 'infant fear code' signals) begin to be sent to the gradually developing conscio, as tutorials of fear and joy. (In a way, it is not unlike the process of loading a computer program onto a computer, except that with the 'infant fear code' the program being loaded is still in the process of being written.)

The 'infant fear code' experiences are not made up of concise memories of actual detailed events, as at that period the infant brain is not sufficiently developed to retain in memory any specific details of its 'infant fear code' index. However the overall experience between comfort/discomfort feelings are stored (as 'infant fear code') in memory, never to be deleted. In this way, all the 'infant fear code' experienced feelings and emotions are categorized in memory as being either pleasant and attractive, or upsetting and repellent. Whichever physical experiences become more prevalent in the infant brain's early life, these then become – more often than not – the prominent characteristics influencing the development of conscio.

The more an infant experiences hunger, anxiety from neglect, physical pain from being mistreated or from illness, or any other discomforts, the higher will be the register of fear in its 'infant fear code' index. A conscio possessing high 'infant fear code' levels on the index of fear, retained in its brain's memory, will have a high probability of being pre-disposed to 'feelings' of pessimism, anxiety, insecurity, fear, and lacking self-confidence.

Alternatively, healthier infants who experience less pain, and are well nourished and fondly looked after, generally develop lower levels of fear in their 'infant fear code' index, reflecting in life a mental state of optimism, self-confidence and wellbeing.

As a young brain grows and develops, the balances of higher and lower comfort levels – registered in its 'infant fear code' index – are gradually transferred onto the new conscio. A human brain retains the index of the 'infant fear codes' for the rest of its life, not openly obvious to the conscio but never completely absent and unfelt. The 'infant fear code' index remains responsible for the sensitivity with which individual conscios react to, and manage their secret 'subconscious' fears (these being 'infant fear codes' maintained by the brain's primary internal consciousness). And while adult conscios with a high 'infant fear code' levels may choose to present themselves as confident, successful and happily in control of their lives (that is, as someone with a low 'infant fear code' levels), unless they are good to their brains, the influences of their high 'infant fear code' shall remain unchanged for the rest of their life.

With growth of an infant body and the infant brain, a noticeable change takes place in the relationship between the developing brain and its developing conscio: the infant conscio begins to recognise its consciousness and to make demands for itself. This is not surprising to the infant brain, actively engaged with expanding its efforts on behalf of its undeveloped conscio. By that period the infant brain has already began its new phase of attempting to manage and control its conscio with another compilation of 'fears' maintained in its 'fear code' index.

A more detailed explanation of 'fear code' follows in the next chapter.

Conscio personality development

Unlike any other brain of any other animal in the existence of this planet – and possibly even this Universe – a human infant brain growing into a child brain increases its ability to develop and maintain its mental conscio (commanding override negotiator, selectively controlling input-output) system. That young human being (a representation of a human brain, its mental conscio system and the body) begins to acquire a mental consciousness of an individual 'self', by referring to and thinking of oneself as 'I' and 'me', and developing its unique identity of personality and ego.

PART 1. WHY HUMANS ARE HUMAN?

The parameters that shape an ego and personality are not just the mental 'infant fear code' and later the 'fear code' indexes, but also the external attributes of other human conscios. But even the limiting restrictions of the 'fear' indexes are not concrete or etched in stone: they are only mental guides a human brain attempts to indicate to its conscio. Therefore, should it want to, a young conscio has no big problems in commanding, overriding or negotiation with its brain. And that is exactly what they do without fail.

This means that depending on how persistent a young conscio is in its demands for its 'wants' (by overriding its young brains 'fear codes') and how obliging other humans (such as parents) are in fulfilling these 'wants' this then converts directly to the development of its (conscio's) ego: its self-worth and self-importance.

The same conscio attributes of wanting or desiring selfish outcomes can also become characteristics of a young conscios human personality. For example, the young conscio from very rich family may develop an ego of high-level self-importance, and follow this with a rude, dismissive and disrespectful personality from knowing that its 'wants' would be met. However a conscio with a similar level of ego may be influenced by 'fear code' of its brain, or 'fear code' of its parents' brains, to develop a courteous, pleasing and a pleasant personality, from learning that such personality outlook is more effective at having 'wants' realised, even for someone who is rich and powerful.

Alternately, a young conscio from a destitute family may develop an ego of low self-esteem, becoming humble, insular, and perhaps even subservient in personality. Or, a young conscio, even with initially low self-esteem, can learn to work well with its brain, developing a genuinely charming personality that can assist it in life. Then again a young conscio with an underprivileged birth may develop an ego that dismisses any setbacks to an ambitious desire for wealth and power, using a flexible personality to achieve that by any means: fair or foul.

(Incidentally, a better method of establishing an assessment of any human is to judge their attitude to all others, rather than what they present as a façade of personality, which they can alter according to their circumstances. This attitude to all others also represents that conscio's attitude to its brain [without it knowing this]. For example: a kind attitude to all others represents a generous and unselfish attitude to all others and the brain. A mean or greedy attitude to others represents an overly selfish conscio attitude towards the brain, irrespective of the personality it presents.)

An individual conscio personality – acting on its perceived status or level of ego – can add to its brain's difficulties with unreasonable demands and wants, because it (conscio) has no idea that in order to facilitate its demands the brain first needs to learn what to do, and how to do that which is requested.

An example of this can be observed when a conscio 'self' or 'I' constitutes a 'want' to ride a bicycle, having never done this before. A brain obliges by positioning its body onto a bicycle. It then takes a while for that brain to learn how to balance its body and to coordinate its limbs for pedaling, before it (the brain) can fulfill conscio's wish of propelling its 'self' on two wheels. So irrespective of how frustrated and cross a conscio may become at itself (actually, at its brain) with its 'want' to instantly ride a bicycle, until its brain can ride a bicycle the body will be unable to do so.

The same applies even to walking. Unless a brain calculates and co-ordinates all the sequential functions and balances required to making walking possible, its body will not be able to respond. An example of this would be a human who had suffered brain damage or paralysis, needing to re-learn how to walk regardless of its conscio's awareness of 'it' or 'I' having been being able to walk in the past.

In many ways a conscio's ego and personality often result in a lot of discomfort for a brain. That is because conscios 'feel' themselves not as a consciousness but as real beings who have physical attributes based on what they perceive as their special traits and unique characteristics of their personalities. By such presumptions conscios appropriate their brains' and bodies' abilities and assign them to themselves, as if those abilities are actually theirs.

Conscios then seek to align the attributes of their brains and their bodies with methods by which they can obtain financial rewards (by means of status and employment), applying to 'themselves' the kind of attitude and behaviour that would, presumably, reflect their ego traits and character personality. This seems to give many of them the impression they need to live out those ego traits and character personalities at the expense of their bodies (and their brains).

For example: a conscio considering 'itself' to be a sports person is capable of subjecting its body (and consequently its brain) to enormous physical stress and fatigue, by overriding cautions of its brain, in order to achieve its selfish want to win a contest.

Actually this kind of obstinate behaviour has always been much admired by humans (conscios) who attribute it to "strength of character" and "will power". Whatever their status, occupation or "purpose in life", humans (conscios) constantly act out a behaviour that they presume should reflect it. So when it comes to "life of danger" or "thrill seeking", depending on how they perceive themselves, be it as 'risk takers', 'dare devils', 'thieves', 'soldiers of fortune', or "rock and roll stars", conscios often choose to risk their lives (actually, the lives of their brains) by participating in multitudes of legal and illegal activities, none of which the brains appreciate. And yet, once conscios become determined to command, override or negotiate the common sense and cautionary advise provided by their brains, all that the brains can then do is to assist and empower their bodies to overcome the risks and burdens placed upon them by the selfishness of their conscios.

Despite all that conscios 'want' and 'desire' of their brains, the brains are not altogether helpless. In tolerating their conscios obstinacy, where constant risk is a prolonged factor, brains can mentally punish their conscios not just with conscience and guilt but also with 'feelings' of fear, doubt, anguish, torment, and mental depression.

That is why humans (conscios) who – voluntarily or under orders – continue to participate for long periods under duress of danger, find themselves suffering from their brains' mental punishment of the so-called post-traumatic stress disorder. This applies especially to those who are often involved in daily threat-of-death situations, such as soldiers deployed in armed conflicts. Human brains do not appreciate such exposures to risk, especially if they are needless and prolonged. Therefore, if a conscio torments its brain by overriding it to chose an exposure to danger, that brain (in overcoming that danger on behalf of itself, its body, and its conscio) will torment its conscio for that.

Senses of brain communication

Throughout its life every human brain collects information (input), through its body's physical senses. Currently it is presumed that there are six senses, whereas there are more than twenty senses that human brains use. Apart from sight, smell, hearing, taste, touch, motion and weight (sense of gravity, which can be experienced simply by having any person putting on a backpack full of rocks), there is the sense of pain, of relief, of anger and other emotions.

Then there are the senses that are shared between a human brain and its conscio. For example there is the sense of déjà vu. There is also the human brains' cautionary mental output known as "sixth sense", which provides sensation of 'foreboding of danger' or of upcoming strange or unusual event. But there are also – unknown to conscios – senses that provide communications between human brains.

The best way to explain these brain-to brain senses is when two human strangers may meet and instantly feel a rapport, a connection between them. Alternatively, two strangers may instantly feel a dislike of each other, because, for example, the brain of highly selfish and nasty conscio will warn the brain of the other human to keep away from its nasty and harmful conscio. These are the senses brains use to communicate with each other, something conscios may notice occasionally but cannot understand.

There is another example of brain-to-brain sense that brains use.
Many conscios 'fall in lust'. This is when a conscio begins to want or desire another human being for their physical attributes. The brain of that conscio is then commanded, or negotiated to find ways and means of attracting that wanted or desired person for a relationship, comprising mainly of sexual attraction and the 'want' or 'desire' for sexual relationship.

Other human beings 'fall in love', which is quite different to that of 'falling in lust'. The 'falling in love' situation is where the sexual attraction is often secondary to the 'feeling' of powerful attraction that can occur even between two humans who may have absolutely nothing in common. This fact had puzzled artists, writers, researchers and scientists for ages.

The explanation to the so-called 'love' has everything to do with brain-to-brain sense of attraction in action. This is when a mutual attraction takes place directly only between human brains. This attraction – which is noticed by conscio but is never fully explained by its brain – occurs due to a sense that can only be described as being a 'boundary of influence' attraction (an explanation of which is provided in Part 3 of the book). This attraction is similar to gravitational attraction that takes place between objects and bodies that are separated by a distance of space. (It is for this reason that couples in 'love' often like to physically put their heads together, as this is as close as two brains can unite.)

By being aware of the occurring 'love' attraction but incapable of knowing how this comes about – because the brains have no intention of explaining the process to their respective conscios – conscios often want to become involved by trying to control this situation, due to their own selfish motives. Considering that a conscio of each brain (involved in a sense of 'love') is used to knowing what kind of 'wants' its brain is usually prepared to fulfill, fears a possible change in this relationship with its brain (without actually being aware of any of this). Therefore, despite the two brains being in 'love', their conscios will insist that their brains continue fulfilling their particular 'wants'. This may inevitably produce a conflict between the two conscios. Even with the brains being committed to tolerating the 'wants' of each other's conscios, this can prove to be a task beyond their (brain) ability.

But even when the brains' tolerance of each other's conscio behaviour may fail, resulting in a separation, those two human brains may retain their sense of 'love' for each other to the end of their lives.

By far the most noticeable examples of 'love' are those between a newborn and its bearer parent.

Apart from being responsible for forming their own bodies, brains – especially those of mammals – have devised a process of altering their bodies from those of newborn to those that finally develop to reflect the evolved DNA in the genes of that body. The purpose for this has been to provide a body camouflage to the newborn brain in a form of a 'cute' appealing body: a large head attached to a disproportionately shortened and plump smaller body. The single purpose of this 'appeal' strategy is to increase the newborn brain's chance of survival: by looking appealing the intent is to prevent its mother from abandoning it, and for other adults of the species to like it enough to avoid hurting it, or even adopting it in case the mother ceases to be available, for whatever reason.

This, of course, does not apply to all newborn, some of whom are classified by conscios to be "so ugly that only its mother could love it." And that is exactly what takes place between the brain of a mother and the newborn brain: 'love'. When the brain of a mother of a newborn views the result of birth using its body's eyes, its direct consciousness assesses the external appearance of the child. Depending on that external appearance it (the brain) calculates what it needs to do to make the newborn appealing to its conscio. The most basic element it has at its disposal is the ability to transfer its conscious fondness ('love') for the new brain of the newborn to the secondary consciousness of its own conscio. The mother then 'feels' love for its child, not realising that the 'love' 'she' senses is actually the bond of attractive fondness between its brain and that of the newborn brain.

The newborn brain, as undeveloped as it is, begins to recognise the facial features of its mother, while receiving reassuring communications from its parent's brain. When a new brain finds itself without a constant reassurance of its mother's brain it begins to fret in fear of abandonment. This is then recorded as part of its 'infant fears'.

Unlike all other primates, human brains had devised a procedure with which they can manage to come physically close to the brains they are fond of: this method is that of kissing. While conscios have learnt to apply this technique for their sexual 'lust', using for this their brains' chemical production, the brains use kissing as means of mutual attraction. This means that the needy brain of a child requires the attention of its mother's brain, with the closer the distance between them the better. That is why in most situations, mothers – and fathers – often kiss their children, with their children being pleased to reciprocate, with all these physical signs of affection providing a function of bringing their heads (brains) close together.

'Love' affection between brains, especially those of mothers' and their off-springs, can remain active for the full duration of lives of parents and their children – remaining with the offspring brains even after the parent brains die.

Us, human conscios

While there remains much more to be explained in the following chapters about the external secondary consciousness, that are conscios, and their relationship with their human brains who maintain them, it is now opportune to reflect on us, humans, as to who we are as conscios, and what it all means to be a conscio.

First of all, there is no escaping the fact that we, humans, comprise of us: the brains; us: the conscios; and us: the bodies. Of this trinity, the entities who look out from the bodies' eyes' are the brains, who are safely encased in the scull of out heads. But that is the only activity that our brains utilise to view the surrounding environment outside us, the bodies. All other activities associated with us (brains) and the outside world, are left by the brains to be conducted by us, conscios.

We, conscios, are able to be aware of the external flesh of our bodies – thanks to our brains' ability to provide us, conscios, with the awareness of the seven senses – and of the physical existence outside our bodies. That however is the limit of our abilities. Our brains may share with us what they observe using the eyes, but they do so without actually explaining the physical reality they observe – which they carefully monitor for any evolutionary necessity – instead they allow us, conscios, to convert these observations into any belief of unreality that we, conscios, may choose to request from us, brains.

We, conscios, are provided by our brains with ability to have a sense of our bodies as bodies, but we, conscios, are deprived of an ability to physically 'feel' our brains, because this sense is out of bounds for us, conscios. These circumstances are what they are because we, conscios, are but mental systems of external secondary consciousness that our brains maintain. We, conscios, are allowed to exist, by our brains, only while our brains keep us maintained, and when they do not we cease to exist, as in sleep, or when a brain is anesthetized and is incapable of maintaining its conscio.

And yet, it is the very same brains that allow us, conscios, to have an infinite ability to 'want' whatever we imagine we want (always using our brains' abilities), and our brains actually attempt to fulfil them, where and when they can.

Needless to say, there shall be those who will refuse to accept this information, preferring to think that 'they' are actually the brains and not some secondary consciousness of conscio. What these doubters should do, sincerely and openly, is to consider the following:

'You' have a need to solve a problem at work within a certain period, while having not the slightest idea of what that solution could be. 'You' are frustrated and worried as the deadline is approaching and no ideas are presenting themselves to your consciousness. Then, finally, an idea becomes consciously apparent to 'you', which 'you' assess as either being workable or not. From that decision 'you' begin to receive adaptations to the original idea, gradually fashioning a practical solution to resolve the problem.

Now then, no matter who 'you' are in life, or whatever station 'you' hold, all notions that arrive to 'your' consciousness do so without 'you' having any direct awareness how the ideas originated. Of course, the only place they could come from is 'your' brain, but 'you' are not privy to know exactly how your brain achieves this, for it does not share this information with 'you'. 'You' only become aware of the ideas in 'your' consciousness when they come to 'you' from your brain.

Therefore, were 'you' to be your brain, then 'you' would know exactly how 'you' – the brain – had come up with solutions to problems. Were 'you' to be the brain, then 'you' would be totally consciously aware of how to produce all the mental and physical internal functions necessary to operate your body. And as a brain, 'you' would know exactly how all brains transform their bodies in the process of evolutionary change.

But this is not what actually happens, is it? As an external secondary consciousness, 'you' are granted output from your brain only when your brain is ready to do so, no matter how much 'you', conscio, may fret and worry. That is because your brain cannot provide a solution to 'your' problem before it can work it out for itself. Even if the solution provided by your brain is just short-term, from being that what 'you' require, your brain still has to present it in such a manner that 'you' can consciously grasp it.

So the fact is: we, conscios, ask and they, our brains, provide. They achieve our wants not with magic or virtual reality of an illusion but physically, in physical reality. They do this by providing for us, conscios, mental solutions to our mental wants, and then have their bodies physically realise these solutions. When the fulfilled wants are beneficial to mankind (brains, conscios and bodies) the brains take no due credit, permitting us, conscios, to accept the accolades, because they also allow us, conscios, to presume that it us, conscios, who are so smart. And when our brains fulfill our, conscio, harmful wants – such as weapons of war or addictive narcotics – that are detrimental to humanity (brains, conscios and bodies) our brains remain silent and non-judgmental, while we, conscios, kill each other (brains, conscios and bodies) for our selfish self-interest.

But whatever our brains do for us, conscios, a single physical fact remains: it all had been devised, implemented and maintained by our brains, and only by our brains.

When we, conscios, wanted a language to communicate with one another, our brains provided that for us. When we wanted many alternate languages, they provided that as well. When we, conscios, wanted religion, they provided religion for us. When we, conscios, wanted many alternate religions, they did this as well. And whenever we, conscios, wanted to form multitudes of societies based on our, conscio, specific principles, customs and traditions, our brains also realised this for us, conscios, without fail. Now we, conscios, want new methods of extending and entertaining our lives. So once again, as always, our brains are doing just that.

This brain-ability of fulfilling our conscio wants applies to everything that is 'manmade' on this planet – and now even off this planet. What this means is that at no period of human (brain, conscio and body) existence did anyone or anything, other than our brains, provide us, conscios, with our fulfillment. It is our brains, and only our brains who had been responsible for everything that constitutes human existence – including human existence. Therefore, no other entity can be given credit for all that human brains have achieved for us, conscios. After all, even the concepts of gods, souls, spirits and supernatural beings had been devised for us, conscios, by our brains.

If ever there was a direct, undeniable proof that there are no gods, souls, spirits and supernatural beings in existence, it is in that we, conscios, owe our brains an incalculable debt of gratitude for our very own conscio existence.

3

Understanding fear

Animals, in general, have no understanding of fear. Some animals have a sense of unease from mistrust of other animals, but this is not an instinct to impending danger. Instead, for most animals their instinct of fight or flight reaction is usually when danger is actually upon them. Take, for example, a herd of Impala, or zebras, or any other herbivore herd on African plains that will calmly graze within the sight of predators, such as a pride of lions. Should the lionesses begin to make a stealthy approach, the herd instantly becomes twitchy and nervous, but will remain where they are. Once the lionesses charge towards the herd, the animals stampede away from the danger of the advancing lionesses. Should one member of the herd be taken by lionesses and stop their pursuit, the herd also halts, and even as the predators drag away their kill the remaining herd begins to graze once again as if nothing had happened.

This shows that the grazing animals do not run off at the first sight of predators, nor do they hide from them. Only when danger is impending do they react, not out of fear but of necessity to evade that threat to their life, and then forgetting all about the danger once the threat had past.

Such is not the case with human beings. While we may not be conscious of this, we are constantly aware of some form of fear – be it stress, worry, doubt or simply concern – which human brains employ in a manner of mental devise for controlling and navigating us, their selfish conscios, towards being more accepting, charitable and tolerant. And to do this is far from easy for human brains.

As a human species whose brains maintain a mental conscio (commanding override negotiator, selectively controlling input-output) system, we have insatiable appetite for possessions, always wanting more and more; never ever being content and satisfied with what we have. Indeed, we actually pride ourselves on the fact that we are driven to gain more and more, considering this to be an improvement to human life and an advancement of humanity. As nations and societies preoccupied with business and commerce, we are determined to do everything possible to exploit all the resources of this planet so as to build and manufacture all that may be needed (and not needed). Consequently, we have become responsible for man-made global warming, which impacts many animal species and shall negatively affect the lives of future human generations.

Do we really care about this? Not really. As individuals we are proud of our constant hunger for that which we don't possess; prepared to take great risks to our body and brain, so as to obtain whatever we want and crave, be that fame, wealth, control over others, or certain 'feelings' obtained through intoxication of thrills, alcohol, or drugs. Even all our acquired knowledge – proclaimed to be an aid to all humanity – is actually directed towards pursuit of financial gain.

Yet despite all the selfish activity we are engaged in, we, as conscios, have been unaware that all of our 'wants' and 'needs' are achieved, if possible, by our brains, who are very often in disagreement with our conscio selfishness. Our brains do not need the ownership of possessions that we, conscios, desire and seek. They are content with much less, and try to instruct us, conscios, that a life with less burdens of possessions can be more satisfying.

Unfortunately for human brains, they have very little ability in directly resisting conscio addictions to selfishness. Human brains are committed to providing – where possible – fulfillments of their conscios' desires. And if, for whatever reasons, this fulfillment is not constantly provided for a conscio by its brain – as it can never be – then the outcome of this inability results in the inevitable mental conflict between conscio and its brain. While conscios have the commanding override negotiator, selectively controlling input-output, the brains have the ability of mental 'fears' projection by which to influence conscios.

Fear, as a component of overall guilt structure employed by human brains, is not simply an element of influence that a human brain uses indiscriminately to subdue its conscio. Instead, fear – as a subtle, mental navigational mechanism – is usually applied very sparingly, so as to enforces the honesty and morality of conscience: the very conscience by which human brains try to direct their conscios to provide them (the brains) with a safe passage to their future.

All the efforts that the brains apply with their use of 'fear' to foster a more harmonious relationship with their conscios, may sound harsh and mean, whereas it is just a mental projection that demands attention and cannot be easily ignored by a conscio. After all, human brains are incapable of any direct malicious or nasty actions, something that their conscios are very capable of instigating by using their brains' abilities to achieve their devious means.

(By the way, how can it be shown that human brains are not malicious and

PART 1. WHY HUMANS ARE HUMAN?

nasty? By the fact that when a human brain – even that of a very deranged murderer – deactivates its support to its highly selfish conscio, so that the conscio is almost non-existent during 'sleep' (while the brain continues to function) the brain itself does not use its body to go around committing crimes. In other words, humans cannot commit selfish actions while they are asleep. But as soon as conscios are re-activated by their brains after 'sleep', all the human selfishness immediately becomes restored.)

All conscio-instigated highly selfish actions do not go unnoticed by human brains, nor are they forgotten, and in some instances not forgiven. That is because human brains had never intended to have their conscios behave as selfishly as conscios do. For that reason, as part of conscience (the mental component that tries to reason with its conscio's negotiator) guilt and fear are functions whose application is to provide leverage for the brain when it has to confront high selfishness of a determined conscio.

Of the two, 'fear' is much more persuasive than guilt. A conscio may ignore guilt, which is a form of mental nagging about selfish deeds (especially those that fail, or go wrong) conducted by a brain. But 'fear' – any fear – cannot be ignored, cannot be avoided, and cannot be negotiated. 'Fear', for conscios, is a mental sense – or a mental feeling – that is learnt rather than inherited.

In order to facilitate its armory of 'fears' to be used against its conscio – so as to use fear as a controlling mechanism – each human brain gathers all the fears that its conscio encounters from birth and compiles these fears into its mental 'fear code' index.

After the mental 'infant fear code' index had been compiled, the growing child begins to physically experience its environment for itself, and in the process unavoidably encounters unpleasant, painful, and dangerous situations. These experiences a human brain compiles into a mental 'fear code' index, on behalf of itself, to be used as a defense and controlling mechanism of its conscio.

To give an example of this, if an infant accidentally burns a finger that baby will experience pain, which it will not understand, apart from a brain's consciousness of a most unpleasant feeling of physical hurt. While the incident itself would not be remembered, the duration of unhappiness would be recorded as 'infant fear code' onto the combined 'infant fear code' index, with the level and the duration of the discomfort being part of the total unhappiness and discontentment experienced by an infant brain, balanced against the total experience of joy and contentment.

From such a physical experience, an infant brain develops for its conscio a reference to an understanding of what pain is and how it 'feels', with the instruction being: "pain is unpleasant, and should be avoided from being experienced". Should the same brain, now being that of a child, sense that its body is coming close to danger of a flame, it would immediately pass onto its conscio a prompt of 'fear' of pain from a burn. This cautionary reminder by a brain is intended to prevent its conscio from exposing the body to potential harm.

Should that conscio dismiss its brain's warning of impending harm and actually come into a physical contact with a flame, suffering a burn, that brain then shall chastise its disobedient conscio not just with mental sensation ('feeling') of pain from a burn but also with addition of guilt for overriding it (the brain). The brain will also record an increased level of overall 'fear' factor by inferring: "disobey your conscience (the brain) and suffer the consequences as pain". This is retained in brain's memory as part of 'fear code' to be used when necessary to remind its conscio to pay attention to 'conscience' and 'guilt'.

Because of an unfortunate experiences, some 'fear code' fears of a young conscio – recorded and retained by its brain – may be of such high levels that they become classified as 'phobias', often remaining active throughout life. As phobias, these fears may comprise of: fear of heights; fear of enclosed or open spaces; fears of other life forms, such as reptiles and spiders; fears of being surrounded by other humans at close proximity; fear of being alone.

Then there are fears that may develop from a personal imagination (mental projections requested by a conscio of its brain), or the imaginations of other humans, in a form of frightening stories and myths, as well as superstitions.

As a child grows, the young brain's conscio begins to be influenced by many sources. Apart from the parents and family members, the influences also come from various friends, peers, educators, and instructors – all those who were also previously influenced by their parents and other adults. Consequently, all these influential conscios end up passing on their own fears such as: prejudices, biases, and discriminations, with all these selfish presumptions presented as instructions for survival; for conformity; and for domination or submission.

There are also the possible fears for a conscio of a youthful brain and body to understand what is occurring with its developing body: these fears being in a form of sexual frustrations and anxieties.

As a human body continues to grow and develop, and as a young conscio accepts the societal and cultural influences passed onto it by others, it (conscio) begins to formulate and exercise its own form of opinions, prejudices, insecurities, and fears, all of them based on influences gathered on its behalf by its brain and stored within memory as conscio's 'selfish fears'. In a way, these con-

PART 1. WHY HUMANS ARE HUMAN?

scio-developed fears are like a list of prompts provided by conscio freely and directly to its brain, with instructions for the brain to switch-on a particular 'fear' for the conscio whenever a specific dislike, loathing, or fear included on the 'fear code' index is physically encountered.

The fears that a conscio has its brain include into its mental 'fear code' index usually comprises of that which may be acquired but is not wanted, and that which may be lost but is wanted. For example, a conscio fears catching a disease; or having pain from damage to its body; or experiencing a financial loss: all the things it does not want and is fearful of acquiring. In a similar way it selfishly fears a loss of its life; a loss of limbs; a loss of its body's attributes, such as a loss of scalp hair or teeth; a loss of its faculties; or a loss of its possessions, including those of its living family members: all the things it does want and is fearful of losing.

While some selfishness is absolutely necessary for all life existence (in the form of self-preservation concerns), the levels of conscio selfishness often go far beyond any reasonable limits by being unrealistic. To counteract this, the brains resort to 'infant fear code' and the general 'fear code', as a leverage to influence conscios to accept and to appreciate lower levels of selfishness. This means that no fears (be they worries, dislikes, hates, revulsions, repulsions, paranoia, or phobias) can ever be erased, nor can they be hidden, without the mutual, sincere and intentional participation of conscios to accept the brain's advice to relinquish their high-leveled selfish 'wants' and 'desires'. So depending on conscio's attitudes to its brain's advice of appreciating rather than accumulating, the levels and the list of fears can be reduced.

For instance, the fears in the later years of human life can often be fewer in comparison to those of the early formative years of life. That is because of a realisation by a conscio (provided by the brain) that the pursuit of possessions had not provided the conscio with its core desire: an experience of happiness and contentment.

'Fear code' also formulates cautionary impulses, passed onto conscio by its brain's thinking (calculation) to counter its conscio's insistence on acting with selfish impetuosity and recklessness. In many ways the 'fear code' is an assistant to a brain's self-preservation efforts.

For instance, one of many reasons in using 'fear code' is to instigate acts of friendship, generosity and sharing, especially with strangers. A brain instructs conscio to be friendly and less selfish with strangers. The best way to be friendly with others is to share something with them, be that just a smile;

an act that the conscios' of strangers would, hopefully, appreciate, and be influenced by their own brains to reciprocate in kind. By applying 'fear code' to its conscio, so that its conscio be charitable, compassionate, respectful and peaceful to others, that brain tries to obtain for its conscio – as well as for itself and its body – some 'insurance' in a form of 'good will' from others, with an instruction to its conscio to treat others as it would have others treat it in return.

When a conscio accepts this advice from its brain and behaves accordingly, that brain rewards its conscio with satisfaction and some happiness. This is interpreted as: 'being pleased with oneself'. It is for this reason that strangers, in strange surroundings (especially where laws do not prevail), usually behave with more tolerance, respect, and friendliness towards all others than they do towards those they know intimately in familiar surroundings.

In reality, fear, which retains a constant presence within a conscio's consciousness, has no actual physical existence. Even at its ultimate level as 'terror', fear remains nothing but a mental projection based on 'infant fear code' overlaid by conscio's personal 'fear codes'. Fear, therefore – whether mild or strong – is actually a sense of mental 'feeling' of overwhelming doom or strong pain passed from the brain to its conscio.

'Fear', while being just a mental projection from the brain to its conscio is never of the same level. It can be in a low-level of precaution, mid-level of direct caution, or of high-level in the face of possible or real physical danger.

When a brain recognizes possible or real physical danger, in projecting high-level 'fear' to its conscio, it (the brain) takes other actions – as if stressing to its conscio that 'fear' is real – by converting the sensation of 'fear' from a mental projection into a physical action affecting the whole body. That brain begins sending instructions along the nervous system to its spinal chord and its body's organs to prepare various chemicals (such as epinephrine for itself), and similar chemicals, such as adrenaline, to its heart and muscles.

Should the threat be real, then the brain activates all the required chemicals, reduces its support to its conscio so as to be more in charge of its body; increases the heart rate; opens the pores for cooling of the body; and primes the whole body to instructions of 'fight or flight'.

However, activating high-level 'fear' for its conscio can have drawbacks for the brain, should its conscio begins to enjoy the effects of the chemicals produced by the brain and the body in preparation of facing danger. These body-produced chemicals can become as habit-forming as any other narcotics or hallucinogens. Usually this occurs when conscios seeks thrills and danger for personal reasons with intent of receiving a 'hit' of body-produced chemicals where high-level 'fear' is involved.

PART 1. WHY HUMANS ARE HUMAN?

It is difficult for a human brain to activate chemicals and pulses for a second rush of 'fear'. Once the mental and physical body-produced chemicals are used up, that brain has to stimulate further body-producing chemicals to counteract those produced to stimulate 'fear', which are required for a brain, its organs, its nervous system and its muscles. This chemical change in the body replaces the 'feeling' of 'fear' with that of 'calm', where any further perceived or actual physical threat is no longer feared, no matter how frightening, due to the brain's takeover of solving the problem of repelling or avoiding the physical treat by excluding inputs or commands from its conscio.

Once reasserted after the brain had traversed and survived a physical risk, a conscio can experience indirect rewards from the brain's instigation of other body-producing chemicals needed to relieve the stress of the body's organs and muscles. These chemicals provide conscio with feelings of relief, pride, and superiority from a presumption that it was 'I', or 'me', who overcame 'fear' and lived to talk about it. Conscios 'feel' that they had shown the courage they are proud off, and which, in their estimation, should be admired or envied by other humans (brains and conscios) who are unable, or unwilling, to face a threat of danger.

The enjoyment of these body-produced chemicals are sought by many conscios, who instruct their brains to pursue activities that expose human bodies to a physical risk and danger, be they legal or illegal; where the outcomes of being caught, or hurt, or killed, are strong possibilities. These kind of adrenaline-stimulated bouts with fear, derived from intentional gambles with risk, apply to nearly all human physical competitions and conflicts, whether these are staged in the wild, in gymnasiums, in sports arenas, or on military battlefields.

But because human brains put up a resistance to all forms of chemical addiction, this reluctance results with a disappointment for those seeking a 'hit from a rush of 'fear''. However, this objection often stimulates conscios to increase the levels of danger to which they are prepared expose their bodies (and their brains). Such conscio 'wants' and 'desires' to expose brains to needless danger, so as to receive chemical delivery to sense ('feel') a 'thrill from fear', are far from being appreciated by the brains. The brains' response comes in forms of mental punishment, which can comprise of 'feeling' of depression, fears (including that of failure), mood swings and other maladies.

While 'flight or fight' instincts of most animals can be applied to human behaviour, it should not be presumed that in critical situations it is the human brains that panic, for they do not. What actually occurs when a conscio is in a situation where it desires not to remain in its present space, or location, due to approaching physical danger, is that it (conscio) may present to 'itself' (that is: to its brain) a demand to alter the present circumstances by either removing 'it' (conscio) from that dangerous circumstance (from that particular space or location), or to physically take action so as to alter the circumstance, even if that is an impossibility. This kind of irrational conscio commanding overrides, requesting for an immediate alteration of the circumstance, can result in conscio behaviour being classified as 'panic'.

Panic is a selfish act. It is often considered to be 'cowardly' – such as that of a conscio of a soldier wanting to avoid being killed and running away from danger – it may also be selfishly 'heroic' (and by that considered as unselfish) when, for instance, a conscio panics in rushing to the aid of someone else in danger, such as parent who is a poor swimmer jumping into water in an attempt to rescue his or her drowning child.

Panic does not assist a brain's chances of survival too well, or too often. Nonetheless, when a conscio panics its brain, it (the brain) has no option but to reduce its support level of its conscio, while still trying to fulfill its conscio's 'want'.

In desperate situations a human brain prefers to be given an opportunity to function at its full capacity, without the needless shrill panic overrides from its conscio. In order to function on a high-awareness level, a threatened brain rapidly deletes the fear signals it had been sending to its conscio, replacing them, for itself, with a sense of acute clarity and awareness of surrounding space. During this period a brain can have the body perform brief feats of incredible strength, agility and speed, with a suppressed awareness of pain. The absence of pain due to absence of conscio's awareness, as only conscio can 'feel' pain and not the body itself. (This can be witnessed in every surgical theatre, where the body is painlessly penetrated by various surgical methods, while the anesthetized brain is incapable of maintaining its conscio. Once the conscio is re-activated after surgery, the 'feeling' of pain in the body become felt.)

While the heightened brain activity continues the 'unconscious' conscio is either deprived of memory or is provided with a feeling of calm, where all the actions occurring in surrounding space seem to proceed in slow motion.

(Should the physical threat appear unstoppable, with death being a certainty, then that conscio would be provided by its brain with a decision to accept the inevitable outcome in serenity, or to allow for selfishness of 'anger' to assist the brain and the body to oppose adversity to the last moment of the brain's and body' ability. Once either decision is made, at that point a brain sends to its conscio an acknowledgement of life's futility, so that the conscio – in most instances – can face its demise with little or no fear.)

More often than not, once a brain restores its conscio to its full level of activity after a panic, that conscio will usually have little, if any, recollections of the past event. In a case of a 'heroic' act, by having no perception that its brain was responsible for all that physically was achieved on its (conscio's) behalf, a conscio presumes that it was it, the 'self', or 'I', who was responsible for any amazing physical deeds, and in all ignorance accepts any forthcoming accolades.

It may seem amazing what a brain is capable of achieving when required to perform with a suppressed conscio. For once unburdened of its conscio's overriding control of placing restrictive limitations on it, a brain can make itself and its body – even if only temporarily – function at a much higher level of physical and mental ability. But this happens only on rare occasions.

The reason why the brains do not function without their fully conscious conscios for long periods is because conscios, as mental system of secondary consciousness, are an integral part of the brain's conscious navigational system. Without maintaining and supporting its conscio mental systems a human brain becomes virtually rudderless. While a brain can provide instructions to its body what to do, the body is unable to return an active response without a conscio's conscious response to the brain on behalf of the body. The brains need the conscio mental system – which they maintain – to be the consciousness involved with awareness of the external presence of their body and of the space surrounding the body.

In other words, every human brain is like an internal engine that requires the steering mechanism of its conscio's observation, as input of all that is occurring with and around the body, so that it can store this information in memory, and act on it so as to navigate them (the brain, conscio and the body) physically in the surrounding environment, which amounts to life.

Each and every young conscio has an awareness of 'fears' from personal experience, as presented by their brain in the form of 'infant fear codes' and 'fear codes', all of which can produce mental levels of pleasant 'feeling' of joy or of pain and discomfort. As a conscios develop with growing brains, they begin to recognise a factor that 'fears' can also be used by them, personally, as means of controlling others.

The 'fears' that conscios learn to take advantage of are those that arise from selfishness and apply to selfishness. These 'fears' become developed from individual conscio's personal likes and dislikes. But considering that all conscios have a similarity in their individual likes and dislikes, conscios have learnt to recognise that if they themselves had these particular fears then so would other humans (conscios). Therefore, they (conscios) came to use these personal selfish fears on each other as ways to worry and frighten others into doing what was expected of them.

But just like 'infant fear codes' and 'fear codes', 'selfish fears' have to be learnt from infancy, because if not personally experienced they make no impact from conscio not knowing that they even exist. For this reason, one human (conscio) generation after another have been teaching the young conscios the 'selfish fears', so as to indoctrinate them in accepting 'selfish fears' as being a natural order of every human life.

The lessons of 'selfish fears' teach that to survive in safety with sufficient sustenance is not enough. There has to be the motivation of selfish 'want' and 'desire' in order to obtain more of everything for each individual, even those who are young and weak. And if this cannot be immediately achieved then there's 'envy' and 'hope' for a chance to fulfill that 'greed'.

By teaching 'selfish fears', conscios have developed large and complex social systems so as to survive not in the natural elements but within human societies, all of which represent conformity to domination by means of submission. These social systems became the social customs of cultures that invented the concept of 'might is right', where the 'mighty are rich', and that high-leveled selfishness of domination is to be admired, respected, envied, and if possible, replicated.

Then there is the instruction dictating that any quest for high rank, for possessions and for power over others achieves happiness; and that alternatively, to have no status is to be disrespected and shunned, because absence power and wealth equates to a failure worthy of social contempt.

These 'selfish fears' lessons that promote selfishness govern all current human mental and physical relationships between themselves and others. With 'selfish fears' conscios fear that they will be somehow let down, disappointed, harmed, compromised or cheated by others, and so they resort to suspicions, jealousies, paranoia, and aggressive behaviour.

Through 'selfish fears' they fear physical pain and verbal abuse. They fear for their possessions, or fear having none. They fear what others have and they do not. They fear their children and they fear for their children. They fear eating too much, or too little, or having nothing to eat. They fear what others may think, and they fear what others may say. They fear appearing publicly before others, and they fear not being given that opportunity. They fear being noticed too often, and fear being ignored. They fear power of others – often justifiably – as much as they fear loosing it. They fear new ideas, and knowledge, for these pose a threat to their established thinking and order. Yet, others fear that new ideas and knowledge will not produce changes, or at least not quickly enough. And they fear losing in any field of conflict, or competitive aggressions, be it in sport, business, or national and international politics.

But of all the 'selfish fears' by far the greatest is that of having to face and tell the truth.

Truth is a very easy concept to grasp, as it comprises of just two factors. The firsts: is to accept a physical fact that that which had physically occurred had occurred, and any claim that it did not is a lie. The second: is to accept a physical fact that that which had not physically occurred had not occurred, and any claim that it did is a lie. It is then possible to apply these criteria to any past and present actions that supposedly have taken place, so as to differentiate between truth and lies. That which is claimed to have happened and had, represents the truth; that which is claimed to have occurred and did not, represents a lie. Similarly, that which is claimed not to have taken place and did not, represents the truth; that which is claimed not to have happened but did, represents a lie. Likewise, this applies to any claim addressing any future intended action, whereby fulfilling any promise to do something or not to do something represents the truth, and failure to fulfill these promises represents a lie.

But then conscio 'selfish fears', with assistance of imagination, become involved with any need to disclose the truth. The two main impediments to telling of truth are that of: 'selfish fear' of losing any personal advantage or gain that may possibly be obtained by withholding, denying, or distorting the truth; and the 'selfish fear' of receiving mental or physical harm at the hands of those for whom – for whatever reason – a revelation of truth is detrimental.

These 'selfish fears' have caused truth to be avoided, distorted, altered, hidden, disguised, neglected, or ignored by most conscios. Presently, it is almost considered a mental deficiency and a sign of physical and moral weakness to speak the truth, because truth is considered to be more damaging than lies. Lies, therefore, have become a more preferred method of communication.

This is nothing new, considering that human societies had always been structured on communications comprising of lies, fabrications, distortions and deceit. Currently – despite unprecedented high levels of present-day communications and broadcasting of information, opinions, and ideas – lies continue to be accepted as facts, not with just reluctance but with glee, or with gratitude or relief.

For an example, most politicians of any political persuasion – in fear that they will not be elected, or not be re-elected – discard their morality and ethics by finding nothing wrong with debasing their opposition with innuendo and accusations based on fabrications and lies. They endeavor to out-do each other's rash, short-term promises to their voters, by proposing to make them better off than ever before; playing up on voters' 'selfish fears' of unemployment, national insecurity, levels of crime, and illegal immigration.

Once in power, after the conversion of votes for promises, the same politicians distance themselves from their original pledges with other lies, excusing their generous promises that were never sustainable, and were contrary to their selfish interests and those of their main financial supporters. They present their new lies accompanied not with humble apologies for their previous lies, but with proud and arrogant claims that place any blame for their previous lies onto anyone else – including the opposition – but never themselves. They then present more lies as assurances that the outcomes of their new promises shall still turn out to be in the best interest of the voters.

In societies based on principles of selfishness, any pledges made regarding financial salvation of all cannot be kept or achieved. This however does not prevent the politicians from continuing to make promises of lies, then lies of promises. Their 'selfish fears' of losing governing power (their control over others) are so strong that enormous amounts of public and private funds are spent by them on propaganda, publicity and advertising campaigns intended to support and promote their selfish lies, and by that secure their election or re-election chances.

And what is the response from the voting masses to such political manipulations? The political lies and half-truths are fed to the voters by media concerned only with instigation of conflicts, divisions, and scandals – as these topics provide sales and profits – to the public who, on the whole, are content to remain ignorantly oblivious of any real facts, preferring instead to treat their political parties in a manner of preferred sporting clubs. This blind attitude leads most voters to disregard the reality that their favorite political party representatives, whom they continue to re-elect, are nothing but self-serving liars.

With all societies relentlessly indoctrinating their young (conscios) to accept the teachings of high-leveled selfishness (based on 'selfish fears'), which is also advocated by many political parties of all nations, it is not surprising that the selfishness of conservatism dominates national outlooks with 'selfish fears'. There are the 'selfish fears' of government spending on social issues; aggressive opposition to those advocating humanity and ecology, from 'selfish fear' of change to power and profits; bigotry from 'selfish fear' of all those who are different in their customs, race and gender. All these 'selfish fears' become ingrained in conscios, especially those who only care about a selfish circumstance of: "what's in it for me?"

In such cycles of deceit, derived from 'selfish fears', majority of humanity retains its fear of any change towards more socially responsible alternatives, having been taught to reject socialism by their educators, governments, and business owners who fear the very thoughts of social solutions, preferring to uphold selfishness and greed for their own interests. This results in humanity accepting the same lies from their ruling elite, over and over again.

To pacify their insecurities derived from 'selfish fears', conscios have always been attempting to obtain glimpses of future, as means of controlling change. In their efforts to alleviate their 'selfish fears' of failing in their enterprises or missing out on opportunities to gain wealth and power over others, the insecure conscios continue – to this day – implementing unreality of guesses and predictions. Conscios classify this as methods of forecasting, by using astrology, tarot cards, palm reading, and fortune telling. They also gather various statistics and compilations of data in attempts to gleam some probabilities, averages, and possibilities. This kind of guesswork has resulted in emergence of various 'risk assessment' experts advocating their opinions and estimations on future of anything: ranging from national economic projections, to movements in stock markets, derivatives and futures, to weather forecasting.

Yet, despite all their computers, projection charts, and database forecasts, the accuracy rates of all these future-predictors are in line with those of gamblers: a 50-50 chance. For that is all forecasting ever is: a gamble – with the results being either that of a win or a loss: success or failure. This means that all the banks, all the investment institutions, and all the businesses operate as gamblers; a description they, not surprisingly, avoid in fear.

This means that no 'business strategies', 'business plans', 'projections', and any other human-devised hopes-on-paper can ever produce any actual insight into future. And because conscios react in panic to any 'feared' unfavorable event, no amount of forecasting can prevent any business institution – or a nation, for that matter – from becoming unprofitable or bankrupt.

Considering that risk is unavoidable in a selfish enterprise, there are those who more than willing to offer – at a price – some protection from risk, while their real function is to be instigators of 'selfish fears', for it is in their financial interests to have their customers and clients be as fearful as possible of potential future misfortunes. Such profit-from-fear operators including all forms of insurance and financial institutions; security and surveillance companies; and even spiritual and religion orders, for they too profess to predict future events from god to their god-fearing believers. After all, they do promise salvation to the faithful.

But of all the current profit-from-fear industries, manufacturers of military equipment are by far the most profitable. These industrialists produce land, water, and air war machines and killing devices for nations that fear being threatened by the aggressive behaviour of other nations, and those that are intent on instilling a fear-of-death onto others. This applies especially to wealthier bully-nations, gloating over their arsenals of atomic bombs and various missiles with which they can threaten other nations. And yet, no matter how big their arsenals become they still crave more deadlier and more destructive devices from fear of losing their weapons superiority to others.

And then there are the leaders of rogue nations who crave destructive devices with which to project threats of their own.

Apart from having 'selfish fears' of other nations, political and business leaders of both rich and poor nations are the biggest instillers of 'selfish fear' onto their own populations.

In guise of "providing security for the citizens", political leaders often intentionally exaggerate the level of threat to their nation. This allows them to exploit the insecurities and paranoia of their public, who, in 'selfish fear' for their safety give them a mandate to spend public money on armed forces and all the necessary machinery and equipment. This is also done for the police and security personnel, who need weapons, surveillance systems, and all else that goes with controlling, monitoring and spying on their own citizens.

Such government 'selfish fears' are much valued by their friends, the business corporations, who supply all the government needs so that the spent public money ends up in their bank accounts. In return, these same business corporations are more than willing to make financial contributions (or bribes) to the re-election funds of their friends, the politicians, as means of reassuring themselves that these actions will dispel their 'selfish fears' of losing any future business. For that is what all business represents: 'selfish fears' of failure to survive.

This applies equally to big or small business enterprise, as all business is based on conscio 'selfish fears'.

When the concepts of mass production were first introduced – as a method of reducing unit purchase prices – the mass production and media industries devised selfish fears' techniques with which they exploited, and continue to exploit the 'selfish fears' insecurities of their consumers. They did this by methods of converting human (conscio) insecurities ('selfish fears'), such as: vanity, envy, and pretentiousness into profits, by advocating benefits of fulfillment.

They presented an image of fulfillment – in a form of a purchase – being

PART 1. WHY HUMANS ARE HUMAN?

directly responsible for acquisition of personal contentment, sexual appeal, career success, ability to impress, gain of status and influence, and attainment of happiness, all of which could supposedly be exponentially broadened with more frequent purchases of all the mass-produced products.

Currently conscios of all ages are encouraged by business to have the 'right' selfish attitude towards materialism in a disposable society: "I consume, therefore I have a purpose in life."

Fearful of being classified as backward, poor, without discernment or refinement, many conscios – especially those with low self-esteem or highly envious – become indoctrinated by marketers, promoters and advertisers, as well as peer pressure, into accepting and embracing the culture of consumerism, where there is a never ending pursuit for the 'right' branding. Many conscios succumb to thinking that in order to be happy and successful it is necessary to consume the 'right' brand of merchandise and products, while being blatantly oblivious to the harm that all this rampant and disposable consumerism brings to their environment and all else living on this planet.

The irony is that the media, producers, manufacturers and retail businesses that preach and promote 'selfish fears' to their consumers are themselves victims to the same affliction. They selfishly fear any loss of profitability caused by competition, fickle consumer tastes, demands of their workforce and their investor expectations, as this affects their business existence. And so, the methods of encouraging and marketing fear, avarice and vanity are constantly recycled, as if they are the only available solutions for human needs: resulting in market saturations and devaluations, producing ongoing trading cycles of 'booms–and–busts'.

What initially had began as a self-assured benefit to all humanity from mass production has now evolved into an ever growing, desperate, greed-based process of waste production, because of human conscios' continuous use of 'selfish fears' for personal advantage. Locked into cycles of escalating needless production requiring increasing quantities of resources, investments, and financial returns (while ignoring the escalating chemical by-products causing toxic waste and the increasing carbon pollution) the profit-from-fear businesses are left fearing any kind of action that would interfere or halt this process.

Instead of finding solutions for preserving natural resources and easing needless overproduction and consumerism, the approach taken by all governments and businesses is to assure each other that: "no matter what, its business as usual" – simply because they can rely on their customers' 'selfish fears' to keep business in business.

Exploitations of 'selfish fears' are not just the domains of human conscios with military, political, industrial or commercial power. It is available to all human individuals, be they males or females with differing political and social ideologies, or religious beliefs; those who may presume to have a grievances against the authority; those who want to oppose that authority; or those who may want to have a share in the power of that authority.

The authorities usually comprise of regressive and repressive conscios of human males, who are, more often than not, selfish and inflexible in their views. They fear any opposition to their self-centred thinking; acting with intolerance to any questions of sharing their control or power over others. Against such autocracy, every human generation give rise to new would-be-leaders, who may hold more liberal views, or those with more radical solutions to social problems, or those with dogmas even more regressive and oppressive than those of current authority.

Not surprisingly, when ignored, suppressed, or forced to be silenced, the new generations of would-be-leaders resort to applying 'selfish fears' against those in authority, presuming this to be the fastest means of overcoming authority. As a way of influencing the social, political or religious outlook of a society, they may become political or religious terrorists, employing kidnapping, assassinations, and bombings of humans important to the state. By destabilizing society and threatening authority with their fear-based chaos that cannot be overcome by conventional military means, such rebels, terrorists, and zealous religious fundamentalists hope (if not expect) that the frightened and anxious population will somehow relent to their uncompromising intentions, and in the process possibly cause a revolution. After all, historically such events have taken place in most countries, including those that call themselves as First World nations.

Most human conscios selfishly fear losing to others what they already control and possess. They also fear being restricted in what they can control and possess. For that reason they often fear and hate – or at least dislike – all who stand between them and that which they desire. Beyond that, conscios fear those who, like them, would desire their possessions and attempt to obtain it by any means. Such 'selfish fears' apply to most conscios, including the very rich and powerful, whose 'wants' and 'desires' are often on a massive scale. While ordinary individuals may covet their neighbor's house, the rich and powerful desire estates, counties, and even countries.

For that is what most human conscios desire most of all: land. Be it a sandy desert, or a swampy quagmire, or prime real estate: as far as conscios are concerned all land is worth owning, defending and killing.

According to most conscios, the size of the land is "the measure of the man". The larger the backyard, or the estate, or the country, or the empire, the more

PART 1. WHY HUMANS ARE HUMAN?

respected, the more envied, and the more feared are those who claim to be the owners. And the more land that they have, the more they desire to obtain more, while selfishly fearing that their ambitions may not be realised.

Such highly selfish attitudes from fearful conscios had brought no great benefits to humanity or to nature at large – only waste. All the former empires – obtained at the point of a spear, or a sword, or a gun of the land-hungry kings, emperors and dictators – have either dispersed into dust or have evolved into new and different societies and nations that have discarded and forgotten their earlier languages and customs. This simply illustrates the futility of individual conscios wanting to own, by military might, all that was never theirs.

Currently, all similar attempts by conscios to forge empires by use of bombs and missiles, or to develop living spaces for humans at the expense of other living creatures, will result in the same pointless futility experienced by the past empires and civilisations. But, needless to say, due to their 'selfish fears' conscios will continue to go on as before. That is because the fear of losing their land – be that of a patch of grass or a country – is more fearful than the fear of never having had any land. That fear drives them to possess, occupy and develop all the lands that they can possibly acquire, disregarding the fact that those lands are so necessary to the lives of all other life forms on this planet, on whose existence their own human existence depends.

While avarice, greed, and envy are all part of 'selfish fears' that conscios have adopted (or rather, had their brains organise) in order to manipulate and dominate each other, there is another factor to 'selfish fears' that had never produced any benefits to anyone. That component of 'selfish fears' is anger.

Anger is an emotion that conscios activate to expresses displeasure with the way that surrounding events are taking place, mainly those that obstruct and prevent the anticipated realisation of a 'want' or 'desire'. This can include dissatisfaction, or a resentment of an abstract perception – something that has nothing to do with any physical event. These abstract activities can be that of a thought or a memory sent to it conscio by its brain, which may disturb and infuriate conscio into anger. An abstract act can be that of an expressed opinion, which can produce anger in those who may be vehemently opposed to that point of view from having a 'selfish fear' of being opposed. Anger can also be the result of a fear or resentment directed at a predicted future event that contradicts the desired outcome, even when this may not even eventuate.

Considering that conscios constantly want something, their wants can be very selfish, as in: wanting no opposition to one's own opinion, or unselfish (and yet still selfish as a 'want'), as in wanting justice and equality for all. When the desired 'wants', be they selfish or less so, do not eventuate or eventuate not as intended, conscios may react with anger from being disappointed at not having obtained their 'want', or in frustration of having to pursue the intangible means of trying to obtain that which they continue to 'want' or 'desire'.

The problem with anger is that it can become addictive. Conscios who are allowed to manipulate others with acts of anger develop a liking for it. Some conscios learn as infants that a tantrum of anger attains a desired result to their 'want'. Unless these early strategies for achieving 'wants' through application of anger are discouraged, that conscio shall continue to rely on anger throughout its life, despite lacking any satisfaction from this. Other conscios grow into anger by having achieved beauty, wealth, rank or a dominance with which they manipulate and oppress others. These 'self-important' conscios consider their own 'wants' to be prime, non-negotiable and a must, no matter how ludicrous and impossible to achieve, and exhibit anger from 'selfish fear' of anything that stands in their way of their 'wants'.

There are also those conscios who learn that with anger it is possible to instill a 'selfish fear' of physical violence and pain into others, and by such threats and actions dominate and control others with offensive anger.

As with any kind of conscio addiction, the next step to unrestricted 'liking' is 'dependence'. And so it is with anger. Once conscio has a dependence on anger – be it reactive or offensive anger – then anger and hostility become a prevalent emotion, used as a responsive approach in most of its relationship behaviors and interactions with other humans. For such angry conscio it makes little difference what its expectations of other humans may be, for they are all met and returned with hostility of anger.

This development occurs because once a human brain is instructed to provide 'anger' for its conscio, it slows down the production of various chemicals by which to maintain contentment, reason, guilt, and the 'fear codes', used as a counter-control of its conscio. One of the chemicals suppressed by anger is serotonin, which is produced by the body for the brain at a brain's instructions, as means of rewarding conscio with contentment, well being, and happiness for not requesting aggression as a solution to every selfish conscio 'want' or 'desire'. By reducing such body-produced chemicals due to conscio's addiction to aggressive behaviour associated with anger, the state of anger gradually becomes more permanent, often resulting in a cycle of anger.

Anger breeds anger. It can become contagious when one angry human may anger others, who, in turn, also anger others. By angry reactions to anger, anger can spread and increase in its scope, especially when a mutual anger inflames a group, a crowd, or a society, who become irrationally determined to obtain their mutual 'want', irrespective of future outcome to themselves and to others. Consequently, conscios addicted to anger have a high chance of inflicting physical harm onto others in a fit of rage (often without premeditation), by that harming themselves, their brain and their body, because they can be punished with incarceration and possibly even with a terminated of life.

Anger, as an emotion, is part of most human (conscios) lives. In many ways it is unavoidable because conscio selfishness lacks tolerance, patience and acceptance. It can be an angry outburst of annoyance from someone seeking solace but is disturbed; a flash of anger from someone unintentionally stepping onto a dog's excrement; or an angry expletive from someone who just missed the bus. Such small-scale angers are accepted and tolerated by the brains as they represent little disturbance to them.

This is not the case with more pronounced and on-going angers. For example: a commander wanting victory in a battle abusing and raging at his officers for being incapable of achieving this for him, would be used to abusing anger as means of avoiding personal responsibility by blaming others, and unreasonably expecting someone else to fulfill his 'wants' irrespective of any reality.

When such conscio continues to desire selfishness by means of anger, both its brain and other conscios and brains disapprove of this. Any brain dislikes the fact that it has to stop treating itself to chemicals that produce for its conscio a 'feeling', or a 'state' of contentment and happiness. And other humans dislike the disturbing noise and the spectacle that often accompanies a display of anger.

For that reason the brains attempt to subdue and pacify anger of their conscios by application of 'fear codes' and devised methods of anger repression and moderation by anger management. This consists of agreeing to hide from direct memory the events and notions that activate anger, and to convince conscio through its conscience that anger actually achieves very little but can cause a lot of trouble.

Unfortunately anger cannot be eradicated. Even in a repressed mental state it is stored in memory, capable of being resurrected should a specific occasion eventuate. But if conscio learns to be kind to its brain, by reducing the scale, the immediacy and quantity of its 'wants' and 'desires', this lessens the need to replace a disappointment with anger.

4

Sex, conscio and human brain

Another component of 'selfish fears' that impacts all conscios involves sex, sexuality, and reproduction. In fact, there so many such 'selfish fears' attributing to 'sexual fears' that it is a wonder that humans (brains, conscios and bodies) manage to reproduce at all.

For example, there are the 'selfish fears' of not being sexually attractive. Fears of genitals being too small or too large. Fears of breasts being too small or too large. Fears of not having hair, and fear of being too hairy. Fears of being rejected and fears of being committed. Fears of domination, fears of submission. Fears of promiscuity and fears of missed opportunity. Fears of faithlessness, and fears of being taken for granted. Fears of lacking sexual performance, premature ejaculation or the inability to orgasm. Fears of unusual sexual preferences, or unusual genders in partners. Fears of forced, non-consensual sex. Fears of impregnation. Fears of pregnancy. Fears of impotence. Fears of sterility. Fears of having too much sex and fears of not having enough. Fears of remaining a virgin. Fears of dying from sex-transmitted diseases.

All such 'selfish fears' applicable to human sexuality are eagerly exploited by the profit-from-fear businesses – these ranging from various clinics to individual psychoanalysts – while being constantly assaulted by religions and intolerant moralist minorities. But despite high levels of anxiety associated with human sexuality, it seems that humans (conscios) have a 'selfish fear' of actually knowing what their sexuality comprises of, and whom it represents.

Once a human sperm and the egg unite, a gender is set for a newly formed physical body within the embryo. In humans, an X chromosome from the female egg and an X chromosome from the male sperm result in a female, while an X chromosome from the female and a Y chromosome from the male result in a male. But while it is known that the XX for female and XY for a male produce DNA to make all the necessary multiplying actions, occurring at appropriate moments in developing a human body of a precise gender, this should not mean that a body cannot possess physical attributes of an opposite gender.

The problem with current human understanding of human sexuality is that the XX and XY combinations are presumed to represent the only possible gender outcomes in reproduction. As is the case with much of human wisdom, this understanding is wrong. Humans (conscios) seem to be adamant to belittle brains. For them brains are simply some organs that do some calculations, a bit of memory retention, and that is all. Whereas, in physical reality, human brains are the very essence of human sexuality.

What currently is unknown but should be understood, is that within each human being there are varying levels of femininity and masculinity present not only in every human body but also in every human brain.
That is correct: every human brain is an instigator of sexuality in its own right, possessing its own gender, of which its conscio is oblivious.
While each body develops its sex organs according to its physical body gender formation, together with the relevant chemical and physical structures in support of them, the human brain is not obliged to represent its body's gender development from the initial X or Y sex chromosomes.

As a human embryo begins to develop, the very first part of the body that has a priority of development is the brain. At the earliest moments of its growth inside the womb, a human brain acquires its gender – which does not, necessarily, imitate the gender of its body. This means that a new human brain begins a new life with a gender that may not match the gender of its body.

Furthermore, the gender of a brain, just as that of a body, is never 100% male or 100% female. Within every body gender there is a balance of genders: in some bodies there are more of male traits and less of female, while in others the very opposite may apply. A male body can have an effeminate structure just as a female can have a masculine shape. The very same gender variances apply to human brains.

The difference between the body and the brain genders is that the body gender is apparent to all, while the brain gender is hidden from view: initially even from its conscio. And because of the invisibility of the brain gender, only the body gender had ever been taken into account and considered to be the true sexual representation and orientation of that person.

This is despite that only the brain gender is responsible for all sexual attractions, and not the body. Whatever the sexuality of the brain is, this then becomes the basic sexuality of that human being.

When a young brain initially instructs its young conscio of its gender, usually, very little notice is taken of this. A child's sexuality has been, and continues to be determined by its body's sex organs. Infants with a vagina are classified to be a 'girl' (a female), while those with a penis are established to be a 'boy' (a male). But in situations where the gender of a brain is different to the gender of a body, that young conscio inevitably becomes confused upon reaching puberty, when it is unable to align 'itself' with that of its body's gender.

Despite its brain's gender instructions to its conscio (in variety of mental communications, especially when the brain-body gender imbalance is in contention), a conscio can become frightened and disturbed from experiencing sexual attraction to those of the same body gender as its own. Such mismatching of a body and brain genders within the same being, result in sexual attraction between same body genders. Currently body attraction between two males is classified as being homosexual, while the same kind of attraction between two females is classified as being lesbian. Instead, all such same body-gender attractions should be called just that: **sa**me **bo**dy-**ge**nder **at**tractions, or **sa**me **bo**dy-**ge**nder **at**tractives, abbreviated to: **sabogeat**.

But while there may be gender mismatching between a body and a brain producing sabogeat (same body-gender attraction) between same body genders, there is another vital influence in sexuality: conscio itself.

A conscio is initially given its gender awareness by the parents and other surrounding humans. This information, however, does not prevent an insecure or highly selfish adult conscio from ignoring even its real (brain) sexuality so as to experience sexual relations with all genders and sexualities. This means that for many reasons a heterosexual conscio (attracted only to the opposite body gender) may participate in sabogeat activities: once, occasionally or continuously – initially, perhaps, because of voluntary curiosity; from voluntary necessity (such as prostitution); from sexual frustration, or from an involuntary subjection to rape.

For example, many heterosexual humans incarcerated in a same-gender prison institutions, may conduct sabogeat relations with other inmates, which they may cease and never repeat upon their release from prison. Other initially heterosexual humans may be introduced to sabogeat behaviour by curiosity or sexual frustration in the same-gender boarding schools, with some of them, possibly, accepting bi-sexuality (receiving sexual pleasure from both genders) for the rest of their lives.

Mismatches between brain genders and their body genders are very common. Even more common are the 'almost' gender mismatches, where there is just a sufficient level of a brain gender to match its body gender. The fact remains that depending on the male-female scale of a brain's gender, and conscio's selfish desires, a conscio may be attracted to the opposite genders, or attracted to the same genders, or to both genders.

PART 1. WHY HUMANS ARE HUMAN?

To illustrate this, the following list indicates some body and brain gender combinations (with conscio intervention), showing the resulting sexual attractions:

- Male body, male brain, (with conscio considering itself to be male); attracted to female bodies = heterosexual attraction.
- Male body, male brain, (with conscio considering itself to be male); attracted to male bodies = no such sexual attraction.
- Male body, male brain, with conscio intervention causing attraction to male and female bodies = bisexual attraction.
- Male body, female brain, (with conscio considering itself to be female); attracted to male bodies = sabogeat attraction.
- Male body, female brain, (with conscio considering itself to be female); attracted to female bodies = no such sexual attraction.
- Male body, female brain, with conscio intervention causing attraction to male and female bodies = bisexual attraction.

- Female body, female brain, (with conscio considering itself to be female); attracted to male bodies = heterosexual attraction.
- Female body, female brain, (with conscio considering itself to be female); attracted to female bodies = no such sexual attraction.
- Female body, female brain, with conscio intervention causing attraction to male and female bodies = bisexual attraction.
- Female body, male brain, (with conscio considering itself to be male); attracted to female bodies = sabogeat attraction.
- Female body, male brain, (with conscio considering itself to be male); attracted to male bodies = no such sexual attraction.
- Female body, male brain, with conscio intervention causing attraction to male and female bodies = bisexual attraction.

What can be seen from the list above is that it is the brain's sexuality that is responsible for the attraction to the opposite body gender, so that, for instance, a male brain, whether with a male or female body, will only be attracted to the opposite body gender (that is: a female body), unless conscio intervenes with its selfishness to alter the fundamental brain to body gender attractions.

Considering that both a body's and a brain's genders are of varying levels, never 100% female or 100% male, the above basic sample of sexual attractions could be used to define the male and female sexuality of individual humans more specifically, if the scale of each gender were to be taken into consideration. For an example: female body-gender 50%, female brain-gender 40%, attracted to male body-gender 50%, male brain-gender 40%; as in a masculine female being attracted to effeminate male.

Apart from the complexity of genders and the varied gender sexual attractions, there are also the multitudes of human sexual stimulants, which are very complex in their own right, having been derived from 'selfish fears'. These can include: age, race, the nature of relationship, types of sexual activity, as well as, innumerable fetishes. The current ignorance of the human body and brain gender mismatches that can result in female and male sabogeat (same body-gender attractives) – together with the ignorance of full scope and variety of conscio sexuality and sexual stimulations – produce a great deal of confusion, anxiety, and anger in many human conscios concerned with imposing their own morality upon other members of their societies.

In their ongoing attempts to dictate their oppressive and puritanically biased notions upon sexual activities of individuals, these self-appointed moralists – often empowered by the support of their local politicians and clergy – presume that by suppressing their own sexual urges they can indefinitely force and shame other humans (conscios) into following their example. But as long as members of societies follow subserviently the dictates of zealous moralists and their champions – be they the rulers, politicians, and clergy – no realistic developments of social structures will ever take place.

While sexual fears and taboos remain in place, these societies shall be prevented from honestly addressing the problems of child molestation, female abuse and rape, and other physical and sexual violence perpetrated against women, children, and sabogeats. These problems shall remain where they currently are: safely hidden behind the closed doors of the victims and their molesters, and behind the turned backs of the knowing. These problems also give support to corruption in societies, stemming from influences of illegal profit-from-fear industries of pornography in its various forms; illegal prostitution; slavery; and illegal abortions.

While all the naive moralist – even those who mean well – presume that on their whim human sexuality and sexual stimulations can be stopped, simply because they wish to eradicate human sexuality and sexual stimulation, this can never be done.

What human conscios can do instead, is to increase their tolerance of others and broaden their perceptions of sexuality and sexual stimulation, and by such means eradicate all the despicable human conscio highly selfish behaviors, including that of child abuse and molestation, which are human-made (conscio-made) problems that have evolved into cycles, where one generation of victims do the same to the next. Instead of openly assessed and assisting such human practices to be altered, so as to be eradicated, societies usually prefer to ignore these issues, despite making occasional pretense of taking some affirmative actions, such as having police make a sensational exposure of a pedophile network. Such efforts by authorities, simply results in other pedophiles becoming more careful and devious, while continuing on regardless.

The reason why such practices continue is because societies are governed and controlled by human males who are taught from early age that they must strive to dominate others, be it nature (environment), other animals, or other humans. If humans (conscios) seriously wanted to solve the problem of child abuse and sexual molestation then their first step would need to be a mutual, conscious, and intentional alteration to their societal behaviors, which encourage a fondness for violence and dominance in males, allowing them to enforce an unjust inequality and subservience upon children and females. It is these male conscio-devised and maintained attitudes that successive generations of human males have continued instigating into their young as if these were aspirational ideals that are to be implemented into their sexuality; constantly desiring to bend anyone to their will, including defenseless children and women.

Throughout the human physical process of change on this planet (their history), all manner of human males had, and continue to have, absolute authority over their female family members and children. They could use them or abuse them – and often did, and do. Male humans continue buy child brides, and visit child sex slaves in prostitution establishments: theirs to be fondled, molested, and penetrated.

In many societies families were the husbands' and fathers' properties, who could sell them, or give them away, or kill them, with no one daring to come to their aid, for they were their master's possessions. Many male humans continue to presume that they still have the same physical and sexual rights over their family members. The reality is that they do not, and never did. They were fooled in believing that they did by accepting the selfish lies and deceptions of their rulers and priests, intent on justifying their own inexcusable behaviors.

For instance, within the Christian faith alone, the attitude to sexuality had been devious. While many Catholic Popes and Cardinals publicly proclaimed their celibacy – a requirement of their religion – they had (historically) given birth to illegitimate offspring's, as did the various nuns and priests. But instead of changing a silly custom of celibacy, so as to openly have families and children, they preferred – and still do – to hide, and cheat, and lie about their sexual practices.

Over many centuries, many priests of various religious denominations have been responsible for sexual abuse of children in their care. Such sexual predatory behaviour is ongoing into the present, and if not openly condoned then certainly disguised by their orders; secreted with 'selfish fear' within the walls of silence of these religious institutions. Unless humans (conscios) acknowledge their sexual 'selfish fears' and desires, and openly address them, their young shall continue to experience inhumanity.

Regardless of body genders, brain genders, and conscio intervention into sexual gratification, the one fundamental purpose for human sexuality is to have human brains reproduce themselves, and by that, navigate into future space (environment). By assessing what kind of future space is possible for them, human brains – as all living brains – have devised, and continue to devise appropriate bodies for themselves, so as to be successfully adapted for future space.

But while the process of reproduction is the method by which the brains of each species alter and refine their bodies, the brains themselves usually choose to remain mostly unchanged. Unless, that is, they feel – as human brains did long ago – that they need altering. The early human brain and body changes had resulted in a different approach to reproduction, which had overturned all previous mammal models.

Most mammals (mammal brains) develop annual (seasonal) 'sex drives' when a female is most fertile, during which period the males and females perform sexual intercourse, the females become pregnant, and the new offsprings are born. Where the process of natural sexual stimulation differs with human brains is in their ability to develop 'sex urges' not once a year but every day, on behalf of their conscios' 'wants'. This came about from human brains' agreement to circumvent any need of seasonal reproduction, in favour of reproduction that follows no specific seasons.

Apart from that, the brains also had acknowledged that to nurture and protect a young human brain for a long period, the male parent needed to obtain regular sexual stimulation to remain interested in the female parent, while the female parent needed to please the male parent with regular sexual stimulation so as to retain him with her and the child, as their protector and food provider.

While allowing for sexual activity to result in impregnation throughout a year, human brains did foresee the need to prevent every sexual intercourse to result in pregnancy, by restricting female ovulation to just a few days in every thirty days. The reason for reducing the probability of impregnation was to limit the birth rate, knowing that every human (brain, conscio, body) child needs a long duration of concentrated adult care to reach maturity, unlike most other animals.

Just because the early human brains implemented such sexual strategy with the best intentions, they did not count upon the influence from their selfish conscios in always wanting 'more'. This 'more of everything' attitude came to apply to the number of offsprings a human couple wanted procreate. Of course, in the early stages of human existence the urge to breed large families was related to the need of human species to survive diseases, famine and misfortunes. But as humans (conscios) became more adapt at maintaining their lives in security, the selfish urge to breed more offsprings by growing numbers of humans (conscios) has led to the current problems of human species overpopulating this planet.

This is not what the early human brains intended. Not that conscios would care about this. As far as conscios are concerned, when they decide that they 'want' of 'desire' a child, that 'want' or 'desire' has to be realised. So when a female body is incapable of reproducing, for whatever reason, they simply seek and accept offers of artificial insemination, surrogate birthing, chemicals, medical machines, and the surgical skills – all of which are provided in exchange of payments in a business transection.

What humans (conscios) fail to see is that human populations have to be carefully managed for the benefit of all life forms on Earth, instead of presuming that humans, with extended lives, can go on increasing in numbers indefinitely; using up more and more of Earth's finite natural resources, while replacing them with a degraded and polluted environment.

The impact of the overpopulation may not be apparent to most conscios, as they cannot witness the overall, combined devastation produced by them all. There is a strong probability that they will only recognise such problems (of their own making) when it is too late to do anything about them; a situation the human brains always wanted to avoid.

5

Living with conscio

The human brains' physical process of managing their conscios is quite amazing. The most impressive feature of this is in how each human brain maintains itself and its body in the instant of the present-moment, while simultaneously constantly fluctuating its conscio's perceived existence between a mental past and mental future. In other words, every human brain retains for its conscio's consciousness a constant mental three-dimensionality of past-future projection.

Most life forms do not perceive the instant of the present-moment. Nor do they do what they do with any awareness of 'yesterday' or 'tomorrow'. Such is not the case with conscios.

Conscios may know that all physical actions occur in the instant of the present-moment, but on the whole they pay little, if any, attention to it as it is so fleeting. Instead, what they are aware of in their present-moment are the memory-ability of mental revisits of the past (provided by their brains), which do not physically exist but in memory maintained by their brains, as well as the mental imagination-ability of future projections (provided by their brains) of what may occur, but which also has no physical existence.

A human brain inputs to memory what its conscio allows it to: recording events in the present-moment, in the same sequence as they occur to its body, in the physical process of change. In doing such memory recordings, a brain, in effect, is being its conscio's biographer and a historian. On a request from its conscio, a brain can retrieve past events as feelings and mental images, in a mental language of a process currently called 'memory'. Memory is not just a physical ability to acquire and retain information, but is also the ability to release it quickly on demand.

In inputting life experience to memory, a brain does not take instant snapshots, like a photographic camera that can capture on film all the details of a frozen instant. Instead, the way a brain records for memory is by placing all events of a single day (from awaking to falling asleep) onto, what can be described as a single mental memory frame (made of memory code for a combined stream of pulses acting as a single unit of consciousness), where all continuous imagery, as physically witnessed by the brain – but occasionally imagined, on conscio instructions – are piled one-on-top-of-another, like a pressed-down concertina. In this way, at the end of a day there are no millions of individual frames but one frame only, comprising of one, compressed, continuous run of all imagery, accompanied by sounds, personal thoughts and observations that the memory code recorded on that day.

When either instigated by it (the brain) or requested by its conscio, a brain locates the relevant information, or any relevant knowledge, then translates that mental memory code back into a language and images that its conscio will understand. This is why a brain communicates to its conscio in a language the brain understands (be that language English, or Chinese, or whatever), because a brain cannot 'think' to its conscio in a language it (the brain) does not know.

The recorded memory code maintains very little detail, and yet when required to be recalled the needed segments are presented as blurry 'blow-ups', with substantial details but not all that was recorded. The only way to maintain more detail to memory input is by constant requirement of its output.
Take for example the learning of an alphabet. The alphabet is retained in memory by the virtue of its constant application to reading, whereas the memory of being taught to read (by whom and where) soon diminishes into vagueness, as this memory has no bearing on the constant memory-recall needed for reading. So unless constantly used, even a learnt function can become faded in memory. That is why a foreign language learnt in youth can become forgotten over the duration of a life, if not practiced.

In making a recall, a brain can often falsify details in its rebuilding of a memory output – especially if requested to do so by its conscio after a long period of being unused – by softening the harshness of reality. Nonetheless with this process of memory recall a brain can reconstruct various feelings, senses, and emotions it had previously recorded.

However, an inability to provide a full memory-recall does not represent a 'wiping of memory' as in a computer's random-access memory (RAM) bank. No human memory is ever deleted: only the paths to those memory frames are diminished from lack of use. So once those paths are re-invigorated by a review of past learning, the memory frames can be quickly restored for recall by a brain's memory-recall function.

A human brain has three components to its memory system, all of which a brain manages as separate divisions. It has a conscio-dedicated memory, which is a brain's ability and speed of response to its conscio's requirement of recalling information that it (conscio) had physically witnessed with or without concentration, and any information that it (conscio) had consciously input to its brain. Then there is a body-dedicated memory component, together with the spinal cord, which is involved with its body attributes and functions. Finally, there is its own, brain-dedicated memory, which is devoted to its ability to function and to the ongoing body development. In providing a memory recall for its conscio, a human brain is absolutely resolute in not disclosing to the conscio-dedicated memory its other memory systems, committed to its other functions.

While a human brain is sound and unaffected by physical damage, diseases and aging malfunctions (derived from DNA deficiencies and DNA hereditary maladjustments) its ability to operate all of the memory components provides a combined and seamless human body operation of a normal, healthy human being. Should a brain be somehow affected in being prevented from managing all three components of memory, the results cause a malfunctioning of the whole mental and physical structure of that total human being.

Perhaps the most common form of this occurrence is when a brain begins failing to maintain the memory component of its conscio and that of its own. This problem comprises of a brain's growing inability to retain in memory not just its conscios requests and wants, but also the memory-recall of its conscios past existence. That is: a brain begins to 'forget' its conscio's wants (classified as dementia) and then goes onto 'forgetting' its ability of maintain its conscio altogether, which then leads to that brain's inability to provide mobility to its body. In such circumstance a brain and its body can be kept alive artificially with assistance of other humans, as otherwise that brain rapidly ceases to live from lack of sustenance needed by its body to feed it (the brain).

The amount of work that a human brain performs, without ever stopping its continuous function, is beyond all current human (conscio) perceptions. But at night it does need to deactivate and reduce its support of maintaining its conscio during its 'sleep' period, so as to provide more dedicated attention to its body and to itself.

A human brain is not a rigid structure where different regions perform specific functions and nothing else, as is currently perceived. While there are regions of a brain performing specific functions, a brain can alter regions to take on different functions. As a mental performer, a brain is a fluid entity, juggling, so to speak, its resources according to its performance needs in response to various conscio requirements, as well as, external physical influences on its body and its internal body needs.

When some brain regions are overburdened, it simply shares the burden with other regions not usually associated with those particular functions, by

PART 1. WHY HUMANS ARE HUMAN?

transferring to them some of the workload in order to fulfill what it has to accomplish. In maintaining its memory systems, especially the conscio-dedicated memory system, the same principle is used. All the conscious experiences, and any other information input by conscio to its brain, are transferred into a mental language by its brain, which it retains in codes of daily frames (segments of conscio consciousness) allocated to a floating syntax. Conscio memory is, therefore, not part of a specific region of a human brain, as none of them are precisely that (unlike what the current scientific wisdom presumes). Instead the memory syntax is a flexible and mobile code library that 'floats', or is moved about by a brain within its functions.

This mobile code library can be compared to workbench trolley, laden with tools, which is wheeled to any location where work needs to be performed: be it in an operating theatre, a mechanical work shop, a dining room, or an artist's studio. Within this floating memory code library, there are other syntaxes not related to conscio, which are relevant to a brain's association with its body, its spinal cord, and its human brain species development, all of which are withheld from conscio's awareness.

Once a brain senses, or is specifically requested by its conscio that a specific recall is required, the ability to do so, and the speed with which this ability is performed is dependent on the brain. Because all human brains have different abilities to function, their abilities to perform recalls for their conscios can vary for multitudes of reasons. The variables include affects of drugs, nutrition, occupation and aging.

Current humans (conscios) associate good memory (rapid recall ability of a brain) with higher intelligence, which really has nothing to do with intelligence. Intelligence is the ability to physically respond to any physical influence with a calculated reason. A rapid recall by a human brain is just a speed with which it is physically capable of locating paths to pulse frames of appropriate information it retains, as memory codes located at its floating memory syntax, then translating and presenting the results to its conscio as information comprising of mental thoughts and mental visuals. Many conscios consider a slow memory recall as a sign of low intellect. This is also incorrect, because the ability to reason with clarity does not depend on a fast memory recall.

Perhaps the best way to view memory-recall is in its fundamental purpose, which is: to assist the brain's reasoning function. And the purpose of a reasoning function is to use its memory-recall and imagination-projection to perceive all surrounding existence, so as to mentally navigate one's life, by constantly moving mentally back-and-forth between the non-existent past, the present-moment, and non-existent future.

By retaining an ongoing record of its present-moments on behalf of its conscio as its (the brain's) memory-recall, a brain allows its conscio to return mentally to its past at will. Similarly, by having the experience of the present-moment in relation to its past, a conscio can also request its brain to project mental images of events and actions that have not physically taken place. With this method of mental projection from its brain, a conscio is able to experience its future – not a physical 'future' which does not physically exist, as only the physical actions taking place in the present-moment exist – but that of an imagined future event.

These kind of past recalls and future projections can be illustrated by a conscio (with its brain and body) enjoying a meal with friends, reminiscing of past events shared by all those present, and discussing mutual or personal aspirations. What would actually take place in such a scenario is that the conscio and the conscios of other diners would be mutually sharing the ongoing physical present-moment, while having their individual brains provide for them mental recalls of the pasts and mental projections for their futures. As they do this, each of their brains is making mental recordings of the present event, in the present-moment, simultaneously with the re-recordings of the past recalls and future projections. In recording what is occurring in the present-moment, the brains of some of the diners may have been instructed by their conscios to devise exaggerations, or lies. These lies then have to be retained in memory of the brains whose conscios had them produce lies, in order to prevent the conscios producing lies from being caught out telling lies at a later occasion, by remembering what lies were told, to whom, and when.

For a conscio, future is an undeniable reality, which it tries to map out for its life-navigation, by using as reference points the progress of its past, based on which it then takes actions in the present-moment, in an attempt to ensure its future existence. By having experience of the past, a conscio has its brain project possible future outcomes as mental visions, which a brain does with amazing speed and detail, but never with accuracy, as most future projections never turn out as expected.

With every moment that a brain maintains its conscio, the fearful conscio may require its brain to project scenarios of doubts and reassurances, in order to have instant comparisons for its decision-making. To achieve this, a conscio has its brain flitter between past events and short-term future projections. Take, for instance, the following hypothetical event.

You (conscio) have agreed to present a speech. As you begin to write your speech, one moment you are searching your memory (having your brain make recalls) of your past experiences to include in your speech; the next moment you are daydreaming, picturing yourself being applauded for your speech (in future). As the writing progresses, you go back and forth between past experiences and the future expectations. Then, even during the presentation of your speech (using your brain to do so), as your body pronounce your words to the audience you are still having your brain make memory recalls and projections for future. There are the instantaneous assessments on how the speech is being received by the audience, comparing this to any similar past experience. You (conscio) may have your brain re-evaluate the speech (seeing past mental image of yourself in the process of writing it), questioning: "Did I cover all the relevant points in the speech? Is my speech effective? What if they don't like it?" But almost simultaneously you may have your brain flash an image of future projection of the audience before you being receptive to your speech, applauding loudly, and with yourself feeling relieved and pleased with yourself.

By such a method – with the efforts of its brain – every human conscio is unconsciously spending its present-moment between where it has been in the past, and where it presumes to be in future. To this end, nearly every physical action produced by its body, results with conscio needing mental past-recalls and future-projections. By providing simultaneously the mental bridge between the past and future, between which the present-moment is present but not often recognised, every human brain is projecting a mental three-dimensionality of the past, present, and future to its conscio's consciousness.

This mental past-present-future projection from a brain, to its conscio, is a unique human brain ability. Extraordinary as it is, it is seldom noticed by conscios, who considering it to be a natural part of their "thinking and remembering ability", worthy of no particular importance.

When a conscio experiences events in its life, its brain's input does not retain just a visual memory but can also retain the emotions and feelings that a brain collects from the body's sensors. When a brain makes a past-recall, the retrieved memory is capable of producing an emotional re-living of feelings. Some mental re-visitations of past actions can be intentional, as they are made on specific request from conscio to its brain, and, if necessary, may be repeated again and again. An example of this would be a person (conscio) being impressed with a character from a movie film, and trying to adopt either the look, or the mannerisms, or the speech of that character, which requires a constant memory revisit of that film. Other recalls, however, may be involuntary, because they occur as a result of an unintentional stimulus or a prompt to a body sensor. In one way, they can be a 'flashback', when, for an example, a fragrance can instantaneously bring a mental image and a resurgence of a past experience or a feeling that occurred in some place long gone. In another way, the unintentional impulse, or stimulus may give a sharp, conscious feeling of déjà vu: a sensation of having previously experienced the same emotion or function, but at a different, vaguely remembered location or period.

Conscios do not like incidents of flashback or déjà vu. For them, such involuntary exposure to past events and unexplained experiences represent an unwelcome mental activity, occurring without their awareness or permission; an unintended disclosure of something that they, perhaps, had chosen (for their brain) to suppress.

The past-recalls and future-projections also represent the method by which a brain communicates with its conscio. While being unknown to its conscio, a brain presents its sentiments for reduced levels of selfishness and greed by presenting its conscio with punishment of conscience when its harmful and highly selfish past deeds are recalled, then projected as images of harmful and painful experiences of the past, or repercussions yet to be endured in future.

In order to alleviate themselves of such feelings of guilt, while choosing not to alter their high levels of selfishness, conscios have devised means by which the messages from their brains can be silenced. One of these is a physical task or a function. Once a brain is set a task or a function – be it simply to have the body dance – it (the brain) will be too busy to further punish its conscio. The other methods are those of noise and visual distractions. For conscios there is no better method to silence – or at least temporarily quieten – a brain's communication of guilt, then by having it input to memory notions based on noise, sound, and visual imagery, such as a memory request by conscio of a noisy, vibrant carnival. By having to facilitate input that may require a responding output, in form of assessed opinions, or calculations, or projections, a brain concentrates on recalling the projected sounds and visual imagery instead of influencing its conscio with conscience.

It is for this reason that most conscios surround themselves with specially produced noises, sounds and visual imagery. By being fearful of remaining on their own in silence with their brains, conscios provide their brains with distractions of songs and music, television and movies, computers, mobile telephones and video games. In fact, they do everything possible, throughout their days and nights, so as to avoid remaining in motionless silence, listening to their brains.

6

Conscio and elements of sleep

A human brain does not work as a computer that calculates from a command to the fulfillment of that command in a progressively sequential order. Instead, it works by producing examination pulses that first assess all that is familiar in the conscio consciousness, and then progressively moving to that which is less familiar. Additionally, being aware of its conscio's 'infant fear code', 'fear code', and 'selfish fears' indexes, a human brain is often unsure as to how honest and sincere its conscio is at wanting solutions based on physical reality, instead of those supporting existing fictions and lies. That is why often, due to a conscio's insincerity and selfishness of purpose, its brain will firstly provide only an obvious and simplistic solution, lacking depth of perception and insight. This kind of dismissive brain attitude can be described as being: "You want rubbish, you get rubbish."

So whichever outcome a conscio truly desires, its brain will provide it with suitable mental code phrases obtained from a series of ongoing compilations and separations of information, which is how a brain compiles and prepares the necessary information prior to passing it onto its conscio. Once these mental code phrases – that can also be grouped into mental code statements – are ready to be presented to conscio, they are then translated (transferred) into mental three-dimensional visuals and pulses of thoughts. When a conscio is presented with such output from its brain, it has the selective control of accepting or rejecting this mental output.

For many reasons, such as: drug influences, or illness, or chemical imbalance, or physical damage, a brain may present to its conscious conscio mental three-dimensional visuals and pulses of thoughts that are disjointed and incoherent. A conscio that is incapable of rejecting and suppressing such an output by its brain is usually not fully activated by its brain, and therefore is in a mentally incapacitated state. Such circumstances are representative of a conscio experiencing hallucinations, flashbacks, or paranoia.

Usually a healthy brain and a well maintained and a fully activated conscio do not find themselves in circumstances of sharing disjointed and incoherent mental images and thoughts. At least not while a conscio is fully activated, and from that being fully conscious. This becomes a different matter when a brain reduces its support to its conscio at night, as it does during a process known as 'sleep' – the presumed period of a brain's and a body's rest.

The fact is, however, that a human brain never switches off from birth to the end of life. And while every human brain requires daily periods of withdrawal from supporting its conscio, this is done not to rest but to conduct concentrated efforts on other requirements. Despite that current scientific monitoring of brain activity during sleep may show a brain to be less heated – and from that assessed to be less active – it is actually in those less heated periods that a brain conducts all the main work for its dormant conscios, while also working on its body's repairs and recoveries, as well as, performing calculations for future human body structure changes. It is during these night hours that brains are most productive: experiencing less interruption, less haste, less waste.

This does not mean that during the night sleep periods a conscio is left unattended by its brain. What actually takes place during the night is a series of brain's gradual deactivations of its conscio before resuming to make gradual activations of conscio. By such a series of deactivations and reactivations a conscio becomes deprived of its consciousness before regaining some of it, then again losing it and then regaining it. There are usually from three to five such repetitive fluctuations conducted in a single period of a night's 'sleep'.

There is another vital factor to a brain's night activity that needs pointing out: the distinct difference of brain behaviour between day and night periods. During the day, when a conscio's consciousness is fully activated by its brain, that brain attempts to facilitate its conscios' bidding. A brain hears and accepts what its conscio 'wants' or 'desires' – be they selfless solutions that would assist others, or selfish solutions concerning personal gains – then tries to provide mental and physical fulfillments to those 'wants' or 'desires'.

At night, however, when conscio anticipates a restful period of unconsciousness – otherwise known as 'sleep' – the brain of that conscio assumes all control of its own efforts. This is the only phase in the day when a human brain is capable of directly rejecting or ignoring its conscio's requests.

This usually occurs at the low level (deep level) of the 'sleep' cycle, when a conscio is virtually unconscious, hardly existing at all. It is at this juncture when a brain takes over all control of its and its body's functions, so that the conscio and its requests temporarily become of no importance to the brain. And even when the 'sleep' fluctuation moves towards the period when a brain has to provide some consciousness to its conscio, that consciousness is not capable of overriding its brain, so that brain either disregards or refuses its conscio, in order to go on with its work. Depending on the relationship be-

PART 1. WHY HUMANS ARE HUMAN?

tween a brain and its conscio, the night dominance by a brain can produce various outcomes for both the brain and its conscio: some bad and some good.

'Deep' sleep is the period when both the body and the brain can benefit. During that period a brain manages to do a lot of uninterrupted work for itself and for its body. In gratitude for being able to accomplish this work a brain may rewards its conscio with a refreshed and joyful feeling once its conscio is activated in the morning. Alternatively, if a brain is preoccupied with a conscio problem, worry, concern or some kind of illness, with all of these diversions preventing it from doing its own private work, that brain will not be able to refresh its body; resulting in the activated conscio feeling tired, drowsy, and possibly irritable (because that is how its brain also feels).

At the high level (shallow concentration level of sleep) of the 'sleep' period fluctuations, a conscio is semi-conscious, almost awake. It is at this level of the 'sleep' period that a conscio's consciousness may receives glimpses of its brain's mental work: most of it relating to conscio. These glimpses are inadvertent crossovers between the conscio's consciousness and that of the brain's consciousness. What conscio may glimpse can be that of conscio's 'selfish fears' of failures; emotional and physical burdens; obligation-pressure fears; and sexual fears: these possibly being sexual, physical and mental frustrations and longings. These glimpses can also represent a brain's 'infant fear code' and 'fear code' prompts. While these mental depictions are often structured within the context of present-moment – where the 'dream' activity is taking place in the present moment of 'now' – this is accompanied with hints and indications of the events being linked to past-recalls or future-projections.

Furthermore, there are the possibilities of unintentional crossovers between the conscio-dedicated memory and the brain-dedicated memories. And while these memory crossovers occur very rarely, some human brains had, and do produce, unintended and accidental mishaps that may cause a conscio to glimpse a bit of brain's work on body development, something that a confused conscio can mistakenly interpret as a depiction of 'reincarnation' of itself.

While this happens very rarely, a brain can provide a meaningful subliminal message to its conscio in a dream. Usually though, most of the mental images and thoughts presented to a semi-conscious conscio comprise of mismatched, abstract, and blended mental images and phrases. And while these may give a semi-conscious conscio a perception of cryptic meanings, containing predictions and hidden messages, they represent no such meaning.

But while most dreams and nightmares are no more than glimpses of a brain at work, on occasions a mental visual or verbal presentation can be that of a specific communication from a brain to its conscio. This may be part of a brain's intentional address its conscio, as its conscience, reprimanding conscio with guilt and foreboding due to that conscio's current self-centred selfishness, or for its past bad deeds. It may also be a directive from a brain, regarding the state of its body, cautioning conscio, for instance, that a tooth is in need of repair, or that the heart needs attention. It may also indicate that the bladder is full and the body is in need of urination. Or else, it may be a work-in-progress of solving problems on behalf of its conscio.

Occasionally, when experiencing a dream or a nightmare, a semi-conscious conscio may request, semi-consciously, for its brain to retain that dream, or nightmare, for when it (conscio) is fully awake. A brain may do this or it may not. Mostly, the brains do not facilitate their conscios with retention of dreams, resulting in most dreams being 'forgotten'.

Alternatively, a brain can make a recall of a dream to its conscio during the day, seemingly for no apparent reason, but to which there may well be a purpose. There are also some dreams that a brain may repeat periodically over many years, with these message-replays always having the same theme but different backgrounds, which is a way of a brain reminding its conscio that its (conscio's) past actions are leading it towards a re-experience (repeat) of a past situation.

On occasions a brain may communicate a solution to a problem in a dream. Though rarely, a brain with a solution may also activate a boost of 'full power' to its semi-conscious conscio, shocking the conscio into wakefulness, to be consciously aware of that solution. A brain can also provide a solution, as a mental statement, or a thought, during a short daytime rest period – a 'cat-nap' or a brief half-sleep during the day – when a powered-down conscio may give its brain more ability to successfully work out a problem.

Human brains are not equal. This has nothing to do with their size. Just because female brains and bodies are usually smaller to those of males, this does not reduce or prevent their ability to function just as well as those of males. Where brains differ is in their ability to maintain their conscios and to provide for their conscio's requests. While some brains are robust and full of confidence, never seeming to tire, other brains are more reserved, or easily fatigued, or less capable of tolerating all the petty demands from their conscios.

Then there are the multitudes of physically hurt or damaged brains, or those affected by some disease, or even those that had been born with deformity. Multitudes of such brains often share one common trait: they have a little tolerance of mental stress, whether that stress is derived from shortcomings of their body or mental stress derived from their conscio's demands. Due to all these reasons there are countless variables to human brain behaviour at night, when they power-down their conscios. The most common of these is that of sleeplessness or insomnia: a brain's inability to reduce its conscio level of consciousness. Or as it is currently understood: the inability to fall sleep or to maintain asleep.

There are many causes that can inflict insomnia. Sleeplessness may occur suddenly or gradually; it may be temporary or of long duration, but irrespective of the length of its persistence it is not beneficial to the brain, or its body, or even to its conscio. And yet, this unhelpful brain behaviour is the only significant constituent in a brain's relationship with its conscio in which a brain can reject not only conscio's directives but even conscio's pleas for sleep (its reduced consciousness), because the brain either does not want to, or is unable to do so.

Currently there are vast numbers of sleep clinics conducting studies into sleep disorders and offering cures for sleeplessness or insomnia. Their business model is based on chemical, cognitive and psychological assistance to their patients, in attempts to induce sleep. Possibly some of these remedies are successful for some patients and not others.

There are many selfish conscios that probably deserve to be tormented by their brains with sleep deprivation – even at their own detriment, due to brains' sentiment of: "You make me suffer, conscio, so we may as well suffer together." There are also those whose brains need sincere assistance to power down their conscios. It is for these humans that the following technique is provided. Admittedly this is done with some reluctance, because brains do not like this intrusion. Therefore, the explanation on how to circumvent sleep deprivation caused by a brain will not go into many details.

The procedure to alleviate sleep disorder cause by a brain requires patience and perseverance over many nights, as it provides no magical instant relief. The intention of this procedure is to obtain a loss of consciousness, which represents sleep, by actually rejecting (overriding) the brain's ongoing activity of keeping its conscio's consciousness active. The method of achieving this is to comfortably lie in bed with eyes shut while making a calm but repetitive mental statement of "No", as a rejection of brain's persistent subtle interjection with mental imagery, ideas and thoughts.

Human brains are so very clever at slipping in their mental input that their conscios often remain oblivious of the mental journeys their brains constantly take them on. So whatever the mental imagery, though, suggestion or ideas that the brain may mentally presents, the constant response should be a calm mental chant of "No."

This process of rejection should continue even when a brain provides a mental feeling of calm – with conscio being aware of mental comfort – because the intention is to 'sleep' (be unconscious) and not be aware of any calm. But even then a brain will attempt to infiltrate some of its mental activity, requiring further use of mental "No".

Initially this mental tug-of-war between the brain and its conscio may last all night, without actual attainment of sleep. However, even this activity results in surprising level of rest.

Continuing this practice over the following nights will achieve some periods of unconsciousness, and then longer and more pronounced periods of 'sleep'. But whenever consciousness is returned, the same rejection of brain's attentiveness needs to be maintained.

Should there be a definite consciousness of wakefulness, it is advisable to leave the bed and drink some warm water with a slice of bread (and not anything sweet or spicy, so as not to stimulate the brain,) and after a short pause return to bed.

Once the brain begins to provide levels of unconsciousness (sleep) to its conscio without any mental overrides, the procedure can be halted.

There is another advisable practice to assist with the "No" technique: the 'warm bed, cool room' facilitation.

Human brains perform better in cooler temperatures because a heated brain becomes stressed and less obliging. The same applies to a human brain during the night, when the head is exposed to cool air of a cool room, while the body can retain comfortable warmth inside the bed. At night a cool brain (not a cold or freezing brain) is more able to reduce its efforts of maintaining its conscio, converting the day period activity into conscio's nightly rhythmic power-down and power-up fluctuations.

There are multitudes of variables to sleep containing dreams and nightmares, but what can be said of them is that they are all the result of directly intentional and unintentional communication between a brain and its semi-conscious conscio. No other entity is involved in this. This mean that despite any human conscio perceptions or beliefs, there had never been, and never shall be, any spiritual or supernatural content or communications passed – as dreams – between the 'self' or 'I' of conscio and its brain and any spirit or a god.

The only exception that may occur between a direct communication between a brain and its conscio is by means of a process currently known as 'hypnosis'.

Currently hypnosis is presumed to represent a sleep-like state of 'mind' of the subject who is directed to obey instructions of the hypnotist, because the hypnotist is thought to penetrate the subject's 'subconscious mind'.

The first flaw with this perception is that in physical existence there is no such entity as a 'subconscious mind', just as there is no 'mind', conscious or otherwise. There is only a conscio mental system and its brain, who maintains this conscio consciousness.

The second flaw is that 'hypnosis' is not a correct description of the procedure. Considering that this is a verbal procedure where a conscio and its brain of one human can communicate directly with the brain of another human by bypassing its conscio, the correct terminology is 'brain-conversing'. The main reason for it being brain-conversing is because the conscio and its brain of the human converser are conversing directly with the brain of another human; a brain that would not accept any suggestion or order to do harm to its body and conscio, but which may accept encouragement to improve the attitude and behaviour of its conscio (if warranted by the brain), and enhance the ability of its body (if that is also warranted by the brain). In other words, a brain may accept counsel and encouragement to solve its problems outside of its own conscio by means of brain-conversing with another human (brain and conscio), but only while its own conscio is not overriding it with its selective input-output control.

The way this usually occurs is by having the conscio and the brain of the brain-converser convince the conscio of the subject to allow it to be bypassed (by being voluntarily deactivated) so as to allow a direct access to its brain. This is not that easy to achieve. Ordinarily, should a fully conscious conscio (that is: a conscio that is fully active thanks to its brain's full support) be requested to allow another human (conscio) a direct access to its brain, this would be refused. However, when a fully conscious conscio (subject of brain-conversing) allows another conscio (that of a brain-converser) to be granted an access to its brain for the purpose of acquiring a personal physical benefit – as in therapy – then this represents a great deal of trust being given to the brain-converser.

Once the permission to the brain's access is granted by the conscio of the subject, the conscio of the brain-converser can speak directly to the subject's brain, because the subject's conscio becomes unable to override the brain-converser's input, nor that of its brains mental output. Nevertheless, the semi-conscious conscio of the subject maintains some vigilance over the proceedings, so that if it does not like the direction of the communication taking place between the brain-converser and its (subject's) brain, it will then demand a full activation of itself from its brain, shutting off the direct communication between its brain and the brain-converser.

When a conscio allows a direct access to its brain, that brain – not shielded by its conscio – shows itself too be incapable of lying and of being devious (as it can be requested to do by its conscio). Therefore, a sincere conversation with a brain can reveal the cause of that brain's dissatisfaction with its conscio. From such revelations a reason can be obtained as to why that brain is punishing its conscio with maladies or other mental and physical problems. By getting to know what bothers a brain, a brain-converser can (possibly) negotiate a truce and an understanding between a brain and its conscio, so that both can reconcile with each other when a conscio obtains a better understanding and appreciation of its brain, and gives it more respect.

(Of course, none of this is understood by current humans and the 'hypnotists'. They simply presume that their suggestions to the 'subconscious mind' of a subject either work or they do not. For that reason, when providing therapy, the hypnotists request for many sessions, so as to be more convincing to the subject's 'mind' [brain].)

But even when a communication between a brain-converser and a subject's brain is successful, the result seldom ends in success. Once a semi-conscious conscio is raised to its full level of consciousness by its brain, that conscio can again override the output from its brain, rejecting that which it does not want to do – such as to stop smoking or drinking – nullifying by that all the brain-conversing suggestions made directly to its brain. And while it is not particularly difficult to 'brainwash' a brain by chemicals and long periods of stress and duress – in the process damaging that brain – brain-conversing cannot make a brain do anything harmful to its conscio and body, or make a conscious conscio do that which it does not want to do. The only way that a conscious conscio will submit to any outside suggestion is by agreeing to sincerely accept it, not because it has to but because it really wants to.

That is why any serious use of so-called hypnotism (brain-conversing) in medical treatment requires many sessions for any positive results to be achieved (even if temporary), unlike the phony hypnotists who amuse paying audiences by sham performances where hypnotic suggestions cannot be resisted, and meant to be amusing.

7

Conscio and brain conflicts and addictions

With the use of its conscio, a brain attempts to overcome any disadvantages to its beginning in life, such as: a lowly birth to poor parents or possessing a weak body. A conscio instigates this change to unfavorable circumstances while being (unknown to itself) supported in this venture by its brain's mental efforts.

From the moment that a conscio is activated by a brain, and with its constant support, conscio begins a journey to gain friends, and, where possible, influence and dominate them in the course of its (brain's) life. Unknown to conscio, this process of self-elevation is intended for achieving an environmental security for its brain, and for breeding of offspring brains, who too may enjoy the gained advantages in their future. But not knowing this, and presuming instead that it is achieving gains 'just by itself for itself', drives many conscios to high levels of selfishness and self-centeredness.

This desire to gain advantages over others is present in all humans (conscios). The intense selfishness can be suppressed or managed, but it will remain. If an intention of self-elevation is agreeable to the brain of an ambitious conscio, because that conscio is reasonable to its brain, then such efforts can result in good outcomes for both conscio and its brain. If, alternatively, a conscio obstinately overrides its brain by producing directives for selfishness far beyond the requirement of self-preservation and self-advancement, this can result in bad outcomes for its brain, and by association, to that conscio.

Some young conscios assessing their self-importance, especially during their adolescence, may find it difficult to influence or to dominate other humans (conscios). Such young conscios may resort to rebellious, unreasonable and over-demanding behavior to test their resolve at dominating others. Later in life, when such openly selfish behavior is deemed inappropriate, many conscios refuse to alter their desire to dominate others. They develop methods to get what they want by means of false charms and servitude, or by flattery, or by lying, or by cheating, or by theft, or by physical violence. But whatever means they choose to employ – fair or foul –most conscios persist in producing change in their lives by which they selfishly want to benefit.

For most brains, their relationships with their conscios result in a relatively well-balanced and satisfactory existence. But there are no guarantees to this.

Depending on how a conscio relates to its 'infant fear code', 'infant code' and 'selfish fears' index and other factors, such as good (selfless) and bad (selfish) influences from surrounding humans (conscios), a conscio can be a good (agreeable) or bad (obstinate) provider for its body and its brain.

A conscio may take care of its body and brain by use of a healthy diet and helpful physical and mental exercise. A conscio may have liberal attitudes towards acquisition of knowledge, allowing its brain to reward it (conscio) for generosity and input-output of unprejudiced views.

But then again, a conscio can abuse and harm its brain and body with an over-stringent diet, equivalent to starvation, or lack of any dietary restrictions, resulting in obesity; with excessive physical or mental exhaustion, or a lack of any exercise; excessive and needless exposure to physical danger; mental damage from legal and illegal drug abuse; participation in deeds unpleasant to a brain, such as: lies, theft, and murder; or repeated overrides of its brain, causing its brain to be a prisoner of its conscio; who is, more often than not, a prisoner of other conscios' opinions – which are, in themselves, derivatives of yet other conscio's opinions. A conscio can also be highly (and boastfully) biased in its perceptions (voiced as its opinions), forcing upon others its attitudes of arrogant disregard of other opinions and its high-leveled self-centeredness.

When a brain is required to suffer any high selfishness from its conscio, especially if that brain's influence of fear has little affect on the rational of its conscio, such a brain can begin a process of rejecting its conscio.

For example, a conscio over-stressing its brain and body (by placing on them unreasonable physical and, or, mental demands) can end up being forced to experience a 'nervous breakdown'. This is currently classified as a psychiatric disorder, but actually it is a situation when a brain refuses to perform its usual tasks for its conscio, while reprimanding its conscio with guilt and various fear anxieties.

Similarly, a brain of an overly selfish (but not necessarily greedy) conscio can cast a feeling of rejection and self-dejection – classified as 'depression' – upon its conscio, even if that conscio's selfishness is based on a high concern for others rather than just for itself. A brain in dispute with its conscio over conscio's strong wants and desires (no matter how selfish or unselfish those wants and desires may be), does not reward its conscio – or itself – with happiness. As far as human brains are concerned, all high levels of selfishness are to be avoided and punished, even if there are moments of temporary elation for a selfish and a competitive conscio, please with itself for getting its own way.

There are other more drastic measures that some brains choose to use against their badly behaving conscios. These can be amnesia and an implementation of multi-conscios (split personalities and multi-personalities).

A human brain, big as it is, can be harmed or damaged very easily. It can be affected by disease or illness, damaged in an accident, contused, shocked, or mentally over-stressed by its own, or another overbearing and insensitive conscio. Under such trying conditions a brain may spasm, going into shock, and reducing immediately the level of its conscio activation support. In such circumstances a brain may continue to support its conscio but deprive it of past-memory from an inability to do so. Such condition is known as 'amnesia'.

In other instances a brain may choose to develop a secondary conscio. This can occur when a brain is ill, or damaged, or becomes dissatisfied with its conscio's disregard of its (brain's) influence; especially when a conscio is not using its negotiator, but constantly overrides its brain, forcing it to be implicated in conscio's selfish acts, which conscio chooses to commit.

While a brain may be required to devise and launch a new conscio due to its experience of sudden trauma or stress, be it physical or mental, a brain may also do this gradually, developing a new conscio over many years, as a version it would prefer to live with.

Whatever the actions may be to cause a brain to launch a substitute conscio, the original conscio is not discarded, but is kept dormant by that brain in mental 'storage'. And in order to alleviate itself of unnecessary burden of maintaining two memory-recalls, one for active and one for dormant conscio, a brain usually concentrates on the active conscio, even if that conscio is the secondary or replacement conscio.

Should a brain be sufficiently recovered to restart maintenance activity of its original conscio, it may deactivate the replacement conscio altogether. The re-launched original conscio may or may not be provided with information of what had taken place during the period of its dormancy. It all depends on the brain's ability, or its intention, to facilitate its conscio with the missing events. Should the brain decide to restore the memory of events, it is possible that it fulfills its original conscio's demands for missing memory by replacing those missing memories with some of its own consciousness memories.

It is possible that a brain may transfer some characteristics of the new conscio – or even from its own consciousness – to its original conscio. This inevitably confuses the original conscio, as it will not understand from where the new (to it) behavioral characteristics, or new abilities, have come from.

A human brain in conflict with its conscio may, for many reasons, become compelled to implement another conscio, while maintaining an original conscio. It may do this with both conscios being consciously aware of each other, or having no idea that the other exists. Or, a brain may structure its dual conscios, where one knows of the other, while the other will have no knowledge of the former. Using such a multiple (split) personalities, a brain may switch its maintenance from one conscio to another. That is, while one conscio is active the other one remains dormant, either aware or unaware of what the active conscio is doing. Often, one of the conscio personalities will be more permanent, irrespective of whether it is more dominant as a personality or not, while the second conscio remains infrequent, brought about by certain conditions and stimulations that may, occasionally, affect that brain.

A brain may intentionally develop a secondary conscio with full consent from the original conscio. This may occur, for instance, when an actor develops a character for a performing role. The intensity of the actor's efforts may be such that the devised character (secondary conscio to the actor's own conscio) may actually become temporarily the primary conscio, taking over from the actor's original conscio during the actor's performance, and, possibly, from then on infiltrate that actor's daily life in unguarded (output) moments.

A human brain does not have to limit the number of conscio's it can develop and maintain. However, it must be understood that implementation of replacement and secondary conscios indicate an overstressed, ill, or a deteriorating brain. By launching and maintaining more than one conscio, an unhealthy brain places an enormous burden upon itself, in the process quickly exhausting itself with such a massive undertaking. The longer a brain continues to maintain multi-conscios, the more certain is its activity towards a total and irreparable collapse and shutdown. In such circumstances such a brain may become unable to retain full control even of its own functions, allowing for uncontrolled and rapid activations and deactivations of its conscios. Currently, humans with such malfunctioning brains are classified as being mentally ill, or more directly, as being 'mad'.

There are other occasions when a damaged, or traumatised, or shocked brain may not only have difficulty of maintaining its conscio but also itself. In such situations a brain will attempt to heal itself while allowing its body to remain passive, and its conscio either completely or partially disabled. This is known as 'coma'.

When a brain in coma manages to provide some level of support to its conscio, such conscio may have some ability to hear, understand, and even remember what is taking place around its body. But because that brain is incapable of giving function to its body, its conscio shall have no ability to do anything about its situation.

PART 1. WHY HUMANS ARE HUMAN?

An old human brain, or one suffering from a deteriorating disease, may keep failing to provide a constant support for its conscio. In the periods when a conscio is active, that human will have the ability of conducting lucid communication, and its normal behavior. When that conscio is not maintained, then that person will remain inactive, presenting a 'vegetative state' of no behavior or emotions, with no self-awareness.

In some instances when a brain cannot successfully maintain its conscio, the mental signals and information between it (brain) and its conscio become distorted and disarranged, resulting in that person being incoherent, hallucinating, and speaking gibberish.

A brain without a conscio can often continue to maintain its body functions; living like that for many years, as long as it receives its nourishment by artificial means, and as long as the heart continues to operate.

The heart, as a muscle, may be induced to function by artificial stimulants and mechanical aids, while its brain lives. However, no body can remain alive once its brain dies. The proof of this is that no amount of artificial stimulation of the heart will keep alive a body with a decapitated head. Once a brain is physically dead, both its heart and its body cease to function, irreversibly, because a dead brain instantly loses its memory of how to operate. So even if it were possible to resurrect a dead brain, that brain would have no knowledge of how to perform any of its activities of maintaining not just its former secondary consciousness of conscio, but even its own primary consciousness.

All human brains are very delicate and sensitive beings. They had designed themselves and their bodies to function in united harmony, where, with assistance of conscios, the bodies were supposed to be fed in order to feed the brains with clean oxygen-rich blood.

While the brains are aware of, and are responsible for their body-produced chemical stimulants, such as adrenaline, they had never envisioned that foreign artificial chemicals would be forced upon them, by their conscios. But that is exactly what human conscios have been doing. Every kind of drug that we, conscios, discovered or invented (using our brains' abilities to do that), such as alcohol depressants, barbiturate stimulants, hallucinogens and other kinds of narcotics, we have been forcing these chemicals upon our brains. We do this because with use of drugs we, conscios, had discovered more ways of controlling our brains for selfish reasons.

There are inevitably responses that occur when drugs are induced upon the brains. To begin with, there are brain malfunctions that may cause distorted vision, surreal sensations, hallucinations, and even distorted mental projections, such as out-of-body experiences. Then there are the responsive feelings of exhilaration and euphoria with boundless energy, or feelings of tranquility. (Or apathy – often changing to paranoia with extended drug usage.) But these are just the kind of mental effects and sensations that we, conscios, often enjoy.

What we, conscios, really like about drugs is that a drugged brain loses its ability to instigate the use of its influence of fear and conscience. When our brains are affected by drugs, we feel free of guilt; protected from rationalism of reality; liberated from restrictions of doubts. Having a release from fear-enforcing restrictions over ourselves allows us, conscios, to be rid of our inhibitions and fears – even if temporarily. We no longer need to be restrained by enforcements of common sense, caution, and conscience directives of our brains' fear codes, which normally nullify our impulsiveness for the ultra-selfish desires and actions. With drugs, all these freedoms are possible to us, the underpowered conscios, while being artificially empowered to presume an ability to achieve anything – including that of our own destiny.

Addiction, therefore, is our repetitive attempts of fulfilling our selfish desire to re-live our first-occasion experience of enjoying the sensation of an instant gratification within our body, which may include or comprise of mental stimulation. But whatever the source of addiction, a brain does not condone it by contesting it.

Unlike what is currently presumed, brains under attack from a drug do not crave the influence of the drug. Instead, they are trying to cleanse themselves and their bodies from the effects of the drug, while being forced by the drug's chemicals to produce functions and emotions for us, its conscio, against their (brains') intent. But despite that the brains may be attacked, over and over, by multitudes of foreign-to-body drug chemicals – or even body-produced chemicals, such as adrenaline – while we, conscios, can become addicted to the effects of drugs the brains do not. Brains can only become accustomed to the repetition of resisting drugs.

Depending on individual conscio's development, including its various fear codes, its brain's fear influences, and its self-developed likes and dislikes (based on fears), all these influences have some bearing on a conscio's self-control levels. And yet, for all these influences with which brains endeavor to control their conscios can often have no effect on highly selfish conscios. This high-leveled selfishness may, but does not have to comprise of greed for wealth and power. It may simply be that of obstinate selfishness based on ignorance and blatant disregard of physical reality – even in the face of physical proof. Such ill-informed conscios may form a firm commitment to some ludicrous notions

PART 1. WHY HUMANS ARE HUMAN?

or opinions, which neither fears nor reason from its brain – or from anyone else ¬– can alter. Often such conscios choose to maintain this highly selfish disconnect to spite others, because they believe themselves to be persecuted by others for being superior, and because they believe that only 'they' know best what is good for them, and that all other opinions – but their own – are irrelevant. But despite the self-induced obstinacy of such conscios, they are easily swayed to acts of rebellion and those that societies deem as inappropriate and illegal. Not surprisingly it is especially these kinds of conscios that are most likely to develop a weakness for habit-forming stimulations.

The other type of conscios that easily develop a liking for habit-forming stimuli, are those with low self-esteem, who are often exploited by other more resolute and dominant conscios, for their personal controlling and financial advantage.

But irrespective of how an addicted conscio blames anyone but itself for becoming, or being an addict, it is just a ploy for refusing to accept the responsibility for its self-centered desires, which it forces its brain to turn into fulfillments.

In any kind of addiction a brain cannot refuse its conscio's demands for particular actions even if they end up being self-destructive, because a conscio has its brain's permission to override it (the brain), with its (conscio's) commanding override control, and its control of the input and output of information. Unless a conscio decides to stop submitting its brain to influences of drugs, that brain is powerless to do anything but to implement its influences of guilt and fear upon its conscio, once it (the brain) recovers from yet another bout with drugs forced upon it by its conscio. A brain, however, unlike its conscio, does not willingly submit to drugs.

Perhaps the easiest way to explain what happens to a brain that is being constantly subjected to a drug abuse by its conscio is to compare it to a pugilist in a boxing ring, forced to fight in an unfair and unequal combat against an unrelenting, constantly reinvigorated foe.

The boxer-brain has no ongoing barrier of defense from drug chemicals, sent in by its conscio, apart from its, and its body's chemical-making ability, and its body's cleansing organs. But while it tries to break down and nullify the drugs with production of its own chemicals, it still has to work at maintaining its body and its selfish conscio, while having to provide a feeling of a thrill for its conscio during the process of being beaten-up by the drugs introduced into the body.

As the brain continues to meet fresh doses of foreign drug, it refuses to provide its conscio with the same feeling it gave at the drug's first encounter.

"Go down for the count!" demands conscio, not having received its usual response from the drug intake, "and give me my usual thrill!"
No response from the brain.

"Alright, we'll see about that!" warns conscio, as a larger dose of the drug is administered to the brain.

And so it goes. The brain refuses to submit, by developing an ability to resist – as it can learn to resist even poisons – with the use of its body's limited short-term chemical adaptability and purification processes.

This, however, does not prevent the brain from being worn down, becoming punch-drunk. Gradually, it can hardly perform, having little ability to support its spinal cord and the body's immune system, which, while trying to help their brain in return, are similarly failing in strength. As the brain's ability to sustain its conscio is diminishing, its conscio persists in demanding that it, the brain, submit to larger and larger doses of drugs – and with that, to more damaging beatings – in order to give its conscio the feelings it craves. And still the brain refuses to provide its conscio with a repetition of the original thrilling experience, because, basically, it cannot even if it wanted to.

On occasions when the anticipated foreign drug does not eventuate, the brain and its body suffer the consequences of their own body-produced chemical antidotes. Therefore, when the usual foreign drug is not provided, the brain and body collapse, but continue to produce their own chemicals in the efforts of opposing their expected drug intake – in no way helped by conscio's continued overriding demand for a repeat of previously enjoyed drug experience.

Often, when a foreign drug begins to wear off, the drugged brain resumes its influences of fear and guilt. Under a barrage of guilt and hope from its brain, the drug-addicted conscio may begin to agree to alter its habit. But as that brain begins to cleans itself from the remnants of the drugs, its conscio either has its brain recall a sublime past experiences with the drugs, or has its brain rationalise the reality and difficulty of living a mundane life without the stimulus and escapism provided by the drugs. With such prospects of joyless reality and self-induced perceptions of hopelessness, conscio falters. It overturns its previous decision not to submit itself (its brain) to any further drug attacks. Once again conscio returns to commanding an override of its brain's fear directives, shutting out the brain's messages of guilt, remorse, and hope, as it reverts to use of drugs as means of gaining a temporary escape from reality.

As with all addictions, unless we, the affected conscio, truthfully chooses not to be addicted – which can be done – our brain is doomed, despite of its valiant efforts of fighting our, conscio's, addiction to the very end.

PART 1. WHY HUMANS ARE HUMAN?

All the mental conflicts that take place between us, conscios, and our brains, can be witnessed currently every day, with millions of us, humans, needing some form of medical assistance from depression or many other similar mental disorders. While this may seem disturbing, it is also not so much as unavoidable as it is a natural event, in view of our current lifestyles.

While, historically, our distant past generations of humans – even those who were poor and destitute – were not prone to mental maladies as current generations of humans (conscios) are, because of their limited ability to obtain their 'wants' and 'desires', we, the conscios of today, consider the accomplishment of our 'wants' and 'desires' as being paramount in certainty.

As younger and younger generations of human conscios grow in affluent and technologically advancing circumstances, both they, and their parents, tend to consider any failure to achieve their 'wants' and 'desires' as being not an options of consideration. In our expectations we, as conscios, 'want', and immediately that 'want' must eventuate. But just like any addiction, this is a form of addiction. And while we, conscios, fail to recognise this, our brains do not.

If we, conscios, choose to expect instant gratification with our every 'want', with no tolerance, acceptance and patience associated with failure and disappointment, and with no conscious humility of ourselves, then our brains shall oppose this, administering their mental punishment on us, their ungrateful, addicted conscios.

8

Conscio and the supernatural

Conscios love mysteries because they are curious. Their curiosity stems from 'selfish fear' of missing out on an opportunity to benefit – even if that benefit is only in solving a riddle (actually, having their brain solve a riddle), and from that experiencing selfish pride of being presumably clever. Many conscios in the distant past were intrigued by natural physical phenomena, such as lightning, which they could not explain. But instead of obtaining solutions to mysteries by physical research, logic and reasoning, they simply had their brains invent superstitions, based on their fears.

When humans, for instance, invented the process of forging tools and weapons, they were physically working with physical properties and physical processes. Yet, it seems as if this was too mundane for them. Instead, magic spells were cast either before or after the metal was, as if this would guarantee a good tool or a weapon.

Conscios appeared to mistrust, or fear, straightforward physical accomplishments. It seems that in all their endeavors – even these days – some form of hocus-pocus blessings always seem to be required to accompany the physical process, be that a preparation of a field for sowing, sending out of a fishing fleet, occupation of a newly built house, or a celebration of becoming an adult. For human conscios, everything physical needed, and needs to have a spiritual, superstitious connection.

This kind of dependence on superstitions began with the early Homo sapiens. All objects and animals in nature, and even human functions and events, were allocated their own dedicated spirits. Behind every rock hid a spirit; every tree had a spirit; the cooking fire had a spirit; even the dark corner in a home-cave had its own spirit. And when a newborn took its first lungful of air and burst forth with a cry, a spirit of some entity was presumed to have entered that infant's body.

Later, when human conscios evolved superstitions into religions, the spirit entering a newborn was no longer permitted to be that of nature, such as an eagle, or a wind, or a running deer. They had to be a spirit of a deity or a god, who, supposedly, had given the infant its soul of life, at the end of which that soul would be reclaimed. And so the physicality and spirituality of human existence became attributed to a human soul.

PART 1. WHY HUMANS ARE HUMAN?

But irrespective of how much acceptance there may be of spirits or souls for conscios, there is only one non-visible entity that any human brain supports on its own behalf, and that consists of its conscio consciousness system: a commanding override negotiator, selectively controlling input-output (over its brain).

A human brain can project mental images from its recall, adapting them to what conscio may be expecting. Human brains actually do this constantly. For instance, when the eyelids blink the eyes miss some activity that takes place before them. A brain, however, can fill in the missing action by using past-recall and blend it with future-projection, making its conscio assume that it has seen something that in reality it did not.

In a similar way, under some influence of physical or mental fatigue or strain, or intoxication from legal or illegal drugs, or even chemicals naturally produced by a body, a brain may present its conscio with hallucinations of actions or events taking place, which, in reality, did not happen. Other natural optical distortions (often described as "eyes playing tricks") may cause physical imagery to be distorted, magnified, or refracted; producing an effect of seeing something which is not really there – such as a mirage.

All such mental influences and natural physical phenomena can cause a conscio with an ill body or ill brain, or an agitated, frightened, or an over-exited conscio to 'see' apparitions or other sights that do not actually exist. A brain can also present mental visual fabrications to its conscio, especially if that conscio is anxious to 'see' what it 'wants' to see.

From the first shamans, priests, warlocks, witch doctors – and their female counterparts – humans (conscios) have been prepared to use 'mind-altering' (brain-affecting) drugs to obtain spiritual or religious experiences. The purpose for this had been to induce a so-called 'insight' when providing prophecies, casting spells, or manufacturing magical cures. These various drugs were brewed and concocted from a variety of plants, herbs, and fungi such as mushrooms. Under the influence of such potions, the spiritual males and females would hallucinate and experience whatever their conscios wanted to 'see', including that of presumed visions of future events, or of presumed spirits or of presumed souls. Actually, what all this means is that their drugged brains would project to them more or less what they 'wanted'.

Over thousands of years conscios had become accustomed to the deceptive notion of a personal invisible spirit or a soul residing inside each individual. Conscios (who are a secondary consciousness maintained by their brains) had ironically credited a soul, and still do, with thinking ability – allocating to this a representation of individual human sense of identity. They presumed their souls to be the immortal: pure essence of a living being, protecting the body from evil and bad deeds of other humans under the influence of evil spirits and evil gods. Most humans still believe that souls leave the bodies upon death, to journey to some celestial 'heaven'. A soul, on occasions, is thought to be able to leave its body and observe it from a distance, or to fly or walk around it. Furthermore, while presumed being dead, conscios have claimed to having experienced 'out-of-body' movements of their souls, visiting heaven through a 'tunnel of light' before returning to life. Considering that conscios have no knowledge that their brains in fact maintain them, such claims are erroneous but not surprising.

Of course, there may be human conscios who will suggest that it is possible to consider conscio to be just another name for a soul. The problem with such supposition is that conscio is only launched when a brain has sufficient ability to maintain its conscio, not at birth. The other reasons why a conscio cannot be a representation of soul is because the same brain can launch a secondary conscio, and by that, maintain multiple personalities. Alternatively, a human brain can lose its ability to maintain a conscio altogether, remaining in a conscio-lacking 'vegetative state' but continuing to function in maintaining its body; living in this manner for many years while being artificially fed and cared for. This means that a human brain maintaining simultaneously more than one conscio, or no conscio at all, can continue to live without having any dependence on any soul.

A conscio, therefore, never was and never can be a soul, just as souls had never physically existed but in imaginations provided for conscios by their brains.

Never having encountered the perception of themselves as conscio systems before in their existence, conscios have underestimated, misunderstood, and degraded their brains – whose mental efforts makes conscio systems possible – to such an extent that they (conscios) had relegated brains to the level of an organ. This now needs to be corrected.

Brains – all brains – are not organs and never had been organs. Brains had devised the organs, and the limbs, and the skeletal structures, and the intestine systems, and the nervous systems – their whole bodies, in fact. They were able to do all this not as organs but as beings in their own right. For all brains of animals are brain bodies: the brains that design and produce their bodies as they – the brains – see fit.

PART 1. WHY HUMANS ARE HUMAN?

In the case of every human brain: each one is capable of the most extraordinary and amazing feats of mental projection on behalf of its conscio, which is the resultant imagination that a conscio experiences. But when necessary it can produce even more amazing feats of imagination and mental visual projection for itself.

When a human brain suffers trauma from exhaustion, concussion, or drug influences, it immediately powers-down its conscio. If the level of reduction of its conscio support is drastic, then its conscio loses all consciousness – classified as a 'faint', or 'unconsciousness'. During this period conscio is usually given no memory support and no ability to make requests of its brain, as it, as a conscio, is virtually non-existent.

If, however, the drastically reduced level of conscio support is such that a conscio still has some consciousness, then that conscio may be semi-conscious, with the brain still providing it with recalls and mental projections. In such circumstances its behaviour can be classified as 'comatose'. Alternatively, it may be 'delirious', when a body is capable of movement, while its brain sporadically continues to sustain its conscio, but unable to present it with an orderly structure of thoughts.

During such a period of low conscio support by a brain – while its semi-conscious conscio helplessly fluctuates between existence and non-existence: reality and non-reality – that brain may choose to conduct a physical examination of its body's exterior surface.

It does this by compiling a mental visual of its body's surface by collecting internal nerve-ending pulses to its skin and forming from this a mental external model of its body. In other words, by mapping from the inside its body surface, a human brain can produce for itself a three-dimensional, external, mental image of its body.

By such method of mental examination and projection of its external surface, a human brain can examine the surface of its body, assessing from that external body examination any signs of damage sustained by its body, which could give clues to the possible internal injuries, as in case of a bad accident where bones are broken. When such external image of its body is glimpsed by a semi-conscious conscio, that conscio, in its powered-down reality-non-reality state, may 'feel' itself outside of its own body, floating over its body, not realizing that what it sees is the process of its brain examining the body. Such experiences can be compared to a space astronaut exiting his space capsule and doing a 'space walk' of floating around it to check for any sustained damage to the space capsule – except that a brain does this mentally.

Should the undertaking of such mental action by the brain be glimpsed by its semi-conscious conscio, that conscio – once it comes to its full senses (recovers its full support from its brain) – will have its brain enhance the memory of this mental experience with imagination. The imagination inevitably redefines the whole experience into an overdramatized and supernatural event, something that is attributed not to the brain but to an invisible and indefinable presence of the soul.

This manner of a brain's mental examination of its body surface is not unusual, occasionally making it possible for its conscio to mentally 'see' the body laying ill, or hurt, or wounded, or running, or fighting, or straining in some desperate physical act. This can occur even to sports participants, when they, as conscios, may 'see' from an elevated distance their own body in action, because they had selfishly over-stressing their bodies and their brains: which has nothing to do with any spiritual or soul connections.

Thanks to its brain's powerful imagination ability, a conscio in such situation can have its brain project a thought, "Oh, look at that! There is my body, and I'm floating over myself! I must be the soul released from my body, over which I'm now hovering! If my body dies I'll fly to heaven, otherwise I'll return to my body."

When an end-of-life moment arrives for human conscios, those conscios in expectation of it – due to a prolonged or terminal illness, or old age – are usually composed in a peaceful resolve of their impending fate, to which their brains respond with no presentations of fear. A totally different behaviour may result for a conscio, if it, or its body are subjected to end-of-life moments under traumatic conditions.

Trauma may occur not just at a scene of an accident, but also originate from a frantic activity that can take place when the nursing staff are attempting to resuscitate a dying patient. In experiencing a sudden, unexpected deceleration of body's life functions, combined with pain, a conscio unable to comprehend what is happening to it panics its brain into action of establishing what is taking place.

As a body's life signs are slowing down, that brain is powering-down its conscio because that brain is also becoming weak, and, therefore, needs to concentrate all its effort to maintain its own functions. Despite such circumstances, a weakening brain may still try to output to its fading conscio a stream of various mental phrases and images, virtually unfolding all of its memory syntax, in efforts to ascertain what is happening to it and its body.

In doing so, a brain may present, at incredible speed, three-dimensional images of familiar, unfamiliar, and even invented humans; various experiences;

PART 1. WHY HUMANS ARE HUMAN?

possibly project imaginary scenes from its memory of conscio's fears, beliefs, and indoctrinations, which a conscio may remember 'seeing', should its brain and the body recover.

While a brain continues to remain alive and support its semi-conscious conscio – with that semi-conscious conscio continuing to make demands of its brain – the conscio may be subjected to an 'out-of-body' experience, in the brain's efforts to view the exterior surface of its body from the inside.

It may also have an 'after-death' experience, when, while presumed dead, a brain may in fact still be alive, but dying. The dying brain may just be managing to maintain its conscio, but can no longer send it any images of communication. From this the conscio may still be aware of a round light surrounded by blackness, which is in fact a visual representation of it – the conscio – as a conscio system.

If that brain and its body are revived, a semi-conscious conscio may once again begin to receive mental phrases and images from its brain, in which the conscio may 'see' a bright light and some recognizable humans from its past. From some of these images that a brain presents to its conscio, that conscio may interpret as being representations of gods, and any other fictitious spiritual characters introduced to conscio in life.

However, these mental depictions of gods and the like, presented to it by its brain, will never be foreign-looking to that conscio, but in images it is accustomed to knowing, or at least being familiar with. For instance, a conscio indoctrinated by European Christian religious customs will not 'see' any black African gods and spiritual representatives; or Chinese gods; or Indian gods – only the images of gods it is accustomed to accepting as likeness of gods: those with pale skins, rosy cheeks, and long golden or brown hair with beards. Likewise, conscios of other races may also have a so-called 'after-death' experience, in which familiar images of gods of their race and colour may appear, but not those that are foreign-looking to them, or in the shape of beings that are devised by imagination for science fiction.

Should a brain or its heart fail to be revived, then the bright light that a semi-unconscious conscio perceives, fades and goes out. At this point conscio stops existing as its brain dies.

Once a human brain is dead, it cannot be revived. Even if it could be restarted, as depicted in tales of science fiction, or by a presumed method of reversing cryogenics (a freezing process to prevent decomposition), a brain would simply be blank. Once dead, nothing can return to it its unique ability of developing, producing, and maintaining its conscio. Once dead, the memory functions cease to exist and cannot be returned, as memory is not stored in some physical compartment of a brain. Once dead, no human brain can resurrect its former conscio, or launch a new one.

Neither the consciousness of 'self' or 'I', nor their presumptive souls that were to have been their controlling life spirits, were ever a sufficient reasons for human conscios to be absolutely grateful for having life. What they 'wanted' – and continue to do so with their 'selfish fears' and greed – is an extension to life in a form of a passage to a life-beyond-death. Because of their highly selfish and disrespectful behaviour to each other, human conscios have felt that life was often unfair and short, so they looked for ways of extending their lives indefinitely.

Conscios came up with a solution: they had their brains invent 'life after death'. This notion became very appealing to most conscios. Whether based on complexity of reincarnation, or simply based on a notion of 'going to heaven', human conscios had their brains devise for them guarantees of an afterlife. By such an invention they came to believe that they can transcend their existence on this planet (in which they produce futile waste with little concern from supposedly there being something far better for them to look forward to after death) into an eternal life full of never-ending happiness and plenty.

Sadly for all conscios who override their brains to think this way, the fact is: one life is all there is per living individual human – as with every other life form – followed by no other conscious life after death. Human death does not open an imaginary doorway, or stairway, to another existence in Heaven or Hell. The physical process of change of every human leads to an end of life moment, which is irreversible and final: at the instant of which the conscio is permanently terminated, while the physical brain, with its body, are influenced to disperse. This occurs by means of cremation, or rotting (decomposing), or being consumed by other life forms, such as maggots, worms, and bacteria.

Therefore, at the end of life there are no further physical options of extending that life. Not even by fantasy derived of human brains' imagination.

9

Human dimensions of thinking

Every increment in reduction of selfishness and bias in the life of a young conscio results in a gain of a substantial improvement in thinking from its brain. This happens because when a conscio lessens its bias with less forceful commanding overrides and more negotiations with its brain, the more comfortable and trusting its brain becomes. And from that: the more affable and agreeable it becomes at providing more important output for its conscio. Such attitude of co-operative harmony between a conscio and its brain results in a development of new perceptions for conscio. It is at this juncture that conscio and its brain begin, in accord, a process of three-dimensional thinking.

One of the fallacies still accepted as being a reality by human conscios, is their concept of 'dimensions'. They take it as factual that there are multiples of dimensions, be it two-dimensional surface without depth, or fourth of fifth or sixth or whatever dimensions.

The fact is that in all physical existence, all physical bodies, no matter where they may be, are completely, totally, and absolutely three-dimensional. That is because in all physical existence there are no entities that are not physical. And therefore, being physical, they all have external surfaces or facings, which – were it physically possible, due to limitations posed by distance and size, etc. – can be viewed from all surrounding sides. Furthermore, as is the case with most objects and bodies, they have internal contents and vacuum spaces, be it an empty room. That internal content or space is as much three-dimensional as the exterior of anything physical, and not a representation of any other dimension.

In fact there is no third-dimension either, as this concept and terms relating to 'dimensions' is just another human-based accepted convention, which has nothing to do with physical reality. All physical objects and bodies have external surfaces on all surrounding sides. This applies to the smallest particles, to every atom, molecule, cell, every brain and every body – including that of every human– and to every blood cell, every tissue and every organ within that body. In retaining the term of 'three-dimensions' has merely been used in this book as convenience from having to devise another term for every shape with a surrounding surface on all sides.

What currently is not known – or even suspected – is that each and every brain of every species is capable of three-dimensional thinking, but only one species of brains can communicate that thinking: us, human beings. This means that had human brains the need to express their three-dimensional thinking, they could, unlike all other brains.

Three-dimensional thinking is quite different to three-dimensional projection that every human brain constantly provides for its conscio. To explain this, consider that every object and body on Earth is three-dimensional. Therefore, every input that our brain receives and stores to memory is based on physical three-dimensionality, as even the so-called abstract notions and ideas are all in some way relate to physical imagery.

For instance, every mathematical equation is based on three-dimensionality because all numbers and symbols represent a convention of accepted values, which are used to give three-dimensional substance abstract notions. Another example could be that of a person (conscio) proclaiming to 'want' world peace. This want may seem like an abstract concept, but for that person and all those who may consider such a concept, they all would have their individual three-dimensional imagery that represents or symbolizes to them the notion of 'world peace'.

In other words, abstract thought cannot exist without being related to perceptions of three-dimensional physical entities, whatever they may be. This can be proven by the fact that it is impossible to share with a brain any abstract concept **without** it containing actions relating to three-dimensional objects and bodies familiar to the brain – those that it can understand and relate to. Just as without knowing a particular foreign language no one can instantly communicate fluently in that language, and be able to, for instance, write in that (foreign to them) language.

So as it is, every human brain gathers three-dimensional information from everything that surrounds its body, and converts this into compressed mental depictions, which are stored in memory. This includes conscio 'wants' or 'desires', which are all three-dimensional, from an ice cream to a kingdom (comprising of physical three-dimensional objects, structures, systems and bodies). Then, when a brain needs to release information as its output, the compressed mental imagery is restored to three-dimensional imagery, be that even as an abstract notion, so that it has a mental three-dimensional shape to it.

Even if a brain is projecting a memory image of a flat photograph, that photograph will appear as being three-dimensional. A brain displays the so-called dreams as active, three-dimensional physical events, including three-dimensional backgrounds, with lighting, sounds, actions, and when applicable,

PART 1. WHY HUMANS ARE HUMAN?

even with music. In all its mental output the presented mental information contains three-dimensional space, within which bodies are presented as being three-dimensional (and not as flat images) capable of walking, running, and even flying.

Where every brain's mental three-dimensional projection is most astonishing is in its ability to maintain its conscio' consciousness of mental past-present-future projection. With every second of its existence a conscio, which (with its brain and body) spends in the present-moment but barely being aware of this, traverses mentally (thanks to its brain's ability) between the memories of its past and the projections of future. Incredibly all these mental projections (from its brain) are in colour, with sounds, smells, emotional 'feelings', and three-dimensional imagery, so that a conscio's present-moment existence mentally blends into a panoramic mental past-future existence of its life. Indeed, this is how all humans (conscios) experience life: in the mental three-dimensional projections from their brains, irrespective of their present-moment existence.

The purpose of three-dimensional thinking is to totally evaluate required physical actions for a far-reaching and wholly embracing **beneficial** outcome, for all brains directly and indirectly involved.
This is nothing like what conscios constantly demand of their brains. While there is no existence of two-dimensions, because conscio 'thinking' comprises of 'wants' that are all tree-dimensional, be it even for knowledge, this can be considered as being flat: a direct trajectory between a 'want' and the fulfillment of that 'want'. Therefore, the term of 'flat thinking' had been incorporates as a descriptive contrast to three-dimensional thinking – both of which are conducted by human brains. It is this that conscios seek and demand of their brains: flat thinking.

So what is 'flat thinking'? Basically, flat thinking represents immediate and short-term selfish demands for "thinking only of yourself". Aside from all such personal demands of individual conscio's placed upon their brains, wanting all that they do for their own benefit with little regard of how the realisation of their want impacts upon anyone else, government decisions – comprising of consolidated numbers of conscios – instigate similar highly selfish short-sighted wants on behalf of their society.

An example of this would be political members of a government being elected to office on a promise to provide the electorate with particular utility, operated under the government's ownership. Once in office, those governing politicians sell that utility to private enterprise for the government's short-term financial gain, with absolute disregard to any consideration (thought) of reduced quality of service and elevated costs the new owners will inevitably place upon the electorate, just as long as they themselves have no more direct responsibility for the operation of that utility.

The other component of flat thinking is when a 'want' has elements of selfishness but for a selfless purpose. An example of this could be, "saving black babies in Africa". A noble quest indeed, and a challenge for the brains to devise a strategy to benefit infant brains. The problem that often arises with such charitable and selfless intentions is that conscios 'wanting' or 'desiring' this, presume that solutions (mental calculations from their brains) can be achieved swiftly, without prolonged and thorough considerations and calculations. These kinds of short-sighted, instant expectations (or instant gratification) of 'wants' and 'desires' also fall into the category of flat thinking, for presuming that best intentions can be achieved with little or no mental effort. As explained previously, the brains deliver what conscios want: "Want mental rubbish, get mental rubbish."

It is for this reason that most human (conscio) efforts to organise beneficial deeds on a large scale end up being either failures or disappointments.

To understand any three-dimensional thinking requires an appreciation of long-term, mental consideration and projection that encompasses most (if not all) outcomes. An example of this would be the simplistic physics of Isaac Newton's laws of motion, where the third law states:

"For every action there is an equal and opposite reaction".

Quick and easy, with no examinations made of all the complexities involved in any reaction to an action. Flat thinking. However, when the process of a 'reaction to an action' is examined in the physical reality of the three-dimensional physical process of change, and expressed in three-dimensional thinking, this is how it would be defined:

"For every action there are direct and indirect reactions, comprising of multitudes of chain-reactions, which result in further unpredictable outcomes, whose consequences will remain as influences of long-duration in the physical process of change."

Should three-dimensional thinking be used to consider a human action, then the expressed definition would be:

"Considering that for every action there are direct and indirect reactions, comprising of multitudes of chain-reactions, which result in further unpredictable outcomes, whose consequences will remain as influences of long-duration in the physical process of change, therefore: before any physical action is taken, it needs to be assess as to who shall benefit from this and in what manner, as well as, who and what may physically be affected in all the reactions to the intended physical action, so as to prepare contingencies to avoid unforeseen outcomes that may threaten or harm them in some way."

In other words, three-dimensional thinking requires that the consideration and calculation of any intention should comprise of solutions that have beneficial outcomes for direct and indirect participants; intention that can only be fulfilled by the following three factors:

1. What width of scope (length) should consideration and calculation of intention undertake? That is: who and what should be included in the consideration.
2. To what level of thoroughness (height) should consideration and calculation of intention be undertaken?
3. What distance (depth) of duration (in numbers of generations) should consideration and calculation of intention be taken into account?

These parameters indicate that the mental three-dimensional thinking does not consider only the point of fulfillment of a human (conscio) desire, but also assesses whether the sought fulfillment of a desire would also be of mutual fulfillment to other life forms, in the present and in future, including those who will be affected directly and indirectly by the outcomes of that initial desire (intention). If the overall result of the desire (intention) does not meet the criteria of providing a mutual fulfillment to all life forms affected directly and indirectly by the outcome of the fulfillment of that initial desire (intention), then the proposed desire (intention) would be considered as being flawed, with that desire (intention) requiring a re-evaluation.

If, for whatever reason, there is no way of assessing whether the progress of the action taken would be beneficial, or harmful, to those involved with the action of fulfilling a desire (intention), then the three-dimensional thinking establishes remedial plans and contingencies, which, in themselves, comprise of the mental three-dimensional thinking considerations.

To phrase this in another way: three-dimensional thinking provides non-sentimental calculations intended to **care** about all the long-term outcomes from intended action, unlike the flat thinking that is only concerned with sentimental short-term solution – with no consideration or care given to long-term repercussions derived from the initial action.

Currently, humans teach each other only flat thinking, where even an unselfish 'want' or 'desire' to provide an unselfish fulfillment often results in flawed and even harmful outcomes.

Take an example of scientists' flat thinking in attempting to provide solutions to national weed and pest problems, by introducing foreign plants and pests, which then become even bigger threat to ecology and environment then any previous weeds or pests. These scientists never gave any thorough and long-term thought or calculation to what they were doing.

Another example can be made of politicians and their generals invading a nation ruled by a dictator, with claims of liberating that country. Consequently, due to their flat thinking, these supposedly well-meaning politicians and generals are forced – many years later – to abandon their intent and return to their own nation, which they had filled with their war dead and financially bankrupted, while leaving behind a nation they invaded devastated and in ruins, and still incapable of preventing an elevation of yet another dictator.

Similarly, when raising charitable funds to aid the starving of some nations, no thought is given to the facts that the distributed funds often fail to reach those in need, and even when they do, those in need become needy again as soon as that provision of charity runs out.

Instead of providing all-embracing three-dimensional solutions to problems, flat thinking attempts to remedy problems with actions that have but one desire: to have those making sentimental efforts of helping others be recognised for doing so, for then they can say with self-pride, "At least we tried to help those less fortunate than us!" This, in reality, is not very helpful as it is merely a waste of efforts and resources.

These assessments may sound harsh, but the fact remains that futile helpful gestures of 'wants' or 'desires' made on behalf of those in need may appear as meaningful solutions (fulfillments) being implemented to solve and eradicate social or economic problems, whereas, this usual flat thinking of throwing a limited amount of money and aid at a problem is futile, often resulting in unchanged or even worsened conditions when no more efforts are made from lack of funds, ability, will, or interest, while increasing doubt and lack of confidence. By refusing or failing to address the very core of any problem with mental three-dimensional thinking, any selfless efforts end up being little more than a sham.

It may well be presumed that three-dimensional thinking is an impossibility. Not so. Just take a look about you at all that lives on this planet and recognise three-dimensional thinking for what it is: evolution of life.

Three-dimensional thinking is part of every body-brain and brain-body. For how else would all these brains develop and implement body changes to

PART 1. WHY HUMANS ARE HUMAN?

themselves, over long periods, and foresee in advance that the body changes they are devising should coincide with the changes in the environment? By merely asking themselves that: "I want a body change!" and presto, it all is accomplished? Not likely. No brain can achieve a body adaptation without three-dimensional thinking: first it is the tree-dimensional thinking, and only then the calculations on how to achieve this takes place.

These brain endeavors are beyond anything that current conscios perceive as advanced technology. Conscios currently involved in developing genetic engineering and artificial intelligence do so without any application of three-dimensional thinking; their only underlining intentions are to obtain financial wealth and eventually gain control (power) over humanity (human brains). In doing this, no thoroughly exhaustive thought, calculation, analysis and consideration had ever been made by them to assess the potential of future environment for such applications, something that all brains do for their body adaptations. So it is this shortcoming that separates human conscios from their human brains: while conscios are driving their brains ahead, they do this blind and therefore in ignorance, having no idea – and in reality, no real intention – of navigating their species to a far reaching longevity, as is the intention of all brains with their three-dimensional thinking.

But even if conscios understood and appreciated the significance of three-dimensional thinking, that ability could not be based on a flippant presumption that three-dimensional thinking is there for the asking. To receive mental three-dimensional thinking from human brains would necessitate conscios to teach their brains to trust their (conscio) intentions, and for conscios to accept from their brains' advice that goes beyond mere fulfillment of wants and desires. It would be necessary to give human brains unreserved permission to make detailed examinations of any intended action to a problem with a three-dimensional scope, requiring a lengthy duration of efforts.

No doubt most humans (conscios) would not acknowledge, or agree, that their flat thinking results in futility and waste, because most of them have always held expedient views of what this planet means to them. They have taught each other to accept selfishness as their right to dominate all other life forms and to consume as much of everything that they can, as quickly as possible, because such avarice is supposed to be healthy for business and economy.

Even in the process of teaching thinking to their young with techniques such as so-called, 'lateral thinking', humans (conscios) insist on inferring that in solving problems it is advisable – if not necessary – to apply selfishness, skewed logic, fantasy, and disregard of physical reality.

Having become so used to their flat thinking, humans (conscios) obviously find it much easier to relate to flat planes, instead of a space with no dimensions. Despite the reality of existing on a small spherical planet, they are only conscious of themselves living and moving on flat ground. They had nominated North Pole as being 'up', despite that from ancient periods of human existence most humans lived, and live, with an awareness of 'up' being 'sunrise' and 'down' being 'sunset'. Or rather, as it should be more correctly known as 'suneast' and 'sunwest', as the Sun neither raises nor sets.

Will humans (conscios) ever alter their current form of thinking? Probably not. Even so, should conscios ever want to obtain from their brains an ability of receiving not just three-dimensional projections but also mental three-dimensional thinking (the true genius level of intelligence), then the only requirement is that they stop ignoring their brains.

10

Conscio fear of living with change

On Earth, as elsewhere in the Universe, no physical entity remains the same forever: sooner or later everything physical experiences physical change. The reason for that is because everything physical comes together in unions before eventually separating in dispersals. This applies to everything in physical existence, as shall be explained in Part 3. Therefore, by moving, vibrating, merging, uniting, colliding and inevitably separating in dispersals, everything physical ends up producing physical change that can be observed and recorded.

This means that the physical process of change – or change – is but a result of all the physical occurrences taking place between, and with, all various physical objects and bodies that experience change, and in doing so, are directly responsible for all the further simultaneously ongoing change everywhere. As such, physical change is unstoppable and irreversible, with no ability to contemplate, reason, or care about what results from physical contacts between anything physical.

This is not how we, conscios, view change. We prefer to consider change as being in a form of events, some clearly visible, such as seasons of the year, and others not so, like birthdays, which are understood to represent aging (change), but which are not distinctly noticeable between one year and the next.

We also categorize events (change) into those we like and others that we do not, where even these likes and dislikes can become changeable. For example, a birthday is liked when young, as this event represents a gay occasion of celebrating with young friends, sweet foods and gifts of presents; whereas a birthday is disliked in old age, for it then represents a reminder of approaching end of life.

Furthermore, we, conscios, fail to view change as a physical process simultaneously taking place everywhere, due to physical actions and reactions occurring between all physical objects and bodies. Instead, we much rather believe in some kind of metaphysical godly entity that supposedly can alter circumstances at a whim known only to itself. This is often expressed as being an 'act of god' (this term still being used by most insurance companies), which is considered to be responsible for all the changes taking place in nature and affecting all human outcomes.

Fate (like god) is also accepted as a metaphysical entity, enveloping everything in existence, and delivering pre-ordained outcomes to human lives, by that instigating change "from what could be to that which shall be".

Then there is the notion of 'time', which continues to be acknowledged by all humans, including those in science, as a real entity, capable of producing direct alteration to all physical actions and bodies as a controlling agent of aging, and which, presumably, can be travelled back and forth.

Therefore, according to the current human conscio perceptions, change can come from many sources, be it god, fate, time, and even 'lady luck'. This means that with so many instigators of change conscios have a choice of whom to blame for events that are not to their benefit, without having to either blame themselves or accept an understanding that most human change is actually the result of human (conscio) interactive activities comprising of actions and reactions.

One day they are healthy, wealthy, and gainfully employed; next... it is all changed: the health falters; wealth stolen or spent; and those in employment not being needed anymore by their employers due to changes in the economy. Who then to blame for this? God? Fate? Time? Luck, or rather bad luck? So the facts that health is a human condition, just as all other events are human conditions, experienced – for better or worse – by those who are directly or even indirectly involved, is either avoided or ignored.

Still, human (conscio) reasoning demands that someone or something has to be at fault for anything that eventuates not in their favour, because then they can seek revenge or retribution on those, or that which shattered their expectation.

There is another factor that conscios fail to take into an account when dealing with change: the instant moment of change. Conscios mostly view change in form of clearly apparent events that take place around them. What they fail to notice is that before anything becomes apparent, the process of building up to that outcome would have been ongoing for a long duration, but in such minute increments that it was not perceived to be happening.

Take an earthquake, for example; an event that is impossible to ignore. The ground tremor seems to come out of the blue and for no apparent reason, despite that the sudden release of energy is the result of seismic activity taking place for a long period in minute increments of distance and duration, out of sight and sound.

Exactly the same correlation can be applied to all events, even those of human activities. The actual reason for such failing of awareness concerns the manner in which human brains present perceptions to their conscios.

PART 1. WHY HUMANS ARE HUMAN?

In order to allow a conscio to be aware of not just its surrounding physical space, but also its position in the constantly changing process of its body's and it's brain's life, the brain constantly presents to its conscio mental images of past recalls and future projections. This allows a conscio to physically navigate Earth's surface but also to mentally navigate its position in life. That is: a conscio can consciously (or not) use its past to plan and take steps to achieve its intentions for its future.

But while the mental past-future projections mentally oscillate between past results and future expectations, a conscio often fails to notice that all change actually occurs only in the instant of the present-moment. It is this oversight to consciously register the instant of the present-moment that result in the difficulty for conscios to have a realistic relationship with physical change, which, in turn, becomes the reason why conscios have a difficulty with their own existence.

By being more conscious of the present-moment would allow conscios to realise that as everything occurs incrementally in the instant of the present moment (even though event take a long duration of such minute incremental change), then no other entity – such as god or fate – can possibly be involved with the process of change. Only the physical actions of physical objects and bodies can produce physical change.

To alleviate the monotony and the sameness of daily life – something that all other animals live with – human conscios had their brains devise means of utilising change, in the form of various entertainments. These activities include all kinds of pageantry, theatrical and gymnastic amusements, gaming and gambling, all of which involve some activity of 'action' (change) taking place.

Indeed, many conscios are enthralled and excited with entertainment that involves deeds and events of violence and risk, where change represents a chance of a human becoming hurt or killed. They are enraptured with change comprising of rapid speed, by which the participants (both humans and animals) intentionally increase the level of risk to their bodies, in their never ending competitions of horse racing, bike racing, car racing, or running, jumping, swimming, and multitude of other disciplines.

In response to such demands for escapism in human lives, based on entertainment, many enterprising humans (conscios) have established vast industries to provide outlets for such conscio desires. In exchange for monetary payments these change-for-sale industries provide arousals in all kinds of racing, computer games, gambling (including stock exchanges) and sporting events. But regardless of entertainment value of these activities, there is always the downside to those who become addicted to some forms of entertainment, and end up experiencing the fearful consequences of these pursuits.

Entertainment aside, there is little for conscios to be fond of, or appreciate about change. It seems to them that just as they become comfortable in their lives along comes change to disrupt everything. So unlike all other species adapting to change, conscios are prepared to control change instead, by doing all they can to avoid it.

One of their most disliked and feared attributes of change is having their opinions challenged by other opinions and knowledge. This is because new knowledge and opinion can bring about change that can threaten everything in their life: be it their personal position, their perceptions, their customs, their faith, their living standards and lifestyle: all established on opinions. After all, opinions dictate all the values in human lives.

Opinions are seldom based on physical reality. Instead they are mostly formed from a human (conscio) dictate – usually that of someone in control of other humans – informing others as to what level of value anything is to be. To give an example of this, when the Spanish conquistadors arrived in South America, the native Incas were incredulous at the Spaniards' lust for gold. Of course gold was admired as a pretty metal by Incas, which could be fashioned into decorative objects, especially for religious ceremonies. But it was not something to be valued above staple food.

Such example confirms how the value of gold is dependent on an opinion and nothing else. This applies to all values in human lives that opinions uphold, including that of knowledge.

For instance, when humans (conscios) are presented with new concepts that have some familiarity to the current notions, they at least make an effort to listen, even if mentally they may have already dismissed the new information. But when presented with innovative thinking, new ideas, or new ideals that challenge existing notions, the rejection is usually made with no attempts to examine the evidence or even listen to the logic. Even if that conscio of the evaluator acknowledges mentally (by its brain's efforts) that the presented new information is sound, it is seldom that this acknowledgement is made public at that point.

(This dismissive attitude to new understanding is contrary to the constant human (conscio) assertions of always welcoming and seeking new knowledge.)

The reason for such dismissive attitude to that which is new, especially to new knowledge, is similar to that of a new opinion: being based on fear that

PART 1. WHY HUMANS ARE HUMAN?

this shall alter and change that which currently stands as a accepted perception of knowledge representing social order, customs and beliefs.

Because conscios are selfish, all of their opinions are based on concepts of selfish endeavors. While some religions and philosophies may preach selflessness, in reality of their lives this is never taken seriously. As far as most conscios are concerned, it is "every man for himself'" in a "dog eat dog world." Therefore anyone in position of having a power of opinion is not about to relinquish voluntarily their position to someone else, even if that new opinion comprises of real knowledge and not a philosophical or religious fabrication.

This fear of "losing what one has" drives humans (conscios) to reject anything that may change their current perceptions (opinions) regarding their existence, no matter how much their understanding, opinions and beliefs remain erroneous and obsolete.

There is no doubt that due to their 'selfish fears' of fearing change, many conscios shall reject that they are in fact conscios, a secondary consciousness maintained by their brains, instead of being a direct representation of a brain, as all other animals are. Or else they shall continue to believe themselves to be a soul, or a spirit from god inhabiting a human body. They would much rather remain being that which they 'want' be, as this would presumably spare them from any need to accept the reasons for their constant overly selfish behaviour. After all, they have been so used to their selfish way of life that they have no wish to alter (change) their behaviour, so as to re-consider what they have been doing to themselves and this planet: their only home – despite all the nagging from their conscience, presented by their brains, warning them of impending catastrophe.

And why? Because they would fear any change occurring to their lives from any undertaken efforts to curb (change) their blindly selfish behaviour.

There is a paradox to conscio reasoning (as requested from their brains): on one hand they allocate the responsibility of change to gods and the like (instead of the physical process of change), and on the other, they hold a presumption of being masters of their own destiny. They do this with false faith that they possess the intelligence to control any change: an ability that, supposedly, shall provide them with an endless life-passage into future. And the way they had decided to achieve this was with materialism.

Although materialism is considered to be a new attitude embraced by modern humans, the fact remains that conscios had learnt materialism a long while ago. Personal possessions became a standard expression of materialism, from which conscios developed an understanding of trading. This led them to an activity that has come to rule all societies on this planet: business: a process where a value (based on an opinion) of goods and services are exchanged for a monetary value (also based on an opinion). Now every individual human – from even before birth, to death – is part of multitudes of interconnected business processes.

This activity has come about from selfishly decisive and persistent efforts of individuals (conscios) determined to introduce change from which they could financially benefit. From the very beginning of human civilization this had always been done with assistance of science. Of course the term 'science' did not exist in the lives of earlier human societies. But be it alchemy or inventions for husbandry, farming, warfare and fortifications, it still involved acquisition of knowledge from conducting of experiments of trial and error.

The application of gained practical knowledge (science) has reached a stage now when every business is using some form of scientific application, while all science has become vast business in its own right. These days all business depends on and supports science, while the business-of-science depends on and supports all business.

In the progress of their development and evolvement, the business and science association had altered (changed) humanity's understanding of itself: they changed self-sufficient survivors to that of producer-dependent consumers, by allowed humans (conscios) to have selfish 'wants' beyond their wildest dreams. Now their selfish desires need not have any limits or boundaries, for everything supposedly is available. After all, this is exactly what science and business had been striving to provide.

Due to the way business and science had changed the structures of all societies, converting every human into a consumer, the governments of all nations had also became dependents of the same business and science activities for their economies. But because all nations have similar dependence on business for their budget profitability, this often results in disasters for them when their economies stagnate from advanced technologies dispensing with workforce and lack of overall business activity and profit contributions, resulting in unemployment and recessions.

The primary principle of any business is to supply the demand. The problem with this occurs when growth of demand cannot be maintained. Then the only methods available in solve this situation is to stimulate demand with new affordable products, and to increase the numbers of consumers. This is where science comes in, to facilitate such requirements by devising new products and making it possible for more consumers to live longer. This then becomes a

PART 1. WHY HUMANS ARE HUMAN?

perpetuating science-business cycle of breeding more human consumers and have them living longer – so that their period of consumption extends – and simultaneously devise methods of keeping growing populations fed with larger quantities of artificially produced food.

There are also other human needs that require business involvement, such as: housing, transportation, education, health and employment. So, in fact, more and more of everything is needed to facilitate the lives of more and more humans (conscios).

This would be all well and good for all concerned, if not for existing limitations and the inevitable repercussions of such economic expansion cycle. One of the limitations is that of market over-saturation, when consumers have more than enough, irrespective of any lowering of price. There is also the limitation of available resources, so that when they are all used up there is no way of replacing them.

There are also the repercussions of all the economic activity in a form of expanding human populations that require more living space and more agricultural land, with these factors imposing a loss of land for wild species. After that there are the problems of vast quantities of waste, pollution, and most devastatingly for humanity the resulting global warming caused by release of carbons into the environment and the atmosphere.

Of course, as always, those in science present themselves as being capable of facing such challenge, which is ironic: From the start, he involvement of science in human materialism has been directly responsible for global warming, and now science is presenting itself as being capable of alleviating the same global warming. Or, put in another way: those responsible for a detrimental change now proposing to revert that change.

Sadly for humanity, once the environment is altered (changed) as it is now, it shall, most probably, continue to deteriorate.

Even if it were still possible to influence global warming with drastic measures of reducing carbon emissions, the leaders of most nations on Earth are fearful of doing so, from concern (fear) that this shall impact badly on their economies: the very same economies that are currently stagnating from the mismanagement caused by avarice and greed of all controlling individuals, including those involved in financial markets.

Instead of all nations defectively and ineffectively chasing economical growth at all costs, many more countries should have been prudently managing their environments, managing their finite natural resources, and managing the growth of their populations.

Having more longer-living populations may have, temporarily, been good for the business, but the continuing uncontrolled population expansion – in no small way accomplished by science – is the real cause of global warming. Science had made it possible for business to use up more of everything in order to facilitate the needs and wants of human expansion, producing in the process so much carbon emissions that nature's processes of storing carbon could not, and cannot cope.

And still more and more carbon continues to be released by industry and human consumers chasing their materialistic ambitions. After all, if all nations had structured their societies on principles of greedy acquisition and consumption for the benefit of their economies, then the populace cannot be blamed for carbon pollution, for they behave as they had been taught to "want it all."

While current humanity is involved in their usual fearful idealistic, patriotic, religious and economic wars, the problems associated with global warming are hardly noticed. Obviously there are distinct changes in global weather patterns and increasing storm activity due to elevating ocean and atmospheric temperatures, but these events, as yet, are not causing substantial hardship for human herds. No doubt the current human populations are becoming edgy about climate change but for now they continue to be more concerned with their personal materialism than the climate.

This shall inevitably change. It is when food production begins to fail that the enormous populations shall turn on their political masters in fear for their lives. Neither will they like it when their daily food insecurity fears begin to be magnified by their employment insecurity fears. Used to getting their way as obedient materialists working in business and nourished by business, these consumers shall react in anger when confronted with an inability to continue their habit of obtaining food and goods for their nourishment and comfort.
So shall begin population discontent.

As more and more inconveniences arise from uncooperative weather and natural disasters, the poorer and underprivileged levels of societies shall convert to anarchy, fighting those with privilege of being wealthy, still capable of living and eating well and being well housed. Violent uprisings, social disobedience and social revolutions will only add further conflicts to those of ongoing wars and insurgencies.

The growing human (conscio) physical conflicts shall result in growing destruction of economies, including those of farming and manufacture, resulting in even more pronounced shortages, where even the wealthy shall find it difficult to sustain their lives.

By that stage billions of destitute refugees of poorer nations, in search of more sustenance and life support, shall barge their way, in huge tides of humanity, to anywhere they think they may fulfil their intent. These human tides shall provoke military conflicts, including those in possession of nuclear weapons.

Right now such calamitous consequences for humans, resulting from global warming, may seem far fetched, because of the faith humans continue to place in science to provide them with a salvation. But what should be understood is that when global warming becomes irreversible, this will not take place in just one region of Earth, but across the whole planet, simultaneously. There shall be nowhere for humanity to escape, just as there shall be no place that will avoid being scorched and devastated. And the science that turned out to do so much damage shall be incapable of stopping this devastating change.

Even if those in science advise a simultaneous discharge of vast numbers of nuclear weapons to produce a 'nuclear winter', by which to cocoon the planet from the Sun and with that drastically reduce the temperature, all of humanity and animal species, as we now know them, shall expire.

Perhaps it is only then that humans (conscios) shall acknowledge Inca's opinion as being right: that the real treasures on Earth were food and water and not the gold, in whose pursuit we, conscios, destroyed ourselves.

11

Conscio making choices

There is no doubt that overpopulation and the ongoing human (conscio) pursuit of materialism – which are directly responsible global warming – have come about from personal choices each and every human (conscio) makes. This is not to say that the choices each and every human (conscio) had made had been done with intent of causing global warming. When making choices, they followed the same pattern of selfishness, simply because they have always lived with great expectations for themselves, as individuals, and with great expectations of their offsprings. If their lives had always been based upon an opinion that selfishness is the most essential component of existence, then it was a natural reaction for them to grasp and enjoy all that their science and business had to offer. However, an excuse of ignorance – from not knowing that their choices would result in a detriment to the climate – no longer can be used. Now, by having a clear understanding of where their individual choices are leading them, they need to consider what choices to make for the benefit of all the future human generations to come.

Despite having great expectations of life, life is not kind to all humans (conscios). The destitute and poor have little ability (financial, social or educational) to alter (change) their circumstances. The seriously ailing or crippled – mentally or physically – often have little ability to alter (change) their situation, even with the best health care. And even those who may seem to have everything are often disturbed with insecurity at the thought of losing what they have, and envying those who have more.

So what do they all do about it? With their expectations, they dream of, and seek the Human Dream. This dream is very familiar to most humans (conscios). In the fable of Human Dream wishes come true, and fate eventually smiles on those who are patient, charming, and just; allowing them to overcome all adversity and receive their deserved rewards; and to live happily ever after.

The heroes and heroines of the Human Dream are discerning and perfect in their appearance, approaching life with optimistic intentions to excel, and end up being fortunate and happy despite any occasional mishaps, trials and tribulations taking place in their lives.

PART 1. WHY HUMANS ARE HUMAN?

As always, a Human Dream begins with a life of Prince Charming and his spouse, Princess Charming – or Mr. and Mrs. Charming, according to nations without royalty.

Irrespective of their initial individual ordeals of hardship and injustice, which they both overcome using their sense of humor, panache, and good manners – the results of impeccable breeding and education – they meet under trying (if not dire) circumstances. Having met, they instantly fall in love with one another (deeply and eternally, of course), after which they end up living happily in perfect accord.

In their wonderful lives underlined by good fortune, these Mr. and Mrs. Charming give life to charming offsprings. The Charming children are sent to charming, exclusive schools, followed by reputable charming universities. At these charming universities the young Charmings meets many charming young people with whom they form charming lifelong friendships based upon their mutually-acquired charming taste: be it in music, clothes, or venues of amusement. Then it is onto a highly sought-after and highly lucrative employment; the work itself challenging and interesting but not taxing, with excellent prospects for advancement, leading up 'the ladder of success', but allowing for many vacations for rest and relaxation. Along the way there are the usual pursuits and rewards of business success, fame, and glory. Of course there are many opportunities to participate and excel in social activities and sporting events, at which it is possible to meet and impress many famous and charming people, including potential spouses.

At the right moments there are the romantic meetings of perfectly charming persons of the opposite genders, with whom strong, endless love ensues. Their whirlwind romances are followed by the most charming, exclusive weddings, attended by the richest and the most important charming guests. Naturally, the second generation of Charming's bare and raise more charming, gifted children, who are given the latest toys and taken to the most popular family amusement parks. Being offsprings to charming parents and grandparents, they also grow up to be supportive, charming, and successful adults.

Meanwhile, all the Charmings go on charming trips and holidays abroad, to the most charming and popular locations around the planet, where they make the usually tasteful acquisitions of exclusive real estate, and the most charming but expensive objects of art – for enjoyment and investment security for the retirement years.

The expanding Charming families continue socialising with the charming people of 'their own class', who enjoy the very best things in life, while constantly being blissfully joyful and happy.

After a long and charmed life, the parent Charming are farewelled into death, one after another, by members of the Charming family, friends, and famous relatives – with tearfully charming eulogies. From there, the new generations of Charming's continue the Charming dynasty into happiness never-ending.

There are many different versions to the Human Dream according to the various human societies. Nevertheless, the basic desire of having a charmed, if not a charming life, in which all kinds of material possessions would be available and easily obtainable – and from which supposedly happiness would be derived – applies to most humans.

With their Human Dream ideal before them – presented to them in all forms of media: from novels to advertising – every year millions of young humans (conscios) join billions of others in the process of earning and spending for the rest of their lives, in their chosen efforts to achieve their Human Dream. Yet, while trying to obtain happiness from wealth and possessions they begin to suspect that they are making choices that trap them in a repetitive cycle of materialistic pursuit that leaves little opportunity to seek any kind of self-development and interests outside of work.

Despite the overall absence of enjoyment, they continue to convince themselves that the best of life is yet to come – their happy Human Dream is just around the corner, as all the books, films and advertising assures them. Not for them is the enjoyment of every moment, be that in poverty or wealth, in illness or health. So they continue to make choices that chase the Human Dream, assisted in this by every business that plays on their fears and insecurities. So they choose to keep on grabbing all that they possibly can, as a security for their future retirements; for surely, that is where happiness is awaiting them.

As they age, they find that their future is no brighter, despite all their possessions. In fact, the possessions turn into burdens and more anxiety; for possessions require guarding and a waste of leisure hours spent on cleaning, polishing, and other maintenance.

On their retirement, not having developed many mental and physical interests outside their work environment and children, they are left to co-exist with other aging humans, most of whom, like them, also lack mentally challenging and stimulating interests.

With deeper aging, such mature humans begin to realise that life has passed them by, and that they have missed real fulfillment and happiness. This leaves them to seek their Human Dreams of happiness in the afterlife, as promised by religions.

PART 1. WHY HUMANS ARE HUMAN?

What turns out for most humans (conscios) choosing to waste their lives in search of the Human Dream, is that their happy endings do not eventuate. While an emotional state of contentment is available to every individual human (conscio) should they want it, this does not mean it can be obtained with any Human Dream. Instead, contentment is obtainable by a choice of accepting and enjoying one's existence, good or bad, which may include some possessions, but not from any possessions themselves. To achieve contentment and happiness actually requires a dismissal of the Human Dream, and acceptance of actions based on personal individual choices that avoid high-leveled selfishness.

As stated before, all choices are selfish, as it is physically impossible for any human to be totally unselfish, or selfless. To be totally selfless would require a mindfulness of all living organisms, on all occasions. In being mindful of all and selfless to all would also mean not being able to harm anything, and by that being unable to consume anything, including water, as water is also full of living organisms. This kind of ultimate non-selfishness would lead to ending of one's life by starvation. But as starvation leads to ending of life, which is harmful to a living body, then that would also represent a selfish act needing avoidance.

Therefore, absolute selflessness is a physical impossibility, as life – all life – requires some levels of selfishness to retain and maintain that process of life. This means any unselfish actions are either an unintended accidental result (misfortune) of an intentional selfish action, or else they are unintentional flukes (coincidental good fortune) of chance.

An example of unintentional accidental selflessness derived from intended selfishness would be that of a person's life being saved when an assassin intent on harm slipping while climbing into a window and falling to the ground, by that preventing harm being done by never completing a selfish intention. An example of unintended selflessness caused by a fluke could be that of a bystander standing next to a person targeted by an assassin, and throwing himself down, supposedly out of harm's way at the sound of gun fire, but as it so happens in front of targeted person, and by that action unintentionally (fluke) saving that person from harm of the assassin's selfish intention with his own body. Were the bystander to throw himself intentionally in front of assassin's victim, then that would be a selfish act. Heroic, yes, but selfish act nonetheless: from intentionally 'wanting' to save that person's life.

With no absolute unselfishness in life being possible, what applies to all life is a selfishness that has levels, by which it can be quantified. The way in which this can be done is by assessing selfishness in a comparison between the 'self' and 'others'.

The selfishness of 'self' comprises of personal wants and desires, consumed with selfish concerns of: "I need...I want..." and, "what's in it for me: what can I get out of it?" The selfishness for 'others' is that of a 'concern' for others, and while this remains a selfishness, it can reduce the levels of selfishness of 'self'. Where this 'others' selfishness reveals itself most often is in the concern of parents for their offsprings when those children are young and helpless. When the young are unable to fend for themselves, the parents fret over them, defend, and feed them. That is: the parent's selfishness of "I want" is to protect their young, which is a selfless act.

Human brains like the selfish concern over 'others', and encourage it in their conscios, influencing their conscios to act in a similar manner to them (brains).

The balance between the 'self' and 'others' is the crucial method by which a human brain tries to safeguard itself from its conscio's reckless selfishness for 'self', as it is unable to prevent its conscio from overriding it. So it is always an oscillating tug-of-war between conscio's 'self' of what it wants and desires, and what the brain's concern for 'others' would have it do.

If this oscillation could be depicted as a physical object, it would look like a measuring device, where at one end there would be 'self' (representing absolute selfishness), while on the opposite end there would be the 'others' (representing absolute unselfishness), and a sliding pointer between the two. And while absolutes of selfishness and unselfishness are impossible, some human conscio, for many reasons, come close to one end of the polarity or the other.

The middle distance of this device would represent conscios who have a balance of selfishness, so that they are neither overly selfish nor too selfless. When the pointer were to be moved towards the 'self', the more it came closer to 'self' this would indicate conscio's greater concerned with its own wants and desires and less concerned for all others. Conscios whose attitudes are very close to 'self' end of the scale are egoists who are nasty and malicious, not averse to killing. They take little or no notice of their brains' fear and guilt instructions, and even when their brains begin to punish them, they still continue to override and oppose them.

When the opposite occurs, and the pointer of the scale is moved towards 'others', the selfishness of that conscio can move past the state of concern for 'others' and enters the realm of overwhelming concern for 'others' with neglect of 'self". This state produces self-deprecation, and self-loathing. This is when a conscio's selfishness of 'self' diminishes to a level of low self-esteem, refusing (by overriding) its brain's instructions to enjoy and appreciate oneself, which impacts the brains requirements for sustenance and self-preservation. When this happens the brain begins to suffer. It begins to lose interest and the ability to maintain its conscio properly, abandoning it to apathy, depression, and even schizophrenia.

While there is no such physical device that records this kind of sliding scale between human brains and their conscios, human brains do constantly monitor their conscio's balances of selfishness and the levels of these balances. Then they give advice to their conscios in a form of choices, by which to avoid – and alter – any extremes, because any extremes of either 'self' or 'others' are damaging to them both.

The brains of all species that have ever existed – or shall ever exist – have an understanding that in order to exist they must make a choice of wanting or not to live as they currently do. While it is perfectly true that many species became extinct due to circumstances beyond their control, many other species became extinct because their brains chose not to live as they lived. These brains chose to end the existence of their present bodies, replacing them with those that would better suit the changing environment. This process is known as evolution.

The brains of these species, such as the sabre tooth tiger, knew that they and their bodies were not compatible to life in a space that was changing. To end their existence these brains chose to end their 'self' for the better adapted bodies of the 'others'.

Evolution is not voluntary: something that the brains decide one fine day. Evolution is undertaken under duress. Evolution, which represents change, is not desired. It is a choice made of necessity, and with uncertainty of success. But still it continues, often in increments that are hardly noticeable.

It is beyond conscio consciousness to comprehend the difficulty for a brain of any species to accept the prospect that they, the brain, and their body not only have to stop existing in their current form, but knowing that the future generations of them will not just look differently but will also behave differently to how they do now.

The early human brains had to make such choices. Not once, but on a few occasions. The human decision to implement and maintain the mental secondary consciousness had not begun with current human species of the so-called "homo sapient, sapient" ("wise, wise man"). The early efforts of mental secondary consciousness were nothing like conscio (commanding override negotiator, selectively controlling input-output). So the earlier efforts were abandoned; meaning that the 'I' of the earlier versions of human brains chose to become extinct, with intention that 'others' would take their place as human brains.

Current studies of early human life find it puzzling why earlier versions of human beings did not live in a parallel existence with their later cousins. They presume that the later humans had killed off the earlier humans. The physical reality is that the earlier versions of humans recognised the fact they were obsolete, and as such chose extinction, as difficult as this would have been for them.

Once having chosen life, the next choice of life is for how long should any individual life form live.

There are amphibian life forms – jellyfish, which are body-brains having no self-contained brain or a heart – that live at vast depth of oceans in absolute darkness, and can revert to juvenile form after reaching maturity. This means that the Turritopsis Dohrnii can presumably live forever.

Most creatures however, maintain a relatively precise lifespan. On the whole, the rule of thumb is that the smaller creatures that reproduce in large numbers have shorter lifespans, while those with large bodies and reproducing in single numbers having long lifespans.

Perhaps the simplest way to explain how any duration of life comes about is by using an analogy of a computing device. The 'brain' of any computer is its central processing unit (CPU), which has an allocated amount of stored memory used by CPU to provide all the necessary functions of computing (doing work). In a way, this is what any brain is: a functioning CPU with a necessary amount of in-built memory. However, a computer has what a brain does not: a connection to random-access memory (RAM) on which to store files and applications. This random-access memory can be enlarged, by having more RAM added to the existing number of RAM, so that a computer can store additional quantities of files and applications. This RAM can also be reduced, if necessary. All brains have no such facility for extra memory: what their brains have of memory is all they have. Therefore, most brains use their memory facility for their function as a brain, leaving very little memory with which to learn anything outside of their function of adapting their bodies and living with their bodies.

Like a computer, brains cannot function without memory, which allows them to 'remember' how to function as brains, and how to respond to experiences that come up in life, which are recorded to memory. But unlike a computer, brains have no ability to clear, or wipe, any memory, so as to reuse it for recording of new experiences. Once memory function is filled, there is no additional memory to be obtained. Once a brain's memory is used up, the brain cannot continue to function to manage life experiences. This memory limitation is the reason for the length of their lives. Discounting any external influences from other animals, acts of nature, and diseases, this length of life is applicable to all representatives of the same species.

Tiny brains, for instance, have developed small bodies so that their bodies will not require too much brain 'computer work', and once they (brains) have used up their available storage of memory in the process of evolving, living and reproducing, this duration of function then become the length of their life.

And when human brains were required, by their conscios, to extend the span of their lives, they (human brains) had to enlarge themselves in order to increase their memory to facilitate more days lived. But in the process of doing this, they had to enlarge the size of their bodies as well, so that their larger bodies could accommodate them with larger quantity of nourishment and longer duration of such feeding. Such developments have allowed human brains to achieve life longevity they could maintain. But irrespective of their brain sizes, once their memories are filled with life experience, human brains know that they have reached the end of their life. That is why no human brain can live beyond its maximum allocation of memory, despite its body being able to function longer, especially with artificial assistance.

There is a factor that applies to brain-bodies that currently is unknown: it is that most species base the duration of their lifespan on days rather than years. For most animal brains a day represents a significant accomplishment. Therefore to maintain a measure of the duration of their process of life they keep score of days from their exit from the egg. While seasons are of importance to many species, even the duration of seasons are counted by days within the overall number of days allocated to any specific life span. Of course any animal may experience a shortened life, due to, for instance, an appetite of other animals. But the average life span of many species is so precise that their brains have devised an end to their life after reproduction, or after a specific number of days of life even if attempts at reproduction fail.

Migrating birds and animals, for example, do not have the ability to reason. They do not look up one day and realise it is autumn, and therefore the need to travel to warmer regions. Instead, their brains know what to do and when, by the number of elapsed days, because their ancestors learnt by trial and error to devise and implement a passage beneficial to their existence – that being a food source – and this direction and duration of the journey (noted in days) had been transferred to the brains of current generations of those species.

What actually prompts them to migrate is the number of days from their day of birth. Their brains know that having been born or hatched on a particular day, they count days to when they are due to depart, counting days of their migration, then counting days for when they need to return, doing this year in and year out (not that they have any notions of what a year is). And when a specific number of days had been lived, they know that they need to die, as that is the completion of their mental memory facility. This actually applies to all animals, including us, humans.

From the beginning as human beings with our currant use of conscio, our human brains had established a life span of 25,000 days, which equates to just over sixty-eight Earth years. While there were always those who had lived much longed, their span of life was an exception rather than the rule. Considering that human bodies were much smaller (even in the eighteenth century), sixty-eight years may seem short, but even in the mid nineteenth century living to sixty years was considered a long life.

For many thousands of years, it was very unlikely for humans to exceed that stipulated lifespan. Currently, however, many humans who are privileged to live with abundance of mass-produced food, artificial life-supporting mechanical devices, medical and chemical assistance and palliative care provided by other humans. All these benefits to human bodies led conscios to have human brains extended their lifespans by over 10,000 days.

With such extension of human lives, together with the growing populations, especially in the poor countries – where the death-rates are high but their birth-rates are exceedingly higher – means that conscios have overridden their brains' initial intent of living and dying in balance with the overall longevity of other animals' lives.

But while human brains have availed themselves to their conscios' wants to live longer, they, as all brains, have never been fearful of death. Unlike their conscios, they understand that an end of 'self' is necessary, so as to give life opportunity to the 'others'. And unlike their conscios, human brains have never been anxious to live forever, which is exactly what those in business-of-science dream of achieving. What human brains are well aware of is that it is the quality of life that is valuable, and not the extent of its duration.

The main problem with trying to convince humans (conscios) to make choices for the benefit of 'others' – which would benefit them as a species – is their brains' ability to provide for them (conscios) with imagination ability. It is the imagination that lulls conscios into a state of mutual optimism. With imagination they believe that all the resources would last forever; that higher global temperature would allow them to grow more of everything, more often; and if everything were to be used up, their science would come to their rescue and invent artificial means of obtaining all the necessities essential to life. And if all else were to fail, they would build spaceships and go off to colonise other planets.

It is also the imaginary fears that delude them into presuming that they have to protect their existing way of life, based on unreasonable selfishness and greed. It is as if missing out on some gain from an advertised bargain in a store, or missing out on promoting oneself online, or missing gossip from some social media, that would forever devastate their lives.

Using imagination as a way of blocking out the physical reality continues to delay conscio decision (choice) that will have to be made, sooner or later: to alter their current attitudes and let future generations live – or not. The problem with 'later' is that it shall be too late for any chance to influence a change to this planet's deteriorating atmospheric and environmental conditions.

The problem is that conscios have grown so accustomed to their high-leveled selfishness that in their imagination any alternatives seem as not worth living, in the same way as drug addicts view any prospects of life without drugs. The reality of lower selfishness is nowhere as bad as they imagine. They can still practice selfishness, which is required by life, just not to the same high level.

Choosing and practicing lower selfishness should be a personal and private choice, and as such it is easy to maintain. In reducing personal selfishness there is no need to promote this stance vocally as being anti-selfish and anti-greedy, or to provoke those who may choose to retain their high-leveled selfishness. There is no need to take on vast or difficult tasks, just to display a less selfish attitude. Irrespective of any social commitments and lifestyle, it is possible to make personal less-selfish choices and to retain that knowledge in privacy: to be shared just between oneself and one's brain. By being generous and caring of others without any expectations of rewards in return, encourages the brain of such conscio to issues its reward of contentment, a feeling that cannot be physically shared with any other human (conscio). It is an unselfish reward from a brain for the unselfish attitude of its conscio.

In choosing to be less selfish doesn't mean depriving oneself of pleasures and functions necessary for normal human life. The only guidance provided for this is that these humans (conscios) modestly observe qualitative and not quantitative pursuits.

To exemplify the directions in life that may influence conscio choices towards less selfishness, here is a list of twelve approaches to life from a brain's perspective, anticipating from its conscio more compromise, more patience, more gratitude, but less desire.

On attitude to life:
- Upon arriving to life: be a sincere guest: demanding nothing while appreciating all that is offered.
- While remaining alive: be a hospitable host: share what you have, including yourself.
- Upon departing from life: leave behind no other footprint but that of a good impression.

On possessions in life:
- In possessing anything: know that it is only a temporary gift in your care.
- In willful taking of anything without permission: know that this is theft, and you are the thief.
- In accepting a gift of anything: know that you are now committed to a bond of reciprocation.

On functions in life:
- Soil only that which only you shall clean.
- Construct only that which you shall use.
- Destroy only that which only you have built.
- Kill only that which you shall eat.
- Produce only that which you need and not desire.
- Give birth to those whom only you shall raise.

Perhaps such guides to achieving lower levels of selfishness may seem as unrealistic, or unattainable, or even silly. After all, in polluting their environment most conscios do not feel obliged to clean up the messes they make. They think nothing of owning many possessions that they do not need, but for the desire of having them. They devour what they choose without consideration or thanks to the fact that that which they have placed in their mouths had previously lived. For them, living in a perpetual state of more-of-more is an everyday normality.

But even for such humans (conscios) there are lessons to be learnt from making less selfish choices, considering that they may selfishly benefit from this. Lesser selfishness is derived from a reality that the essentials necessary to life do not require all the unnecessary and superfluous baggage, requiring teams of servants or porters. So if there are fewer burdens of possessions in life, there is less to worry about. And every human brain loves having less to worry about. Less worry means a pleased brain.

One of the rewards of a pleased brain is a provision of so-called happiness for its conscio. Maybe this kind of benefit should not be ignored even by highly selfish conscios.

12

Conscio pursuit of happiness

There is one elusive entity that humans (conscios) had sought to find for as long as they had existed: happiness. Yet, while happiness seems to be something easy to understand, those who never experienced it confuse happiness with other feelings such as: joy, elation, euphoria, exhilaration, contentment, and pleasure. After experiencing those temporary emotions and still remaining dissatisfied, they conclude that there must be a secret to happiness. Well, there is.

The only secret to happiness is that happiness does not exist. As least not in the sense of how conscios perceive it. They believe it to be a singularity of one distinct feeling that can be consciously felt, whereas happiness is nothing like that at all.

The state of happiness is not some form of singular emotional ecstasy. Happiness has no singularity. It is not a singular emotion or a singular feeling. Instead, happiness is a composite of: contentment (satisfaction with having what one has, both physically and mentally); peace (lack of envy to obtain that which others have, by violent means); joy (thrill at being able to live); and goodwill (unselfish acceptance of all). This composite of mental emotional states is not combined in some single happiness, but remains as independent mental feelings where contentment, peace, joy, and goodwill swirl and flow into one another.

Happiness does not move consciously like a direct stream, but rather as a swirling and drifting waves of different states that flow in and out of each other, never constant, never identical. Therefore the state of happiness is never definite and because of that cannot be distinctly defined as a constant sense of one particular emotion.

The reason that brains can produce emotions, in the first place, is to achieve the means of external communications with their own species and other animals. Because the brains – aside from humans – have no direct verbal communication ability with anyone outside their body – so as to explain exactly how they feel and what their intentions are – they have developed some limited methods of doing just that by using a display of emotion exhibited by their bodies.

For instance, a dog actively jumping around with its tongue out and wagging its tail is an exhibition of the dog's brain communicating pleasure at seeing a human who feeds it and takes care of its body, and whom it trusts. Having made that communication to its master, the dog's brain then resumes its normal placid attitude. Alternatively, the same dog will growl threateningly, its fur standing on edge, at other people in an exhibition of the dog's brain expressing its displeasure at other humans it does not know, and thereby does not trust.

(In a way, this is what humans [conscios] do these days when they add emoji [tiny smiley symbolic faces that suppose to emote a fundamental explanation of their intent] to their written electronic communications, so that the presumed obtuse essence of the written message is not misunderstood.)

With human brain development, this form of brain's emotional communication with the outside world had become more advanced when the early human species had implemented their secondary consciousness of conscio (commanding override negotiator, selectively controlling input-output). These human brains had to communicate without words not only with other humans and other animals, but also, more importantly, with their own conscios.

By means of instigating a particular feeling of an emotion a brain could let its conscio know what it (the brain) was reasoning. For example: a calm feeling represents a contented brain. An uneasy feeling means the brain is warning its conscio to be vigilant and cautious. A joyful feeling represents a brain that is pleased to reward its conscio with a short burst of life enjoyment, which simultaneously displays little concern for its safety, so that this could be used as means of influencing others to relax and lower their guard.

Such method of letting the conscio know its (the brain's) position on issues that resulted from an observance of conscio's attitudes included the possibility of mental rewards of happiness.

Acquisition of happiness (or a levels of it) is not dependent on any human's (conscio's) opinion on how to attain happiness, but on the physical and mental attitudes to life in life.

By understanding and accepting this attitude all notions of power and possessions pale into insignificance, giving way to an ongoing mental celebration of all, with full understanding of the futility of all, but without any nasty or unfair criticism or condemnation. Any criticism raised, has to be based on physical truth of physical reality. And despite the realisation of futility of all, the awareness of all in existence being so complex and so simple – and so beautiful – transpires into a single wish for all that exists, if possible, to last for as long as possible.

During this experience that constitutes happiness, a conscio receives the necessary chemical stimulation with which it can consciously 'feel' and physically display with its body the mental communication from its brain that makes it possible for it (conscio) to experience a bond with all that surrounds it.

Of course happiness is not just the domain of conscio. When conscio receives the conscious mental sensation of happiness from its brain, the brain simultaneously rewards itself with chemicals that give it a similar feeling to that of its conscio. This is why a happy conscio represents a happy brain.

There is another element to the state of happiness: a presentation of happiness from a brain to its conscio is usually conducted in isolation. This means that it is virtually impossible to experience happiness in the company of others or in a crowd. Some of the elements of happiness (contentment, peace, joy, and goodwill) are possible. But not as a complete composite. It is only when a brain can have an undivided attention of its conscio that it (the brain) rewards a compliant and unselfish conscio with happiness. For the conscio (without being aware it this) needs to be totally focused on what is occurring with its (conscio's) awareness of happiness, putting all other consciousness of mundane, daily activities aside.

And that is the reason why most humans are incapable of experiencing happiness. With their daily wants, and ambitions, and living concern, humans (conscios) are too busy engaging their brains with their 'wants' to have their brains present them with an experience of happiness. They may feel joy, or contentment, or goodwill, but at that period they can never experience happiness.

It is inevitable that selfish pursuits and selfish actions distance conscios from experience of happiness. This includes attitudes that utilise high-leveled selfishness, such as anger, envy, nastiness, bitterness, malice and displeasure. That is why humans involved in less-selfish pursuits, as part of their working life, can receive levels of happiness more easily to those who are more vexed in working life, especially in professions that demand highly selfish activities and involvement of falsehood and dishonesty.

The sensation of happiness is not permanent. Each of the mental sensations of contentment, peace, joy, and goodwill do not comprise of a standard value, because like everything that human brains maintain they have levels that can, and do, oscillate between being stronger and weaker.

Once the initial sensation of happiness has been experienced the brain adjusts the individual levels of the mental sensations of contentment, peace, joy, and goodwill to what it (the brain) chooses them to be. This means that the initial overall sensation of happiness can be adjusted by the brain to a mental sensation that has, for example, high level of contentment, slightly less peace, even less joy, but a high level of goodwill. Such adjusted levels of individual components, that blend into a whole combination, is a function of happiness, altering and transforming in strength and structure depending on the ongoing relationship between the brain and its conscio.

Happiness, or rather the levels of happiness, are important to human brains because in sharing that experience with their conscios they benefit with a satisfaction that their conscios are in accord with them (brains). This stimulates the brains to share more of their knowledge with their conscios, expecting a reciprocating attitude from their conscios.

Unfortunately for the human brains such brain-conscio cooperation do not happen too often. Being pragmatic, most human brains accept this situation. However, human brains are not pleased when their conscios present them with either very high levels of 'self' or very high levels of 'others', which constitutes a high disregard of 'self'.

In such situations human brains cannot reward themselves with happiness because they cannot share this with their undeserving conscios. When locked into such predicaments human brains make their conscios experience the misery of unhappiness, while still trying to alter the situation with all the 'fear' codes and conscience.

A mental state of 'depression' is one example of a human brain being deprived for too long of any level of happiness. This applies equally to all conscios, be they billionaires or paupers; physical sport stars or cripples; the successful and the failures.

Currently, 'depression' – often described as a 'black dog' – is diagnosed as an illness where the mood of the sufferer experiences overwhelming despair, and an inability to come to terms with disappointments. The feeling of depression and the levels of depression vary with each individual. The commonality, however, is a state of dejection, which is a rejection of any joy of living, and an acceptance of overwhelming hopelessness. Being oblivious to the reasons for maladies such as 'depression', humans (conscios) presume that medication is the answer. They employ various 'anti-depressant' drugs, which basically stupor the brain and isolate it from continuing to punish its unhelpful conscio. The real solutions to such problems are not in the brains but in conscios. Teach conscios to love their brains and the brains shall reciprocate in equal measure.

PART 1. WHY HUMANS ARE HUMAN?

Humans experiencing happiness, or a version of happiness, are not extraordinary, nor do they live on some elevated plane. They are human conscios like all others. They can experience exhilaration, and pain, and grief. What differs with them is the awareness that all emotions and feelings are temporary, including happiness. If the happiness component is to be retained for as long as possible, then all actions providing other distracting emotions and sensations (be it even physical ones) need to be suppressed, or at least ignored, with a brain's assistance.

In order to obtain happiness, or a level of it, from a brain does not necessitate giving all away and becoming a pauper. What is required, though, is the understanding that everything in the physical process of change has no permanency. This means that all efforts to maintain permanency of anything, including possessions, are absolutely futile. Therefore those who have possessions need to know that they are only temporary custodians of their possessions, which do not actually belong to them, but to all.

Once possessions become meaningless to a human conscio, miserliness, avarice and greed cease to be of any importance. What does remain is the uncluttered appreciation of all that exists without the desire, or fear: for both the desire and fear are ever-present in all forms of ownership. The essence of possessions becomes not the ownership but the care and appreciation of those objects, or bodies, regardless of their value, as all are equally valuable and worthless. It is then that possessions may be kept or given away, with no regret. Just as with possessions, which become of no value but precious (even in memories), all life also ceases to be of value by becoming priceless: beyond and beneath any human worth.

There is no reason for wealthy humans (conscios) not to gain an experience of happiness, though it is unlikely. The difficulty is that the burden of being responsible for wealth – for its expected growth or fear of its loss– involves ultra-selfishness. Worry, concern and fear over possessions are not congenial for experience of happiness. Were this not so, bankers would be walking around with happy smiles on their faces, instead of smug expressions.

An attainment of happiness had often been associated with religious or spiritual ideals. Despite such presumptions, in reality, happiness cannot be received from a brain as reward for a regime of prayers, mysticism and meditation. Some religions, such as Buddhism and Zen Buddhism (Chinese Buddhism), infer that their religious beliefs and religious practices can provide happiness within an individual. This is nonsense, as no prayer can achieve happiness. Furthermore, their notions of nirvana and reincarnation are nothing but fiction.

What the Buddhist religion has in its favour are teachings (which human brains appreciate) where conscios are directed towards humility, less selfishness and the appreciation of harmony in both nature and within oneself. It is these elements alone that can contribute to experience a level of happiness obtained from a brain, without any religious dogmas and teachings. After all, any human (conscio) can attain a level happiness without the slightest knowledge of any religion.

But at least some religions, like Buddhism, permit a pursuit of happiness for their faithful. This is the very opposite to other religions, such as Christianity, where the concept of happiness is frowned upon. For the Christian priests, monks, hermits, and even those seriously committed to that religion, a mere thought of happiness is considered to be an inappropriate emotion. While it had been permissible for the members of Christianity to experience a 'spiritual bliss' of pious ecstasy – and to hallucinate images of saints – this was only acceptable after substantial periods of self-inflicted physical discomfort and pain.

Up to the present day, many religions regard feelings of joy and exhilaration to be inappropriate attitudes for those who are committed to their god, while sternness, glumness, torment, anguish, pain, and misery are looked upon with admiration, for, presumably, these kind of emotions provide a path to a blessed salvation in the afterlife.

While some conscios have the need to avoid happiness, as in case of some Christians, there are others who would like to obtain it with a minimum of effort. These human conscios would like to retain their ultra-selfishness and greed and still experience a state of happiness, by pointing towards the business-of-science and demanding, "give us instant happiness!"

In response, scientists – who in their guise of helping humanity are always on a lookout for some remedy by which to make money – are prepared to oblige these selfish more-of-more individuals. These flat thinking humans (conscios) presume that by mere chemicals they can produce an artificial form of happiness.

No doubt, while happiness cannot be fabricated it is not difficult to narcotise a brain. All legal and illegal drug manufacturers know this fact. Just as they know that many legal drugs and all the illegal drugs can, and do, become habit-forming for conscios.

So then, what kind of drug will the 'happiness pill' be? Will one pill be sufficient to last for a week? A month? A year? Not likely. More like a few hours. After the effect wears off there will be a desire for another 'happiness pill', then another, and another, and another, until the 'happiness pill' produces another addict, requiring other medications and therapy for depression and suicidal tendencies.

Just like the politicians, many scientists often promise a great deal without being able to provide it. And when they do, these chemicals inevitably produce harmful after-effects and bad repercussions. Not that they care about that, just as long as their products fulfill personal 'wants' of those who desire them.

But no matter how humans (conscios) attempt to obtain happiness by artificial means it shall not work. Happiness cannot be found, purchased, won, financed, borrowed, dictated, persuaded, ordered, manipulated, devised, replicated, cloned, implanted, or swallowed. It can only be freely given by a human brain to its deserving conscio, as a reward for being less selfish and less overriding, while being more accepting and loving of all. As a confirmation of this, simply ask your brain.

13

Talking to your brain

Because we (human brains) prefer anonymity, we, conscios), who represent us (the brains), are neither consciously aware of our brains' presence nor aware of the level of control we have over our brains. In fact, majority of us, humans (conscios), never think of our brains, presuming that a brain is just an organ that handles the memory and calculations, while we: the id, or ego, or soul, or spirit, or simply the consciousness of 'I' or 'me', who actually operate the whole living being.

Meanwhile, hidden in our bodies' craniums, we (human brains) remain unacknowledged: never objecting to such arrangement. We do all the mental work required of us by our conscios, and organise all the functions performed by the bodies, also on behalf of our conscios' requirements.

Once in a while, however, we, conscios – you and I – do what is known as: "talk to ourselves." This represents a situation when we, conscios, consciously presume that we are intentionally communicating with ourselves. On these occasions we may be chastising 'ourselves', or asking for a solution to a difficult problem "of our own making", or praying to god, or asking for forgiveness. We, conscios, 'think' that we are conversing with ourselves, whereas we (conscios) are in fact addressing our brains – without knowing that this is what we are doing.

For those who have done this, they would have felt a sense of relief, or a sense of confidence afterwards. That is because when conversing with a brain in a moment of genuine need we, conscios, do so truthfully, as there is no point in lying at a critical moment. Therefore, when such conversations take place between us (conscios) and us (the brains), we (the brains) appreciate the sincerity and honesty. We (the brains) then calm and soothe us (conscios) while proceeding to find a solution to that problem.

While it may seem that we (the brains) are just a brain, what we can actually do is astonishing. When asked honestly, with no deceitful selfish motives, we (the brains) can assist both our conscios' mental state and the bodies.

Those conscios who suffer our (brain) punishments, be that even 'depression', can ask us (the brains) for a way out of that mental state and we (the brains) will do so. Perhaps not immediately, but if the request is sincere we (the brains) will definitely help.

PART 1. WHY HUMANS ARE HUMAN?

From the moment of its formation after conception, a brain begins to take care of its body, and continues this labor, with assistance of its backbone 'brain', to the last present-moment of its life. This means that from the instant of conception to the end of life there is no other entity that knows its own body like the brain. It is the human brain that directs the formation of chemicals for itself and its body, with which to nourish its body; to sooth its body; to heat its body; to excite its body; to calm its body; to cool its body; and – most importantly – to heal its body. Heal it from diseases; heal it from physical abuse (self-inflicted by demands of conscio to take risks, or accidental); and, when possible, heal itself from mental abuse. When there are obvious problems with a body, there is no one better to talk to about this, and to ask for assistance, than a brain itself.

This in no way implies that a brain can instantly heal all the damages and ills that may befall its body. It cannot cure by itself incurable diseases or heal mortal wounds. It cannot simply alter any of its body's physical deformities or shortcomings, or provide by itself any required intensive care, and the requirements for surgical and medical treatment procedures. And yet, it is surprising what physical beneficial changes a brain can actually provide for its body, including what is known as 'miracles'.

Given a sufficient duration, and depending on a brain and its circumstances, when asked, a brain can, and does, more than just activate a faster recovery for its body: a brain can influence physical changes to its body, including a gradual removal of unnecessary growth occurring to parts of its body, such as that of arthritic growth, blemishes and lumps.

The only way an individual human (conscio) can find out what its brain can do, is to talk to it (without nagging). The procedure is to converse mentally with it, at least once a day, inquiring how it is and relate to it about oneself, sincerely and without lies. If a conscio is conscious of any physical or mental problems, these can be specified to its brain – as to a doctor – without lies, exaggerations, or excuses. After all, every brain knows all of its conscio's fears, guilt, and lies.
Should one become ill, promptly seek professional advice, but keep the brain informed of your condition.

In deciding to talk to one's brain, remember to regularly thank the brain for its constantly outstanding efforts. Having a friend in one's brain results in the brain providing much help and relief, perhaps not in the way expected, but in the way a brain chooses to, or can. And while it can, it will not give up trying. For how can it fail to do so, when it (the brain) together with its body and its conscio are all part of the same being?

Incidentally – while they may not openly admit this – no brain appreciates being abused by others, and especially by its own conscio. It does not like to be physically struck on its cranium or its face. It can take offence to being mentally or verbally degraded by its conscio by being called "stupid" or any other offensive degradation. It also does not like to be physically mistreated by too many artificial stimulants and depressants.

In addressing one's own brain it is preferable that a brain is not referred to as "Brain." They dislike pet names, but do not object to human names. An essential aspect of communicating with one's own brain is the requirement of sincere respect, which should also apply to the name. Once engaged, a brain can actually let it be known what name it would like. And if it does not like a given name, in one way or another it can let this be known to its conscio.

Human brains do not consider themselves as brains but as 'space-navigating intelligence', which is what they are. They are independent units of intelligence, who have devised, adapted and operate the bodies of their own design, which can provide them with mobility of movement anywhere on this planet. In addition, they had used their intelligence to devise, adapt and operate their secondary consciousness in the form of conscios (commanding override negotiator, selectively controlling input-output mental system), which act as a façade representations of their brains, and assist them by providing a three-dimensional awareness of space that surrounds their human bodies.

While human bodies and conscios are involved with giving human brains their mobility, it is the brains that use their reasoning ability to employ that mobility to locate the best advantage for themselves: a good environment for their sustenance. This ability incorporates not just the requirement to navigate around many obstacles in space of the present-moment, but to assess, calculate and establish navigation for future life-sustainment for themselves and their future generations.

So irrespective of any conscio ambitions – using human brains' imagination and invention ability – to devise and implement machinery for any outer space navigation, the main concern for human brains is to remain dedicated to this planet for as long as possible. That is: if their conscios allow them to do so. With their love of life, human brains respond to kindness and thrive with generous attitudes. Every brain loves to work, even if its conscio does not. A brain needs exercise and challenges just like a body, in order to remain healthy and in good condition. If this is done, then conscio has a good and helpful companion in its brain. One part of its reward to its conscio for being a kind, honest, active, supportive and unselfish can be happiness, or a level of happiness. The other part is a mentally stimulated and fulfilled life, despite any difficulties presented in life by any of its body's deficiencies and other human conscios.

PART 2

DISCLOSING THE SHAM OF HUMAN CONSCIO PERCEPTIONS

No matter when and where on Earth, all human events – classified as history – have a similarity of outcome: they all end up being waste: nothing but ashes, dust and rubble. But despite that we, conscios, are aware of this, pretending that "lessons can be learnt from history," we inevitably continue to repeat the same behaviour – with the same outcomes – over and over. Or, as we call it, "repeating history."

The obvious reason for such repetitious cycle of rise and demise in human development is that while circumstances of life may vary, the opinions on which we base the traditions and customs of our cultures remains unchanged.

Part 2 of this book examines what the opinions behind all the cultures have actually represented and continue to represent: disregard of any environmental damage done in pursuit of personal wealth; selfish control of others by any means; and ongoing inhumanity of humanity.

14

Imagination:
the path to deception

In classifying their species as "wise, wise man" (homo sapiens, sapiens), humans (conscios) had erred (as is the case with most of their notions) in their perception of who and what they actually are. The single factor that allowed them to achieve a Human Age on this planet is totally owing to no one and nothing but their amazing brains. So instead of naming themselves 'homo sapiens, sapience' they should have named themselves Cerebrum Glorificus (Glorious Brain), in honour of their glorious brains.

After all, while maintaining their external secondary consciousnesses of conscios (commanding override negotiator, selectively controlling input-output), this advanced awareness-ability was only made possible because of the glorious brains' three basic mental functions: enhanced memory, ability to imagine, and an ability to reason. But what all the human brains never anticipated was that their conscios would choose to depend more on the imagination than on the ability to reason.

Humans (conscios) have become so accustomed to some notions concocted by their brains at their request by use of imagination, that these fabrications are no longer questioned, but accepted as facts of life.

The reason why imagination had become so liked is because it provided an exciting alternative to physical reality in the mundane and boring human lives. Imagination allowed for a presumption that anything is possible if physical reality were to be disregarded, including hope of overcoming any deterrents with use of beliefs derived of imagination. So no matter how erroneous and unrealistic the beliefs may be, this had always been good enough for conscios to base all their cultural principles on them, to this very day.

One of the erring notions that humans (conscios) had devised and adopted is that of the 'right to rule' others. This blindly accepted presumption is upheld by, and for, the benefit of individual humans (conscios) who had always aspired to dominate others. Such individuals have been classified as: 'Dominant Males'.

Projections of mental imagination by the human brains, on behalf of

IMAGINATION: THE THE PATH TO DECEPTION

conscios, have no restrictions on how conscios, including those of 'Dominant Males', choose to use that imagination. For instance, it is only with use of imagination that the 'selfish fears' – with which human conscios attempt to control each other – are devised and implemented. Envy, greed, and jealousy: they are all products of 'fertile' imagination, often driving humans (conscios) to commit awful acts, simply because they are incapable of controlling their out-of-control selfishness by use of their brains' rational reasoning and logic.

Having no limitation on how imagination can be used by conscios for their own selfish 'wants' had proved, over and over, to be a useful implement of suppression. The proof of this is in all the human societies that had been, and remain, based upon erroneous notions (derived by imagination) that gods – be they Egyptian gods, Greek gods, Roman gods, Chinese gods, African gods, Jewish god, Christian god, Islamic god, or any other gods owing their invention to human imagination – indorse the supposed right of the 'Dominant Males' to rule others.

(For anyone considering that evocation of god by 'Dominant Male is out-dated and in no way applies to rule of democracy, should know that even today, all leaders elected to be heads of democracies, continue to give a pledge which usually ends with words of: "...so help me God." This statement has but one meaning: "By evoking God, I, the ruler – by whatever title – announce that I am in charge, and therefore I must be obeyed, for God is directly on MY side and is directly looking after ME.)

Yet, just because such fantasies became the structural norms of human societies, even a miniscule level of reason would suggest that these customs came about from a propagation of imagination rather than being based on physical facts. After all, the physical facts are in evidence everywhere that all those who choose to obtain control over others, do this by lies; for no god had ever appeared in person before an assembly and appointed in front of them a particular human to be their leader.

Most species of societal animals, living in family groups, have adapted to having one leading male member, who, by means of physical size and strength maintains physical superiority over all other members of that family. These so-called 'alpha males' take on the role of the head of the family group, for as long as they possibly can.

Most animals living in family groups, including primates, such as apes, accept alpha male behaviour with a degree of reverence, but only while that alpha male remains able-bodied. Within such groups an alpha male has a total dominance over all females, and other members. An alpha male may tolerate young males, but on their maturity they are forced to leave the family. Any sexual activity in the family group can only be conducted by an alpha male, who tolerates no sexual advances towards his females from any other males.

An old or injured alpha male – incapable of retaining his status by means of physical strength – will be replaced by other challenging males. The victor becomes the new alpha male, always vigilant to control his family group, and to defend his position from all other males.

When the early humans began to evolve from their primate ancestors, they retained in their makeup the prerequisites for an alpha male head-of-family, or a tribe leader. An alpha human male, while not a sexually dominant male, had the right to establish a recognisable pecking order for other members. An alpha male needed to be the strongest, the most able and the bravest of the group. He had to be a leader who would be obeyed not from being the most feared but for being trustworthy, caring, and hard working in protecting his members.

With the engagement of the mental conscio system (commanding override negotiator, selectively controlling input-output) a branch of human species obtained both, a beneficial and a detrimental characteristics that forever separated their behaviour from that of all other animal species.

On one hand, the conscio system with consciousness of 'self' or 'I', allowed these early humans to achieved a security of a superior survival ability, by means of their brains' provision of 'thinking ahead' projections as a preparedness for any adversity. With reason and imagination the early human conscios reduced their physical insecurity caused by unpredictability of nature, by devising the use of storage; use of fire; mutual planning of hunting trips and other resource-sharing arrangements; and the use of tools and weapons.

On the other hand, where the initial reason with lack of imagination helped their survival in nature, a continued dependence on imagination diminished reason. This directly attributed to the increase of their fears of the very same nature they always knew. While early humans lacked imagination they, like other wild animals, felt no fear of their environment. Once their imagination ability became more developed they began to have fears of non-existent threats, evoked by their imagination, just as many humans (conscios) still do.

The more they became reliant on their imaginations the more physically defenceless humans began to imagine themselves to be: not from any real danger of other animals but from the imagined threats of imagined beings.

While an animal, such as a wolf, having no imagination and no knowledge of superstitious fears, would inquisitively venture into a dark cave or a forest in search of any possible food and then calmly move on, for humans such places – while presenting no actual physical threat – appeared mysterious, foreboding, dangerous, and frightening because something mysterious, foreboding, dangerous, and frightening could, supposedly, live or lurk in there. Instead of accepting a dark space as just that, a space without light, the imaginations of early human conscios presented such dark spaces as containers of magical creatures and spirits, whom their imaginations invented.

It did not take long for the lives of early humans (conscios) to be ruled by their imaginations that distorted logic and reasoning. This led them to associate positive events in their lives (a successful hunt or a birth of a child), or negative events (an unsuccessful hunt, or a death), with some controlling spiritual mechanisms that could influence an outcome of a future event. With use of imagination also came the presumption that there are means by which these spiritual mechanisms could be directed towards a (selfish) positive future outcome:

"What if those 'somebody' or 'something' are given an offering of our food before the hunt? Would that not appease them, ensuring that the hunt is safe and successful?"

"And what if those 'somebody' or 'something' were again given an offering as a token of gratitude after a successful hunt? Would this not count as some insurance for a positive outcome in the next hunt?"

"And what if for the next hunt we pretend to be a fast, savage beasts? Would that not aid us, if we acquired some of animal attributes, in the form of their spirits, and gave them offerings as our gratitude?"

" 'Somebody' or 'something' that cannot be seen, must occupy that dark place and, therefore, it should be feared! Spirits must live there! We should bring offerings to this place, so the spirits will be kind to us!"

By such imaginative presumptions the early human conscios invented superstitions.

As superstitions spread through the early human tribes, some astute female conscios within the tribes recognised that they could do very well by claiming to have an affinity with that 'something' or 'somebody': be it the spirits, demons or monsters. By convincing others that they, and they alone could communicate with such supernatural beings and have some control over them, would give them respect and power of influence over other members of their tribes.

From there it would not have taken long for some male conscios to realise that a great deal of wealth and power could be had from being in charge of the spirits, and took over from the females already involved in this practice. Establishing themselves as shamans, witch doctors, priests, or 'Spiritual Males' by any other name, these male conscios gradually had their brains invented various rituals and techniques to impress their audience. The 'Spiritual Males' began brewing potions, casting spells, reading signs and making predictions, as well as inventing spirits presumably of greater importance and power; these often comprising of two opposing attributes: one being supposedly benevolent (unselfish) and therefore good; the other being nasty (selfish) and therefore evil.

As more early human conscios were drawn in by the mysteries of the supernatural to worship spirits, others were drawn to the thought of wielding the power of spiritualism and mysticism – wanting to join the service of the head 'Spiritual Males'. As those assistants came along to act as priests, their numbers elevated the importance of the head 'Spiritual Males', who were beginning to devise and present large ceremonies.

While the early mystical gatherings permitted all participants to dance and frolic together, the 'Spiritual Males' and their supporting priests began gradually to partition themselves and the performance of their spiritual rites away from the worshippers. Eventually the 'Spiritual Males' had invented an elevation of a stage, to represent a distinct physical divide between themselves and all other worshippers. The worshippers could chant along, but only if they chanted what they were taught. No improvisation from the worshippers was to be tolerated. To participate in the event the worshippers had to obey the 'Spiritual Male'.

In due course some 'Spiritual Males' conscios had their brains come up with a notion of mighty gods to replace the trivial mystic spirits.

Throughout all these 'spiritual' development, the 'Spiritual Males' in charge of performing these rites expected, or demanded, offerings to the spirits and gods, which they would use as they saw fit. As the beliefs took hold of the societies, the requirements of donations towards permanent temples gave religions a fundamental purpose to exist.

Such system of granting hope and solace in the name of fictitious gods invented by human brains – while collecting contributions in return – had become the basic principle on which all religions are based.

No doubt, some conscios of the early 'Dominant Male' leaders would have been facing opposition to their absolute control within their own groups, or tribes, or clan, from conscios of 'Spiritual Males'. The 'Spiritual Males' (conscios), however, were generally astute enough to realise the difficulties and dangers that 'Dominant Males' had to face on daily basis. By far the better option for 'Spiritual Males' was to support their 'Dominant Male' leaders. This way, their own positions would not be threatened.

With the 'Dominant Males' (conscios) being as superstitious as other humans, when given helpful and encouraging blessings from 'Spiritual Males' – in the name of the gods – the 'Dominant Males' were grateful and often generous to the 'Spiritual Males'. By supporting their 'Dominant Male' leaders, the 'Spiritual Males' faced no personal dangers, while retaining secondary power and wealth, as well as, the respect and gratitude from their society.

Besides, whenever 'Dominant Males' were challenged by other males, irrespective of who won the 'Spiritual Male' usually remained unharmed and in control of spiritual worship. From such beginnings religions, with their 'Spiritual Males', became bonded with the 'Dominant Male' rulers.

During the last ice age a great quantity of the water was retained in ice and snow. This caused ocean beds to be exposed, allowing the continents to be connected by land bridges, including those between Siberia and Alaska, and between Asia and Australia. Towards the end of that period, while humans were moving about establishing their presence in all regions of this planet, they were already worshipping spirits.

The dispersal of humans across the planet was due to their nomadic life-style of following animal herds and game. To live, humans were compelled to follow their hoofed food-supply of wild animal herds that followed the grass pastures amongst the snows. With the coldest period of the ice age coming to an end, the planet's weather began to warm up, producing a change in the human nomadic behaviour.

By 9,000 years ago, the melting ice and snow began causing massive floods around the planet, raising the water levels of seas and oceans; cutting off many landforms and human populations from each other. These frightening events of nature greatly altered the life-patterns of most human hunter-gatherers; engraving in their memories the raising waters and floods, which they retained in their folklores and legends.

As land became released from the cover of snow and ice along the Tropic of Cancer, the moist soil propelled vast growth of trees and vegetation, which in turn allowed for an explosive growth of animal life. This abundance of food allowed humans to rapidly increase in numbers. About 8,500 years ago, humans began to exploit the rivers and seas by fishing with fishing nets and spears from the shore. Then came the human (brain) invention of fishing from reed and log boats.

The incentive to follow wild herds in their migratory treks had subsided for many tribes and clans, as there were always the game and birds and fish to be had by the rivers and the seas. In the next thousand years more and more human tribes recognized that good living conditions could be maintained throughout all seasons, by remaining permanently in vicinity of water.

Without anticipating it, around 8,000 years ago, humans came across another food source, which was to change, totally and irrevocably, the manner of human life upon this planet. They had discovered that some grasses (grains) contained seeds that were good to eat, with the same seeds reproducing if planted into fertile soil. The problem with seeds, though, was that it required a lot of them just to make a single meal. That meant the cereals had to be grown in large numbers, which called for cultivation of large areas of land.

In their efforts to grow grain and cereals in large quantities, those doing so had to decide as to how they would spend their lives. The growing crops required regular attention, preventing the growers from leaving their fields for long periods, in order to go hunting or fishing. They also needed to live in close proximity to their fields, so as to cultivate and protect their properties. These demands of agriculture, unlike the first domestication of animals, where nomadic tribes could move domesticated animal flocks across the countryside while still following the wild migratory herds, turned many migrating hunter-gatherers into domesticated settlers.

Agriculture produced for humans a new concept: possession of land. Instead of all the land being a vacant space freely accessible to all life forms, land now became viewed as a personal property, belonging to a single individual, or to a specific group. With a more permanent and secure lifestyles – based on stable, renewable food supplies of grain – many human tribes began to congregate into permanently residing communities.

With such social development, the role of the early protector and hunter alpha males also began to change, taking on more prominent significance. The alpha males were no longer just the heads of small human groups, but representatives of whole stable communities. They became involved in resolving regional boundaries, military expeditions, and protection of trade.

These leaders took on the authority of becoming unchangeable rulers, controlling their societies with support of underlings they appointed for the task. And so, with the transformation of human dependence from hunting to that of agriculture the human alpha male hunters converted into 'Dominant Males' rulers.

The invention of a settled lifestyle appealed to many 'Dominant Male' rulers. By remaining in one place and claiming surrounding lands – in their own names of course – they had discovered that they were gaining enormous power of control over other human males. This power was not in just keeping the land for themselves, but in allocating portions of it to their followers. By giving to their male subordinates parcels of land, the 'Dominant Males' were in fact handing out pride, identity, conformity, security, and means of income – in return for their submission and loyalty. This meant that the 'Dominant Males' made a discovery of a new form of power – a power more persuasive than that of death: the threat of loss of possessions.

The 'Dominant Males' could now command their male followers, and those orders would be carried out, or else any disobedient follower would be dispossessed of his property, becoming an outcast. 'Dominant Males' could also force their male subordinates to fight for them – offensively or defensively – for in either situation they would be fighting for their own property.

Furthermore, because a human loaded with land and property cannot easily – and without regret – leave to another place, the landowners became an asset to the 'Dominant Males', as they could always be depended upon to obey orders, tend their land, as well as pay taxes. The gathering of taxes is another 'Dominant Male' invention, devised for the purpose of constantly maintaining the level of their personal wealth, which, in turn, was constantly used for purchase of respect and loyalty from their subordinates.

(From that period, land and property owners came to be treated with respect – even those who were, and are, thieves and liars – while those honest but having no land or property have been either despised or looked down-upon for being presumably either too poor or incompetent of obtaining land.)

The 'Dominant Male' rulers quite rapidly became despots: their word becoming the law. Continued possession of land by a farmer depended on the permission being granted by the master, the 'Dominant Male' ruler. If in any way offended by a farmer, the 'Dominant male' ruler could deprive that farmer of all his possessions and simply give them away to those favoured. There was no justice for any one, apart from the 'Dominant Male' ruler.

As the lands and properties of 'Dominant Male' rulers came to require constant attention and maintenance, they needed more humans to oversee and perform all the required tasks, but without having to financially remunerate them for their work. The solution was simple: oppression and slavery. Raid another community, kill those in the way, capture the rest – together with their possessions – and make them all your property.

This became a convenient method of combining business with pleasure: in an exciting blood-thirsty raid the warriors of 'Dominant Male' rulers could obtain for their 'lord' cheap labour and sex slaves; sex slaves who could be bartered or sold. Otherwise, the slaves and sex slaves would be used by the 'Dominant Males' as a method of rewarding the obedience and loyalty of their followers and supporters.

Having gained power over other male members in their communities, the 'Dominant Males' were now in need of resolving how to prevent any young males challenging their authority, and by that restriction allowing the old 'Dominant Male' rulers to retain power and control indefinitely. This is where the 'Spiritual Males' came to be most useful to 'Dominant Males'.

The solution was to have the 'Spiritual Male' priests proclaim their 'Dominant Male' rulers as appointees to their permanent position of rule by a decree of a certain spirit, or deity, or god; with any challenges to this right to rule being forbidden in the name of that supreme being. The rational was that very few humans would dare object to the word of a spirit, or deity, or god. And if anyone dared, they could be accused of being heretics and ungodly, and therefore could be executed or banished.

There was another imaginative task that the 'Spiritual Males' performed for their 'Dominant Males'. In the name of the their spirit, or deity, or god, the 'Dominant Males' would be ordained as supreme rulers and owners of all on their lands, allowing for their titles and all their ownerships to be transferred onto their offspring in posterity and in perpetuity. In return, such cunning falsehoods allowed the 'Spiritual Males' to spend their lives in luxury, governing their own possessions, and being provided with everything by the contributions from their faithful worshippers, as well as being financially supported by their grateful 'Dominant Male' rulers or kings. After all, these 'Dominant Males' – former tribal leaders – were now elevated above their fellow humans to the ranks of 'royalty', appointed with

IMAGINATION: THE THE PATH TO DECEPTION

'the right to rule' by 'Spiritual Males' on behalf of a celestial power; served and protected by their servants: the clergy and the nobility.

With these developments came the formation of a class systems; a social system intended to retain an oppressive power of few over many. By binding human conscios to 'selfish fear' and greed, the conscios of 'Dominant Male' rulers had invented a method by which they could maintain an on-going and unchallenged lie of the 'right to rule in the name of god'.

Some 6,000 years ago, substantial civilizations – already based on class systems – were forming around the planet. In regions surrounding flood plains of major rivers, human conscios practicing agriculture were gathering and living along the 'tropic of civilization' in increasing communities, under their self-appointed rulers.

In vicinity of the thirty-fifth parallel, traversing westwards from China's Yellow River towards the Hindus River in India, then onto Tigris and Euphrates rivers in Mesopotamia, and continuing from there to the Mediterranean and the enormous basin of the Egyptian Nile River in Africa, humans were cultivating crops and developing other industries helpful to their life.

As these centres of civilization increased in size, they began to attract various foreign 'Dominant Males' usurpers from distant regions. These 'Dominant Males' (conscios) used various invented claims as pretexts for their violent theft of lands belonging to others; aggressively thrusting themselves in rivers of blood upon those living by the rivers of life.

15

Greed with no boundaries

Around 3200 BCE (Before Common Era), while most 'Dominant Males' along the 'tropic of civilization' were content in controlling their individual kingdoms and city-states, a more ambitious 'Dominant Male', Pharaoh Menes, was forming a single nation of Egypt. From that action, under successive dynasties of 'Dominant Male' kings, the Egyptian civilization had retained supremacy of its region for thousands of years.

The actual unification of Egypt was achieved later, by several reigns of Menes offsprings and relatives waging wars against other nearby 'Dominant Male' kings. Nonetheless, the unification of the Upper and Lower Egypt can be symbolically assigned to Menes, for being the first 'Dominant Male' ruler to invent a concept of a single nationhood ruled by a single supreme ruler.

Menes came up with another 'Dominant Male' invention. He had himself proclaimed a deity: a god. The intention of presenting himself as a god-king was to prevent any other 'Dominant Males' or 'Spiritual Males' lording over him. As a god-king he could – and did – re-arrange the order of importance of all Egyptian gods, invented in previous centuries by the brains of Egyptian 'Spiritual Male' priests. By becoming a god-king, he and the future generation of god-king pharaohs sought to be regarded as superior to any mere mortal earthly kings.

By such deeds Menes had raised the level of expectation for all future generations of 'Dominant Males'. After him, no up-and-coming young conscio of 'Dominant Male' ruler – in or outside of Egypt – would contemplate anything less than an empire of his own.

To the 'Dominant Males' rulers surrounding the Egyptian empire, the strength of Egypt was not to be ignored but grudgingly respected. Not for too long though, as all 'Dominant Males' believe that no matter who controls the world, the world is still theirs for the taking. So while at that period Egypt was temporarily outside the bounds of their ambitions, any lands outside of Egyptian control were theirs for the taking. That is, if they had the physical means. Some did.

In order to illustrate how the 'Dominant Males' of the early African and Middle East civilisations evolved into 'Dominant Males' of the western cultures, here is a brief history lesson.

To the north by north east of Egypt, heading north along the east coast of the Mediterranean Sea was the land of Canaan; north of it was Phoenicia; and beyond that, Syria. Further to the east of these three countries were regions of Mesopotamia: vast plains cut by two rivers of Euphrates and Tigris, running from the mountains in the north, down to the Persian Gulf in the south. (Currently, it is the country of Iraq.) The lands of the "Two Rivers", as the Euphrates and Tigris were then known (at that period, both rivers entered the Persian Gulf independently), had been considered as a prized symbols of imperial achievement by all the 'Dominant Males' of their day.

(While the region is currently valued for its large reserves of oil, back then the region's value stemmed from rich grass soils of the 'fertile crescent'. The region of the rich soil [extending from Canaan up onto Syrian plains, then east, and from there down onto the flooding plains between the two rivers] could be said to resemble a crescent, or an arch, or a rainbow shape. It is for this reason that many current Islamic countries have adapted a green crescent into their national flag designs.)

Being part of the fertile crescent, the countryside of "Two Rivers" was a natural magnet of attraction to 'Dominant Males', wanting to possess not only a status of power but also a source of agriculture, animal husbandry, and trade. These wide plains were open and accessible from all sides. This indefensible openness allowed the region to be continually infiltrated, invaded, and conquered by different racial groups and their 'Dominant Males', which continues to this day.

Initially the land that was flooded by the two rivers in the rainy seasons was separated between Akkadians living on the relatively barren northern region of Akkad, and those living on the very fertile lower half of the Sumer valley. Sumer was occupied by Sumerians long before 3,000 BCE (Before Common Era), and well before the arrival of the northern Akkad tribes. The Sumerians learned to control floods by irrigation, and built walled towns and ziggurat temple towers. As a well-developed society – whose males shaved their heads and faces – they used copper, devised their own form of writing, and invented accounting. Their 'Dominant Male' kings – those ruling over their townships – were often at war with each other, invading of each other's grain fields.

The religious beliefs of Sumerians were based on the nature gods. There was the creator and ruler of Earth and storm weather; a god of the sky; a benevolent healing god of water; the moon god; the sun god, and his wife goddess of love and fertility. Sumer also had myths regarding creation of the cosmos, gods, and men, all of whom, supposedly, had emerged from an initial watery chaos of the Great Flood. The gods then supposedly created the Earth and all that was on it, including seven pairs of humans.

Simultaneously with the Sumerian culture and civilization development, humans of Canaan (along the coast of the Mediterranean Sea) were also developing and prospering. The region of Canaan (now called Israel) – being the thin corridor of land separating Egypt and the vast deserts of Arabia – had become an important trade link between prosperous Egyptian and Babylonian merchants, as well as, Hittites, Phoenicians, and Aegeans in the north. As early as 2,500 BCE, during the period of Egypt's old kingdom, Canaanites – whose origins dated back to the first humans who settled in the region after the last ice age – were building cities with protective walls, one of which was the city of Jericho.

Unfortunately for Canaanites, the advantage of being a strategic midpoint between Egypt and other surrounding nations also carried with it disadvantages. For many centuries Canaanites had to tolerate different foreign armies fighting each other on their soil, or suffer from those in pursuit of conquest. Not only did Canaanites had to content with invaders from the northern regions, they had to undergo penetrations of Semitic Arab and Hebrew tribes.

Hebrews: dark, bearded and longhaired herdsmen wearing animal skins, were not at all averse to opportunistic robbing and looting. Coming from north-eastern and eastern deserts, these Semite nomadic tribesmen spoke Afro-Asiatic family of languages, comprising a mixture of Akkadian, Canaanite, Aramaic, Hebrew, Phoenician, and Moabite dialects.

As former hunter-gatherers who followed migrating herds during the ice ages, they retained their nomadic lifestyle long after ice sheets had melted: roaming countryside in search of pastures for their vast animal herds. These Semites were a law onto themselves. As is the normal attitude of all 'Dominant Males', finding territories already occupied by other humans did not discourage them from forcing their way onto those lands.

One such large Semite Akkad clan from the northern region of Mesopotamia, comprising of Hapiru (Hebrew) – a multi-ethnic group of Hebrews

– set their ambitions on Canaan. With stubborn perseverance, tribe after tribe of these marauding Hebrew invaders travelled to Canaan regions, attempting to capture few of the weaker Canaanite towns.

Over many years of infiltration, many Hebrews in Canaan managed to settle and assimilate with Canaanite communities. Hebrews liked the fifteen hundred year old Canaanite culture. The comfortable Canaanite houses, their competent government, their various trades and industries, their written language and embracing religions had influenced the Hebrew shepherds to abandon their tents, discard their sheepskins and adopt the Canaanite garments, culture and gods.

Unfortunately for Canaanites, their generosity to all who entered and assimilated into their societies would neither be acknowledged nor rewarded by the later generations of Hebrew marauding invaders, the tides of whom just kept on coming because they could. And so they did.

One of human conscio characteristics is to blame anyone else but themselves for their selfishness and greed. For that reason, when conscios are prepared to steal, be it someone's land, they always make an effort to devise a ploy of blaming someone else for their selfishness, which they presume would justify their actions in hope that the retribution of conscience would not befall them.

The early Hebrew usurpers and thieves concocted the blame for their selfish desire of Canaan on the very people of Canaan, who had graciously accepted their unwanted infiltration. The Hebrews' excuse for usurping that which was not theirs, was a notion that: the meek must forfeit to the assertive. And since the attitudes of acceptance and tolerance were considered by them to be those of weakness, the Hebrew invaders felt justified in maintaining their nasty, greedy and unscrupulous behaviour towards their Canaanite hosts.

Therefore, the tolerance and acceptance shown to the earlier Hebrew tribes in Canaan had not only made the later emigrating Hebrew tribes more determined to penetrate, infiltrate, and dominate all they could of Canaan. And as an excuse for their despicable actions they actually devised a myth by which to claim the whole region for themselves, in the name of their god, Yahweh: a god they basically devised with appropriation of an Akkadian god, Marduk, combined with a Canaanite god, Baal.

Their intended use of Yahweh was to have that god supposedly proclaim that the lands of Canaan actually belonged to them, despite knowing full

PART 2. DISCLOSING THE SHAM OF HUMAN CONSCIO PERCEPTIONS

well that their actions were dishonest. (This is something that the Israelis and Zionists – the latest generations of the early Hebrews, Israelites and Jews – continue to dismiss, and thereby to defend to this day.)

In the period when the earlier generations of Hebrews were assimilating in Canaan, the nomadic Semite tribes of Akkad in Mesopotamia – who originally possessed no knowledge but that of survival in the steppes – gradually usurped from Sumerians a great deal of their knowledge. Akkadians learned from Sumerians how to write in their own language, using Sumerian writing method. They adopted Sumerian calendar and their system of weights, measures, and numbers. Akkadians also adopted Sumerian business and military models, as well as their gods. With this development, the Akkadians managed to build their own cities, including Babilla (later to be known as Babylon).

At around 2334 BCE (Before Common Era), an Akkadian Semite 'Dominant Male' chieftain named, Sargon (Sharru-kin: Rightful king), overpowered the Mesopotamian city-states of Sumerians, and established a Semitic dynasty. According to Akkadian folklore, Sargon was a 'self-made man': a gardener, who had found himself as a baby floating in a basket on the river, and raising himself from a child into adulthood.

No sooner had one lot of 'Dominant Male' usurpers been comfortably installed in their kingdom, that a new generation of up-and-coming 'Dominant Males' come challenging. Around 2200 BCE, Semite Amorite tribes began to assert their ownership of the Euphrates region, just as the earlier nomads of Akkad had done previously under Sargon. For centuries these new kings of Babylon fought for the leadership of Sumer and Akkad and the supremacy of the region.

In 1792 BCE, Hammurapi (or Hammurabi) ascended the Babylonian throne. He made Babylon his capital and managed to subject the whole of Mesopotamia to his rule until 1750 BCE – while constantly battling his opponents of various city-states. Contrary to the custom of other 'Dominant Male' rulers of his day, he declined to be deified, believing that real rulers need not be gods to rule, nor do they need gods to appoint them to be rulers.

Hammurapi may have temporarily regained the attention of the then world, by making the name of Babylon famous, with its – as legends have it – hanging gardens of Babylon, but 150 years later his dynasty was destroyed by an invasion of new 'Dominant Males': the Kassites.

The Kassites were pre-Indo-Europeans (from what is now Western Iran), drifting into the Mesopotamian valley with a new weapon, comprising of horses and light war chariots. After conquering Babylon, the Kassites had at first removed the idol statue of Marduk, a local Babylonian deity worshipped by the Babylonians.

Marduk was initially an ancient Akkadian god, a version of which was later to evolve into a Hapiru (Hebrew) god YAHU, later to become YHWH, then Yahweh, and later still, Jehovah. Marduk was a 'sun within a sun' god, comprising of two other gods: Ea: water (representing life), and Bel: 'lord'. These two gods within Marduk represented a godly trinity. The Kassites, who were polytheistic (worshippers of many gods), later returned the idol of Marduk to Babylon, and renewed its cult.

Before long, the dominance of the Kassites was overtaken by yet another set of 'Dominant Males'. These were the Hurrians of northwest Iran.

Next, by about 1590 BCE, came the Hittites of Anatolia, comprising of a loose union of Turkish mountain tribes, who, for a while, were a considerable force in the region, just before the Assyrian 'Dominant Males' took control of Mesopotamia, forming their own empire.

But before the Assyrian military dominance descended on Mesopotamia, the latest of the migrating tribes of Semite Hebrews advanced to Canaan. These Hebrew usurpers were a clan of tribes called Israel, who broke away from other multi-ethnic Semite Hebrew tribes. The claims of ownership that these Israel tribesmen made on Canaan were now based on their new form of politico-religious dogma – despite that the lands of Canaan were occupied by Canaanites from the last ice age.

It is important to notice how the Israelite 'Dominant Males' incorporated the dogma of religion into a self-serving propaganda by which to convince their own society that their claims on someone's land was justified.

This technique was often used by later 'Dominant Males' to this day, who, to achieve their objectives, would invent their own rules. If they had no military might, then they resorted to spirituality, in order to gain an appearance of having a celestial support to justify their cause. And when they faced opposition claiming similar spirituality, they then invented newer and supposedly even more powerful gods to be on their side. For the ambitious 'Dominant Males', any ideology or religion that impedes the progress of their desires had to be either eradicated or incorporated – with some changes – into becoming their spiritual possession: to be implemented for their needs as required. It did not matter – as it still does not – which version of religion the young 'Dominant Males' choose to employ, just as long as it is based on their opinions, and fulfills their dictates alone.

And that is exactly what was done by the members of the Israelite tribes, in taking control of all other Hebrews.

The nomadic Semite Hebrews clans comprised of many tribes of hundreds of large families traveling with their herds (made up of tens of thousands of animals), not unlike small towns on the move in search of pastures. The heads of larger families, just like the leader of the tribe, would have numerous wives.

Seeing in their wanderings how humans who had taken possession of land living more comfortably and securely with permanent landholdings, caused Hebrews to develop envious desires to possess their own land.

With no Hebrew tribe having any military knowledge to weld other Hebrew clans into a united military structure, the male heads of one particular family developed a clandestine plan by which to cast governance over all other Hebrews, without having to expose themselves to any danger or even criticism.

These male headsmen of the Levi "enveloping pasture" tribe (a name based the Fertile Crescent) resolved to develop and implement a method by which they could achieve their own controlling ambitions, but which needed the involvement of other Hebrew tribesmen desiring land possession.

To achieve all this, the heads of Levi tribes devised a new religion with which to control other Hebrew clans. But to achieve these intentions they knew that they would need to be cunning. Levi tribesmen could never openly present themselves as rulers, because in Hebrew nomadic societies each head of tribe was a power onto himself in his tribe. As such, he, the 'Dominant Male' tribe leader, would never share this power with others, nor would he obey any rival tribal leader; and would certainly not submit to a central, singular 'Dominant Male' ruler. Besides, despite most tribes and clans comprising of hundreds of families, these tribes and clans would seldom come across one another, so it would be futile to select one single 'Dominant Male' ruler whom many would never meet.

The only way that Levis could implement their ambition of acquiring control over other Hebrews – and eventually their kingdom – was by applying a unique new concept to their new religion: exclusivity.

All religions of the period welcomed any worshippers and converts, reasoning that the more of dedicated and incidental worshippers they had attracted the more wealth they would obtain from donations, and the more vital and influential their religion would become. Using this approach, the gods of all religions were presented as unbiased and unprejudiced gods, welcoming all worshippers of any faiths.

What the Levi's did, which was so astonishing, was to present a concept that rejected such embracing religious egalitarianism. Instead, both their god and their religion would be strictly exclusive to Hebrew tribes, excluding even those Semite tribes that were not Hebrew.

The Hebrew tribes, invited to join this exclusive alliance, could only do so as a member of Hebrew tribes amalgamated under a name of Israel. The binding conditions to join the alliance were: only those Semite Hebrew tribes who had no Arab or other ethnic lineage; and all those joining would be forbidden to worship any other god but that which came with the membership: the Israelite god, Yahweh. Furthermore, only the members of the Levi tribe could be priests, managing the religious practices. (Women, of course, were excluded from acting as priests, with Levi 'Spiritual Males' considering that their religion was strictly their masculine domain.)

As to the new Levi god, no god before had such attributes. Yahweh was to be the most singular, fearsome, exalted and the most powerful god. More importantly, though, he was to be exclusively dedicated to the Hebrews of the tribes of Israel, and as such, was to be intently concerned with obtaining for them their own kingdom: that being, not surprisingly, the land of Canaan.

This god, Yahweh, was meant to convince the Hebrew tribes joining Israel that their days of wandering the deserts and other foreign lands were coming to an end. With Yahweh they could expect their own rich and productive land, belonging to them, protected from all others by their dedicated god: a god who would answer no one's prayers but theirs.

Having worked out their strategy, the self-appointed Levi priests ('Spiritual Males') who devised the scheme, were astute enough to present their god as an event derived from the past. In this way, no Hebrew tribesman (living in travelling isolation, with no knowledge of events beyond their lives) could possibly contradict the Levi's presentation of Yahweh, claiming that no such event ever happened. After all, as far as most humans (conscios) are concerned to this day, anything supposedly could have occurred in the past, even the presence of gods. Knowing that their god was a sham, they presented their god, Yahweh, in a way of not having to present him in person, by claiming that their god, Yahweh, had made it abundantly clear in the distant past that he would not speak directly to ordinary humans but only to his prophets. But as the supposed prophets (who supposedly nominated Levi priests to be their representatives to god) were also from the past and were no longer living (and by that also could not show themselves in person), this meant that no one could disprove their existence or their instructions. That left the Levi priests in charge as representative go-betweens between god and the faithful, with no alternatives to such decrees being possible.

The intention of the Levi 'Spiritual Male' priests was, of course, to present a direct link from god to the prophets, and then from the prophets to them, the Levi priests, without needing to provide any proof to any of this.

Cleverly, as god's priests, they made no claims to having directly seen Yahweh for themselves; but they could and did claim that they strictly adhered to the instructions of the no-longer-existing prophets, who had directly witnessed god and had directly passed onto them, the Levi priest, the precise instructions received from their god. In avoiding a direct physical link between themselves and their god they could prevent other Hebrews from expecting a personal appearance from Yahweh. And as expected by them, the Levi priests, being in charge of their god's instructions and laws, they would administer control over all Hebrew-Israelites, and collect their offerings to their god in perpetuity.

Not surprisingly for that age – but surprisingly for the present age – the Levi fiction was believed (and continues to be believed). Hebrews accepted the notion of an exclusive Israelite god, considering that the exclusively 'their god' was going to provide them, 'his chosen people', with an Israelite 'promised land'.

And by being god's priests the Levi priests achieved their desire to have the ruling control of other Hebrew-Israelites.

There would have been little shame felt by the 'Spiritual Male' Levi priests, in adapting for their own purpose the fictitious figures from Canaanite and Mesopotamian primeval myths, which they turned into their god's go-between prophets, such as: Moses, Noah and Abraham, and later Adam and Eve.

By presenting themselves as having a god who was biased and racist towards all who were not part of the Israelite tribes, the original Levi 'Spiritual Males' had, very intentionally, structured the first antagonistic, biased and racially prejudiced human society, as before then gods were presumed to be all-welcoming. By alienating their faithful from other humans with differing religious beliefs, the Levi Israelite 'Spiritual Males' had introduced an attitude of "divine privilege" to hate, despise, and envy all those around them, because that was, supposedly, expected of them by their hateful, biased and racist god, Yahweh.

Given half-the-chance, most human brains love to accept and propagate hateful attitudes towards others. This is exactly the kind of attitude Israelite 'Spirituous Males' encouraged in all Hebrews-Israelites. So from then on, to this day, those of Hebrew faith looked, and look, at all other humans not of their faith, with animosity and dislike.

To make sure that their authority would not wane, the Israelite 'Spiritual Males' needed to preserve the laws they concocted in the name of their god. These were eventually written as records of their beliefs, comprising of fiction and distorted history. These writings became considered to be sacred books of a *Hebrew Bible*, as the canon (from Hebrew-Greek, meaning a measuring rod), by which Hebrews were supposedly given 'measures' to ascertain their life in regard to their laws of religion. *TaNaKh* (*Hebrew Bible*) stands for abbreviations of: *Torah* (Instruction and Law), *Nevi'im* (The Prophets), and *Ketuvim* (The Writings). *The Hebrew Bible* is equivalent to the books of *The Old Testament*.

What can be said of these writings is that the historical truth had never mattered to the Israelite 'Spiritual Males', because for them the historical events, as depicted in *TaNaKh*, were whatever they wanted them to be and not what they actually were. This meant that the nameless and secretive Hebrew-Israelite writers credited their fictitious, or reconfigured and distorted historical events to their fictitious Israelite patriarchs, prophets, and other Israelite heroes (such as king David), as if those events and persons had really existed.

What mattered to the Hebrew-Israelite 'Spiritual Males' was that their rendition of history depicted events according to their vision and their rules. They wrote of Israelite military and human failures as an intended punishment from their Yahweh god, as means of excusing their failure when any real historical occurrence did take place. They produced religious propaganda portraying Israelites as the most worthy race to acquire dominance over non-Israelites. They proclaimed their god, Yahweh, as the most demanding and the most powerful in existence – the creator of everything – while believing that the Sun and Moon were flat disks which moved over a flat Earth, whose overall area extended from regions around the Mediterranean Sea to Media (present Iran).

Also written into the *TaNaKh*, was a very resolute intention of the Israelite 'Spiritual Males' to obtain the possession of their own kingdom by any physical means. While using the 'promised land' as an incentive stimulus, they made sure that the notion of this was fully integrated into the concept of their god. Any mention of Yahweh had to conjure-up the thought of the 'promised land' in the psyche of every Israelite. The intention was to make every Israelite feel guiltless, by presuming that their god would absolve them from their crimes and the inhumane deeds and thoughts they committed.

To paraphrase the instructions of Yahweh, as presented by the Israelite 'Spiritual Male' priests, the meaning is precise (even if they are lies made up of wishful thinking):

"I give support only to Israelites. In providing them with land I make a promise of personal intervention on their behalf, and their behalf alone. I give them my permission to use any methods of ethnic-cleansing, and any manner of extermination in ridding Canaan of its original inhabitants, the Canaanites."

This theme is carried throughout the *Hebrew Bible* and *The Old Testament*, clearly illustrating the heartlessness of the Israelite intent. The following two brief samples of text from *The Old Testament* depict how Israelites were expected to treat other humans who already lived on the 'promised land' of Canaan:

[Deuteronomy 2:34] *"And we captured all his cities at the time and utterly destroyed every city, men, women, and children: we left none remaining; only the cattle we took as spoils for ourselves, with the booty of the cities which we captured."*

[Deuteronomy 20:16-18] *"But in the cities of these people that the Lord your God gives you for an inheritance, you shall save alive nothing that breathes, but you shall utterly destroy them, the Hittites and the Amorites, the Canaanites and the Perizzites, the Havites and the Jebusites, as the Lord your God has commanded."*

Indeed, for many ages the Israelite's 'promised land' – in fact, the 'coveted land' as it was and still remains – was Canaan. The Israelite 'Spiritual Males' obviously knew that the choice of Canaan would be acceptable to all Israelites. After all, what did it matter that Canaan had been occupied by Canaanites for thousands of years. Canaan was a good target. All other kingdoms in the region had armies, which could repel and destroy the Israelites had they dared attacking them. Not so with Canaan. The Canaanites were basically farmers and traders. They were definitely not 'War Male' warriors, even if some of their towns did have formidable defences. So, the peaceful and fertile region of Canaan became a focal point and a rallying cry for the envious and avaricious Israelite plunderers.

But it was not just Canaan that Israelites made claims upon. If the Israelites could not conquer other nations militarily, then they would lay claims to them by asserting lies that it was their Israelite god who had created all – supposedly out of nothing by just willing it so – and that they, Israelites ('his people') were the first humans he created in 'his own image', from whom all other races developed.

The first human, Adam, had to have a direct lineage to Israelites, who had supposedly populated the Earth and founded all the major cities. As claimed in *The Old Testament*, Noah of the Ark, a direct descendent of Adam (the supposed original male), had three sons:

[Genesis 9:19] *"These three were the sons of Noah; and from these the whole earth was peopled."*

According to The Old Testament, these three sons, in turn, had multitude of sons who built cities and civilizations of the whole known world, including Egypt, and even that of Canaan. By pretending to have been responsible for the formation of Canaan, they dared to make physical claims on lands never previously theirs.

(As mentioned earlier, this is a typical 'Dominant Male' ploy to lay claim to land belonging to someone else by inferring a prior ownership through ancestry.)

A great deal of such falsehood and propaganda was, and is, presented in the writings of the *TaNaKh* – the *Hebrew Bible* (*The Old Testament*). These writings are full of assertions made by unknown scribes on behalf of non-existent prophets, in stories exaggerating the physical abilities and powers of the Israelite patriarchal 'Spiritual Males', like Moses, and 'Dominant Males', such as Joshua, David, and Solomon. These writings exalt the kingdoms they were supposed to have established; the cities they had supposedly built; the glories they had presumably gained.

But what can be made of these claims, should they be examined with rational perception of present-day physical reality, supported by historical and physical archaeological data?

16

Monuments built to lies

The most common features of what the past kingdoms and nations have left behind as a solid evidence of their former existence, is their waste, their ruins and their dead. But interned in the rubbish, relics, debris, and the tombs, there are also all kinds of written and pictorial records.

Historical assessments are based on all forms of surviving physical remnants and records. There are the written records ranging from incantations and proclamations on tombs, to instructions and replies between a monarch and his courtiers and army generals, or sets of accounts between merchants and their customers. The old records can also comprise of official decrees, commemorative announcements, treasury lists, dispatches between bureaucrats, and personal messages and letters between citizens.

Therefore, when assessing human lifestyles of a period, all these written and pictorial records are taken into account, alongside the archaeological research of ruins and other physical objects, such as sculptures, kitchen implements and utensils.

Humans who ruled nations and empires did so by communications and accounts, and not just by military might. They built palaces for themselves, and temples for worship of gods, for which taxes were raised, materials purchased, services of craftsmen and artisans paid by the treasury, for which accounts were made and kept. Taxes also would be raised for payment of armies, and records kept of armed conflicts, by both warring sides. The rulers would have encouraged their merchants to travel into other countries to trade with other nations, leaving in those foreign countries evidence of their own country's presence, and its ruler's existence. The monarchs would also leave records of communication between their states. They would also form bonds with their neighbouring countries, often through marriage, with these being recorded. The same would apply to ordinary humans. By developing and constructing towns and cities in which they lived, cleaned, cooked, ate, entertained, were born in – and later buried – all these physical activities leave behind remnants of precise physical evidence of specific past human existence at exact regions on Earth.

This means that wherever humans may have physically existed, the space they occupied had also existed.

The remnants of records and ruins of Egypt, and other surrounding ancient nations and empires of the Middle East, are constantly researched and assessed by historians and archaeologists. Their findings from each region are then compared against their former neighbouring nations. Records of one ancient nation corroborate the records of another, and by such comparisons establish the true relationships that respective ancient 'Dominant Male' pharaohs, kings, and emperors had with their subjects of their own domains, and with their counterparts – the 'Dominant Male' leaders of other nations.

What such research has conclusively shown is that despite the claims made by unnamed writers of the *Hebrew Bible* (*The Old Testament*), none of the events mentioned in those works can be physically found, or corroborated by any records of other surrounding ancient nations. While a great deal of efforts and resources have been – and continue to be – spent by many international research teams, eager to establish proof of the Israelite ancient culture existing for the periods claimed in the *Hebrew Bible*, none had been found.

This absence of physical proof of early Israelite existence (claiming to be where it was supposed to have happened) makes it definite that most events depicted in the texts of the *Hebrew Bible* simply never happened. This signifies that the ancient Israelite heroes and rulers were not just historical distortions but total fabrications, because they had never physically existed, and all they supposed to have achieved simply did not occur.

Here is a short list of Hebrew and Israelite deceptions from the *Hebrew Bible*, or *The Old Testament*.

Moses: fiction
Supposedly Moses was to have been sent down a river in a basket as a newborn; raised as part of a royal family; freed the Israelite slaves from an unnamed Egyptian pharaoh after having god cause plagues, and then the death of the pharaoh's firstborn and all other firstborn across Egypt; parted the Red Sea so that the Israelites could walk across; delivered to Israelites the testimony from god, and who, with his tribe of Levi, from the clan of Israel, took charge of controlling all other Israelites; left everyone, and walked into the desert never to be seen again.

Moses: reality

The Egyptians were a nation of record keepers. From 18th century CE (Current Era) and onwards, the Egyptologists and archaeologists have searched all the available Egyptian records and sources for any hint of Hebrew presence there, and any hint of Israelite or Hebrew exodus, irrespective of them being under the guidance of Moses and Aaron, or anyone else. To this day, there is not a shred of evidence that anything even remotely similar had taken place in the whole history of Egypt.

The episode of infant Moses, floating on the river in the basket, had been appropriated from an ancient folktale about Sargon, the Akkadian, being a self-raised man, who had found himself as a baby floating in a basket on the river, then raised himself into an adult.

The so-called, pestilence, which Moses was to have brought onto Egypt and the unnamed pharaoh, were not unusual natural phenomena; the rest is pure fiction. Nile floods the Nile delta. On occasions fungus can discolor its waters. Similarly, on occasions, floods bursting their banks can drive reptiles and amphibians inland.

But unlike natural calamities caused by weather, what the Egyptian scholars and officials would never had ignored, and recording it, was the supposed event of the Passover, were it to have actually taken place.

The Passover was supposed to have occurred in Egypt at Yahweh's personal intervention on behalf of Moses and the Israelites. For a god to have killed all the first borne of Egypt – including the son of a pharaoh – would have meant a simultaneous murder of around half a million first born Egyptians, and maybe two million animals, all of them dying across all of Egypt in one night. Were such an unimaginable devastation to have physically taken place, such a calamity would not have gone unnoticed in Egypt, and would definitely been noted and recorded, both inside Egypt and other neighbouring nations as well. Yet no such records exist, nor could it, as logistics of murdering such a number of humans and animals in a few hours, across a nation, is even beyond physical capability of plagues. Besides, were this to actually happen, then the Egyptians themselves would have converted to such a powerful god.

In actual fact, Passover is nothing but a vicious tale that clearly depicts the genocidal desires of the Israelites, and the nasty, hateful, and despicable nature of their religion and of their presumed god.

[Exodus 12:26-27] *And when your children say to you, 'What do you mean by this service? You shall say, 'It is a sacrifice of the LORD's Passover, for he passed over the houses of the people of Israel and Egypt, when he slew the Egyptians but spared our houses.'"*

[Exodus 12:29-30] *"At midnight the LORD smote all the first-borne in the land of Egypt, from the first-borne of Pharaoh who sat on his throne to the first-borne of the captive who was in the dungeon, and all the first-borne of the cattle. And Pharaoh rose up in the night, he, and all his servants, and all the Egyptians; and there was a great cry in Egypt, for there was not a house where one was not dead."*

How filled with wishful hatred and selfish intolerance all Jewish conscios must be to continue celebrating each year, for over three thousand years, a cold-blooded mass murder of Passover; an event that – even though fictitious – represents a desired murder of vast number of humans and animals, carried out by their god on their selfish behalf. How filled with vicious hate all Jewish conscios must be towards other races, nations, and religions, to continue celebrating with glee such a vile, despicable, spiteful, and horrid extermination of other humans and life forms, instead of being ashamed of their fanatical blood-lust, envy and hatred. Even as fiction, Passover celebration should be detested and abhorred by any human (conscio) possessing even a miniscule sense of humanity.

If such nasty attitude was insufficient, it is further exacerbated when upon, presumably, leaving Egypt, the Israelites were depicted robbing and cheating the Egyptians along the way, with the help of their god.

[Exodus 12:35] *"... for they asked of the Egyptians jewelry of silver and of gold, and of clothing; and the Lord had given the people favor in the sight of the Egyptians, so that they let them have what they asked. Thus they despoiled the Egyptians."*

And while this, too, is but fiction, by being annually celebrated, these intentions clearly expose the lack of decency, the lack of integrity, and the lack of morality of Jewish attitudes towards possessions of those who are foreign to them.

As to the miracle at the Red Sea, there is no way that a parting of the waters of the Red Sea, or any other large body of water can be achieved, without mechanical and physical constructions taking place. The absurdity of such inference becomes even more pronounced on reflecting that in order for a large group of humans to travel on foot across the seabed of the Red Sea, even at its narrowest point between shores, would require at least a week of walking along a soggy uneven surface of a seabed, with overnight encampments, while the waters would need to be kept apart for the full duration. Only in myths, fairy-tales, the Bible, and the motion picture films could such a ludicrous notion be given a serious consideration.

The breaking of the first tablets by Moses has also been appropriated from another earlier myth. Where the Israelite 'Spiritual Male' priests changed the tale, was in having Moses bring the tablets of Commandments not once but twice, with the tablets remaining intact on the second occasion.

[Exodus 34:28-29] *"And he was there with the Lord forty days and forty nights; he neither ate bread nor drank water. And he wrote upon the tablets the words of the covenant, the ten commandments."*

But if Moses came down, again, from Mount Sinai, with a second set of stone tablets (having smashed in rage the first set, upon seeing his followers worshipping a golden calf, and then supposedly killing thousands of those worshippers), the question is, what had became of those tables, which would have been very large to hold all the information, and requiring forty days and forty nights for Moses to write with no food or drink? These – presumably the most important relics – are never mentioned again in the Hebrew Bible. Why? Because never having ever existed, they were best quickly forgotten.

Israelite spiritual heroes are often portrayed conversing with their god, directly and privately; yet, that communication always seems to have been recorded by a third unknown party. This form of reporting had been taken to extremes in describing Moses, who, having departed forever from 'his people' is supposedly all by himself in the desert, having his last conversation with 'his god':

[Deuteronomy 34: 4-6] *"And the Lord said to him, "This is the land of which I swore to Abraham, to Isaac, and to Jacob, 'I will give it to your descendants,' I have let you see it with your eyes, but you shall not go over there." So Moses the servant of the Lord died there...; but no man knows the place of his burial to this day."*

Even with a limited ability to reason, the impossibility of the described circumstances can be exposed by the following questions: When a book supposedly written by Moses records his private and final conversation with god, which he does not write down, and with no surrounding bystanders to record their word-for-word conversation either, then who recorded that conversation? Or would Moses and god fail to notice a scribe recording their conversation? But then how could Moses be alone with god if there was a scribe present? How would a book written by Moses – which he was supposed to have with him – end up in possession of 'his people' when he was never seen again? And if the book was not with him but in possession of 'his people' then how did his final solitary conversation with god get into the book? And if he has never been seen again, then how can it be known that he was buried and not gone off back to Egypt to live as a prince? Who saw him buried to know that he was buried? And who saw him buried (by whom?) And how did that witness relate this information to the priests so that they knew that Moses was buried, without exposing himself as a witness, and without disclosing where that burial had taken place?

These are reasonable questions that have no answers, because there can be no answers to that which had never taken place, from never having existed. Such writing simply confirms that the conscios of Israelite scribes were using their brains to invent make-belief events in order to fool the conscios of the faithful into accepting that which had never happened.

The simple truth of the matter, beginning with the infant floating in a basket, to the flaming bush, to the acquisition of tablets with written commandment, all these events attributed to Moses were taken from other nations' beliefs and fables. Moses never existed but in the Bible. Based on a Mesopotamian myth and tales, he was devised as a mechanism to launch the Hebrew-Israelite religion. Moses, together with his fictitious patriarchal predecessors, including Adam, were mere inventions of Hebrew-Israelite 'Spiritual Males' brains, their conscios prepared to sell the conscios of Hebrew tribesmen an 'Israelite dream'.

Israelites did not come as slaves from Egypt and conquered Canaan. As nomads, the Israelites followed the footsteps of their Hebrew forefathers, traveling wherever life was more favorable for them and their animals, or wherever opportunities to pillage were available. As many other foreigners, Hebrews had been entering and leaving Egypt, but not in any exodus. The Hebrew tribes also had been entering Canaan for tens of hundreds of years, some assimilating with Canaanites. The later Hebrew tribes, the Israelites, were also entering Canaan as nomad marauders, unruly and difficult – no different in their savage attitudes to their forebears. During that period, Canaan was a province of Egypt.

When Hittites from Anatolia (Turkey) began their push south, they forced the Israelites from the coastal plains, and other regions, to the south-eastern, hilly regions of Canaan – later to become known as Judah. The Egyptian military force that entered Canaan from Egypt, around 1207 BCE (Before Common Era), was not that of Israelites led by Moses, but an Egyptian army of pharaoh Merenptah. The Egyptians crushed the unrest in Canaan, in the process dispersing the Israelite tribes living in the hills. There is archaeological evidence of a scant Israelite presence in that area, as indicated by Egyptian records.

There is also physical evidence that those Israelites worshipped several gods, one of which was Baal: the local town 'lord'. The other deity worshipped by Israelites was Ashera, the goddess of fertility, the altered versions of Babylonian Bel and Ea, who originally were part of Marduk, the Babylonian idol. (Many Israelites of that region and period had also accepted Ashera to be the consort of Yahweh, so there goes the notion of a singular god.)

This was the first occasion that Israelites were mentioned in the Egyptian records. There was no further mention of Israelites in any other Egyptian records for another three and a half centuries. This is totally contrary to the events described in the Bible, when the victorious Israelites were supposed to have over-run Canaan, creating their great kingdom under their hero kings: Saul, David and Solomon.

Not long after Merenptah, his successor, Ramses III, had to battle with invaders from beyond the sea, including the Philistines, who settled on the coastal regions of Canaan. There is a great deal of archaeological evidence relating to those invaders, who established their city-states on Canaan soil. In contrast, there is no evidence of even a modest Israelite military, or civil presence at Canaan between 1200 to some 900 BCE, as it should have existed according to the Bible.

Joshua: fiction
Joshua, the successor to Moses, had supposedly: crossed the Jordan River by parting its waters; conquered Canaan; orchestrated the collapse of the walls of Jericho.

Joshua: reality
Like Moses, Joshua is an inspirational figure for the Israelites. Unfortunately for them, historically he is in the same league with Moses: a result of Israelite priests' imagination.

Not only was there no parting of waters of Jordan, but also there are absolutely no traces of Israelite occupation at the ruins of Jericho. Of the 12 cities that Joshua had supposedly destroyed, according to the Bible, archaeological research and historical records confirms that any destruction that did take place, occurred at a different age – hundreds of years later.

David: fiction
The Israelite King David, who was supposed to have: reined in 1000 – 962 BCE; killed Goliath in a duel, captured Jerusalem and made it his capital; built new palaces; fortified the city; established a harem; ruled an empire stretching from Egypt to Lebanon and from Mediterranean across to Arabian Desert; had a son, Solomon.

David: reality
An empire cannot exist without communications, contacts and trade with other nations and their rulers. Unless David ruled in a vacuum, there are no records or archaeological evidence from neighbouring kingdoms surrounding David's supposed 'empire', or any communications, or contacts, or trade with Israel. There are also no references to Jerusalem becoming David's capital, outside of *The Old Testament*. Nor is there any evidence, outside the Bible, to show that he even existed. There is no archaeological evidence of David's palace in Jerusalem, or its fortifications. By not having existed, it would have been very difficult for David to kill a non-existent Goliath.

Solomon: fiction
King Solomon, the son of King David, was supposed to have: reined during 962-922 BCE; built a Temple in Jerusalem, as well as other lavish palaces in a vast building program; built a port on the Red Sea; impregnated Queen of Sheba; built a large fleet; extended Israel's influence through commercial treaties acquired by marriages to a number of foreign princesses who were part of his 1000 females harem.

Solomon: reality
Like David, outside the Bible there are no other sources to indicate that Solomon had ever existed. No other nations in the region had ever known of King Solomon, despite of his supposedly large fleet, which, were it to exist, would have had an impact on all the neighbouring nations of the period. There are also no records to indicate that any regional royal houses of the period had made unions with Israel through marriage. The cities that Solomon was supposed to have built were constructed at a later periods by different rulers, like Ahab, the king of northern region of Israel, around the middle ninth century, BCE. There is no evidence of a Solomon's Temple in vicinity of Jerusalem. Jerusalem, itself, became an important trade centre

PART 2. DISCLOSING THE SHAM OF HUMAN CONSCIO PERCEPTIONS

well after the so-called King Solomon, as confirmed by records of neighbouring countries. Other sites attributed to being constructed in the period of Solomon, were results of different cultures altogether, as indicated by their archaeological contents.

All-in-all, there are absolutely no records or archaeological proof to indicate that Egyptian pharaohs assisted Solomon in Canaan, by winning cities for him, or of him marrying an Egyptian princess, or to indicate that there ever been an Israelite empire in existence of this planet, even though they still desire it.

Based on regional records and physical evidence of debris and ruins, a small kingdom of Israel was established by the Israelites, but at a much later date, and not by any of the supposed Israelite hero kings. This lack of evidence to support the claims made in the *Hebrew Bible*, or *The Old Testament*, leads to a conclusion not appreciated by the current occupiers of the former lands of Canaan – the descendants of the former occupiers of the lands of Canaan – who now call themselves Israelis.

By being unable to substantiate their past origins to the land that they had always claimed as their own, shows that their present occupation of the former Canaan has no justification, as their claim to it had always been based on lies and fiction depicted in *Hebrew Bible*. Canaan was never theirs, then or now.

The unfortunate Canaanites, whose only fault was to occupy a region before other humans, were besmirched by the lies of the *Hebrew Bible* and *The Old Testament*, and continue to be into the present; lies that had convinced other humans (conscios) to presume that the land of Canaan – later to be renamed Palestine – was a Hebrew-Israelite-Judaic-Israeli possession from their ancient links of ownership, and officially granted to them in 1948 CE (Common Era). These links of ownership had never existed but in fabrications of the *Hebrew Bible (The Old Testament)*.

The unfortunate Canaanites, who accepted and integrated the marauding Hebrews tribes into their culture – as they did with other want-to-be invaders, such as the Greek refugees, the Philistines – did not deserve the debasing of their existence, the kind they received at the hands of such callous, greedy and hateful liars, the Hebrew-Israelite-Jew-Israeli humans (conscios).

But if nothing else, then the unfortunate Canaanites, who wanted nothing from their neighbors but fair trade, and who showed no aggression towards other nations but tried to defend themselves, should at least have their honor restored – as late as it is – by having the physical truth confirm that they were innocent victims of unscrupulous and hateful lies that are the *Hebrew Bible* and *The Old Testament*.

Humans (conscios) are often prepared to believe anything, simply because they want to believe. This is why it is not surprising that otherwise rational humans attempt to seriously and earnestly explain fabrications that are religions, by discarding logic and physical reality. What these humans have to recognize – in their zeal to believe – is that that which is fiction but pretends to be a physical reality can never produce any physical proof of that fiction ever having existed. Therefore, the fabrications that make intentional and unrepentant assertions to past physical existence of that which never took place, are not merely fiction but are lies. By applying logic to illogical depictions recorded in *TaNaKh* – the *Hebrew Bible* – the logic easily discloses the intentional lies intended to impress the ignorant and the gullible.

The process of generational physical changes, as it applies to human existence on this planet, known to humans as 'history', is incapable of being permanently altered or hidden. Even when physical efforts are made to obliterate or falsify physical events in history – as too often attempted by conscios of 'Dominant Males' and 'Spiritual Males' – their physical efforts eventually reveal themselves for what they are: lies; and by that disclose what they have tried to destroy or conceal.

Lies become apparent from having no physical substance and no physical presence. Were the claims made in all the religious writing to be true, they would have had a physical substance and presence. But they do not, because lies cannot build anything physical.

Throughout human history humans (conscios) used their brains and bodies to build monuments – be they temples, churches, or cathedrals – to their fictitious gods; or rather, in fact, to their lies. These physically erected monuments are a testaments to human physical efforts of trying to falsify physical facts that their gods do not exist, by trying to give non-existent gods some physical substance and presence, by erecting buildings dedicated to these non-existent gods. These physical monuments, however, are governed not by gods but by the physical process of change – and so they age, turning to rubble and ruins, with no fictitious god ever retaining its monument in pristine condition. No human (conscio) notion of any gods can prevent all that humans erect to their gods to crumble, including the

155

notions themselves, for lies cannot suppress the physical truth forever.

While humans (conscios) can physically construct physical monuments to their lies, so as to delude themselves with notions that gods exist, their gods made of lies cannot do the same, for they do not exist. That is why all fictitious gods, for all their assumed powers of being responsible for 'all creation', have never, ever, left a single physical structure, or a memento, as a testament or a memorial of themselves to humans. Nor shall they ever do so. The reason is simple: that which does not exist, cannot construct a physical structure. That is why, while monuments may be built to lies, lies build no monuments.

Just as no god had ever erected a monument, no god had ever come forward to make a bald head hairy; a toothless mouth into one full of teeth, a castration into new genitals, an absence of an eye into a new seeing eye, and an absence of a limb into a new limb. It is these physical impossibilities that physically prove that religious lies do have their limits.

17

The ugly truth

In the year 922 BCE (the year that Solomon was to have died), an Egyptian pharaoh, Sheshonk, sacked the region of Canaan. Despite that, Philistines (after whom the name of Palestine was derived from the Greek, "the land of Philistines") persisted to enforce their influence there, until Assyrians re-asserted themselves.

The Assyrians kings, among them: Sargon II, Sennacherib, Asarhaddon, and Asharbanipal, began a new period of expansion in the 9th century BCE, lasting for over two centuries. The Assyrians – with their long hair flowing from under their tall hats, and with long, squarely trimmed beards – were renowned as much for their cruelty and fighting abilities as for their monumental building programs. Their 'Dominant Male' kings had united most of the Middle East, from Egypt to the Persian Gulf. In 701 BCE, the Assyrians forced Hezekiah, king of Judah (by now, former region of Canaan in possession of Israelites), to pay tribute, while northern Israelite tribes were forced to flee and disperse to other regions under the Assyrian onslaught.

By 625 BCE, the Assyrian dominance was being replaced by that of Chaldean 'Dominant Males'. Nabopolassar became a king of Babylon, establishing a Chaldean dynasty that lasted up to the Persian invasion. By 612 BCE, two new kingdoms were established: the Medo-Persian, on the eastern plateaus, and the Chaldean, in the south of Mesopotamia.

Like most 'Dominant Males', Nebuchadnezzar II, the eldest son and successor to Nabopolassar, had imperial aspirations. He intentionally pursued a policy of expansion, claiming a universal right to do so, given to him, as he claimed, by Marduk, the main god of Babylon.

During his life, Nebuchadnezzar made a number of expeditions into Palestine: attacking Judah; capturing Jerusalem; sending thousands of resisting Israelites to Babylon as a deterrent to the remaining population. These deportations of Israelites to Babylon were called by Israelites: the "Babylonian Exile".

Those Israelites from Judah, who retained their faith in exile, became known as Jews. On their return to Judah from Babylon these Jews became considered as deeply committed to their religion. So as not to be accused of

PART 2. DISCLOSING THE SHAM OF HUMAN CONSCIO PERCEPTIONS

possessing any less faith, soon all Israelites were calling themselves Jews. Just as Hebrews changed their identity to become Israelites, so did the Israelites easily discarded their identity to that of Jews. In mid twentieth century CE [Common Era], the Jews had changed their brand again, to become Israelis.

Meanwhile, the plight of Israelites, at the hands of Nebuchadnezzar, had given them a great wealth of new subject matter for their religious propaganda.

For instance, secretive authors writing in the name of prophet Jeremiah, were virtually rejoicing – with a great deal of hindsight – in placing the blame for the Israelite misfortunes on their sins. (These authors called upon Israelites to stop resisting Nebuchadnezzar, and to give themselves up to their punishment of exile to Babylon, because, presumably, god had ordained it.)

Alternatively, other authors – writing as prophet Daniel – were busy producing propaganda that discredited the abilities of their Babylonian captors and their deity Marduk, worshipped by the Babylonians. In their writings – while accepting that Marduk is an actual living god – they, nevertheless, belittled Marduk as being merely a subordinate and subservient god to their own god, Yahweh.

Of course, such ploys of spiritual one-upmanship had been used from the day that human conscios began to request the invention of gods from their brains. Hebrew and Jewish did this; Christians did; Mohammedans did; most 'Dominant Males' did – and many still do.

"Look," they profess to each other, " I know you probably will not like hearing this, but it's a known fact that your god prays to, and gives homage to my God."

Eleven years after the death of Nebuchadnezzar, new 'Dominant Males' began to take over his former empire. On this occasion it was the turn of the Persians to dominate the Middle East. About 550 BCE, Cyrus II, a Persian-appointed caretaker ruler, acting on behalf of Median Empire, had revolted and swept over western Asia with his archers and horsemen. Babylon fell to him in 539 BCE, Egypt in 525 BCE. By 490 BCE, in the reign of Darius, the Persian Empire extended from the Hindus river (because Darius just said so) to the Danube river. Only the successful resistance of the Greeks prevented them from being absorbed into his empire.

The Persian Empire (currently Iran) was thoroughly organized as a governing body and a military force under their kings, primarily: Cyrus II,

550-523 BCE; his nephew, Darius, 523-485 BCE (who secretly assassinated Cyrus); and Darius' son Xerxes, 485-465 BCE. Although the Persian 'Dominant Male' kings expected to be respected as gods, they rejected being worshipped as such; those kings cared more about power and control than religion.

The fundamental Persian religion was devised by an Iranian religious reformer, Zoroaster (628-551 BCE). The importance of Zoroaster was that it was he who first introduced the concepts of a 'good and righteous' god (Ahura-Mazda, Ormazd), and an 'evil' god (Ahriman [Satan]), and the concepts of 'immortality' and 'bliss' – with all outcomes depending presumably upon a single, comprehending, and accepting god, Ahura-Mazda. The exposure of Jews to the Persian Zoroastrianism during their stay in Babylon (from which they were released by Cyrus II, in 537 BCE), introduced them to all these new religious concepts, which they, until then, had never even contemplated. But once having been exposed to them, the Jews lost no opportunities to appropriated most of these new notions and ideas.

In faiths based on a premise of a single god, the difference between Yahweh and Ahura-Mazda was that the supreme god Ahura-Mazda had supposedly ruled his creations with kindness not malice, and had a spiritual antagonist, Satan – who could sway the faithful from the path of righteousness – while the initial Yahweh had no antagonists but 'his people' who had erred. That is why it is only later that Adam and Eve, and the Serpent (as the Devil) were inserted into the *Hebrew Bible* (*The Old Testament*).

Furthermore, according to Zoroaster, the righteous – by following a life of good deeds – could attain a reward of an after-life in heaven, while the wicked were to be punished in hell. At that period the Jews had no such notions in their religion. Such attractive concept – inviting because it implied that appropriate justice would be dealt to both the pious and the wicked – had captured the imagination of some Jews. These concepts, however, were retained outside the Hebrew Bible, and were incorporated centuries later into writings made to promote a so-called human god by the name of Jesus.

Following their release by Darius II, the Jews continued to add more religious fabrications suitable for inclusion into the Hebrew Bible for another two hundred years, until, quite suddenly, everything changed for them. The cause of this was the coming of the Greek 'Dominant Males' from the west, in 332 BCE, bringing with them a new culture with its own fictions and new ideas.

It is at this point that the lies of the *Hebrew Bible* came to an end.

18

Converting a Jewish political concept into a new religion

When the Jewish lands were taken over by the latest invaders, the Hellenes (the Greeks), headed by their all-conquering 'Dominant Male' leader, Alexander the Great, the Jewish priests and their scribes did not actually stop writing their distortions of historical events. What altered the course of such religious inventiveness was the fact that these writings were not allowed to be included into the TaNaKh. The reason for that was simple enough.

The Greeks were accomplished liars, with formidable imaginations. Their culture was already based on fiction, comprising of a whole society of gods, whom they had invented and worshipped. From the moment that Greeks came, it was inevitable that, as masters of fiction, they would take interest in the Jewish religion and would intrude upon it with their own inventiveness.

And that is exactly what happened. The Greeks made an impact on all Jews – whether they liked it or not – many of whom were impressed by Greek knowledge, ideas and culture (religion), and an expressive language – much broader and more precise then Hebrew. With the Hellenization of Palestine, the Jewish writers began to develop new literary styles, one of which was apocalyptic – from a Greek term, apokalypsis, meaning: "revelation of divine mysteries" about god and the resolutions for mankind. Such literature was intended for Judaism, but because they were primarily writings of Hellenic Jews, these works of fiction were rejected by Jewish orthodox priests, as not being authentically Jewish.

Therefore, in order to prevent any foreign influences infiltrating TaNaKh – the Hebrew Bible – all religious texts written after the Greek invasion were forbidden from being included into the existing Judaic testaments. This rejection affected many writings, including the four Maccabees books written in a biblical form, of which the first two books recorded the history of the successful Maccabean uprising against Hellenization (166-142 BCE), led by a Jewish priest, Mattathias the Hasmonean, and his five sons. Due to this rejection by Jewish religious authorities, many of the apocalyptic writings ended up in the Christian *New Testament* instead, such as the book of The Revelation.

CONVERTING A JEWISH POLITICAL CONCEPT INTO A NEW RELIGION

The Judaic orthodoxy did more than reject new Jewish writings. They also opposed all Hellenic influences, and rejected any advances from many gentiles living in Palestine wishing to participate in Judaic religion. The hardline Jews stubbornly maintained that their god, Yahweh, belonging exclusively to the Jews, and to no other humans who were not born Jewish. These issues soon developed into hostilities and unrest in the multi-cultural societies of Palestine and Judea.

With the death of Alexander in Babylon, 323 BCE, his friends, generals and their heirs began to break up his empire by feuding and fighting amongst themselves. The more successful of these, the ambitious Antiochus III, sought to extend his empire into Greece. In trying to do so, he was defeated by a new 'Dominant Male' western power of Rome, to whom he was forced to relinquish (in 188 BCE) all of his territories along the coast of Mediterranean. During the term of Antiochus IV, the entire region suffered from internal dissensions, family feuds and Parthinian, Armenian, and various Arab invasions.

Antiochus IV tried to unite the Jewish and gentile populations of the region by fostering Hellenization. While this was accepted by many Jews, Antiochus IV turned out to be his own worst enemy. By plundering Jerusalem and forcefully converting the Jewish Temple into a temple dedicated to a Greek god, Zeus Olympus, all his earlier efforts turned out to be futile with the Jewish Maccabean rebellion of 166 BCE.

After years of armed conflict, the sons of a Jewish priest Mattathias of Hasmonean's achieved a Jewish rule over most of Palestine. After a century of Hasmonean rule, with a succession of sons and male blood relatives taking upon themselves the combined mantles of kings and high priests, the dynasty began to fracture. By re-introducing Hellenization – the very element that their forebears fought against in the first place – they inevitably invited opposition from Jewish religious groups, each with their own 'Spiritual Males' to attack them. Apart from this discord, a civil war erupted between 'Dominant Male' heirs of the Hasmonean dynasty.

With all the revolts causing deteriorating social stagnation in Palestine and Syria, Roman general, Pompey, restored order in Syria then captured Jerusalem in 63 BCE. The Jews were now under Roman control. The Romans combined all of the western Syria and Palestine into a new province of Syria, with its capital at Antioch.

In 47 BCE, Romans had appointed Herod as the ruler of Judea. This did not prevent continued conflicts flaring up between 'Dominant Males' in

Palestine and elsewhere. The invading armies of Parthian kings of Persia and Mesopotamia fought and won against the Romans, while two Roman generals, Julius Caesar and Pompey, fought each other in a civil war. In 37 BCE, Mark Antony restored Roman dominance in the eastern provinces, regaining Jerusalem for Herod.

Herod (the Great) was a skilful politician, managing for a long period to maintain tranquillity between the Jews and Hellenistic gentiles in Palestine. This didn't prevent a constantly opposition from Pharisees and the Zealots – the conservative factions of Judaism. To them Herod was a foreigner, despite being a protector of Jewry within and outside his kingdom, and the re-builder of their Temple. Towards the end of his life, due to his illness, Herod became increasingly mentally unstable and intolerant: even executing two of his sons.

After his death, Herod was vilified and slandered by nameless Jewish scribes, who accused him of being the slayer of all the male infants of Bethlehem. The spread of this lie was done intentionally, as means of distabilising Roman influence.

On Herod's death his kingdom was divided between his three remaining sons, of whom Archelaus was deposed by Rome in 6 CE (Common Era). He was replaced by a series of Roman prefects, including one Pontius Pilate. Judea was absorbed into the Roman dominion as a minor province.

The closure to new entries for the Hebrew Bible did not mean that new literature, intended to stimulate and inspire Jews to some particular action, was not written. It went on being written. Even the same biblical style of writing was used. The difference between the new writing and the old scriptures was that many of the new writers wanted to present new religious, political, and social notions, learned from the cultures of Persians and Hellenic gentiles. Such new ideas had many critics, leading to disagreements and physical conflicts.

In the period just before and after Herod's death, different Jewish religious factions were trying to attract participants to their cause, so as to enrage them to the level of an uprising. In striving to acquire influence and importance for themselves, the main Jewish religious factions were at odds with each other, not so much due to the differences in their religious dogmas but because of their constant agitation of each other.

The Pharisees were progressive in application of religious law, but sticklers for tradition: rejecting all things considered unclean and unholy. Their greatest triumph was in launching and maintaining the Maccabean

CONVERTING A JEWISH POLITICAL CONCEPT INTO A NEW RELIGION

revolt. They were cautiously anti-Roman. They believed in the resurrection of the dead.

The Sadducees comprised of conservative priesthood and land-owning aristocracy. They collaborated with the Roman authorities. They rejected the notion of the resurrection of the dead, because they found no mention of such a doctrine in the Hebrew Bible.

The Zealots were active revolutionary plotters against the Roman occupation and oppression.

The Essenes, whose copies of religious texts had been found as the so-called Dead Sea Scrolls, lived in secluded communities. Their founder, known as the 'Rabbi of Righteousness', believed that he knew the interpretations of the prophets in a way that this was obscured even to the prophets themselves. The Essenes were expecting two messiahs: one, as a royal representative of David, the other as a priestly representative of Aaron. They believed that when the end comes they alone would be saved.

While these religious rivals agitated and misbehaved, the Romans continued to maintain their occupation of Jewish territories. With no one party trusting the other – everyone relied on spies. The main Temple in Jerusalem remained the centre of religious, civil, legal, educational, and business operations; all of these activities controlled in some manner by the priests of the Temple. The Temple was big business, and the priests were adamant that this remains so.

In such squalid politico-religious atmosphere of insecurity, suspicion, and bedlam, there were the educated Hellenistic Jews – who belonged to no particular factions, or to more than one faction – and who were developing messianic/utopian/socialist views as future direction for Jews. The ideas that these Hellenistic Jews were devising included unique notions of Jewish equality, for the purpose of subtly organising an accord between the various religious factions, so as to unite Jews into a patriotic Jewish front against the Roman occupation. This was one of the early attempts in human history to apply a concept of communism for a nationalist purpose. Not that such idea hadn't been voiced before.

For instance, Yeishu ben Pandeira, a Jewish rabbi preached utopian-revolutionary ideology. He was stoned to death by Jews, around 100 BCE, for treason and supposedly casting of spells.

With the continued Jewish social unrest after Herod's death, the devastated and unsettled Jewish population had begun to anticipate the prophecy of the coming of a Jewish messiah: a liberator. They wanted a heavenly warrior who would win them a Jewish state – a Jewish kingdom; who would provide them with freedom to worship, yet who would not be a dictator by benevolently sharing power with other Jews.

The notion of a messiah, derived from Hebrew 'mashiah' (anointed) is not unique to Jews alone, as many religions have expectations of a spiritual redeemer from god. By presuming 'time' to have a physical existence, a messiah was – and for many still is – expected to arrive at a particular moment from future, when that future 'time' comes in alignment with the present. The expected Jewish messiah was to be heralded by a period of human turmoil, which he would then convert to better and happier days in human lives.

The new Judo-Hellenistic religious doctrines experienced by the Jews, gave more relevance and fervour to the Jewish messianic religio-political movements throughout those periods. There were a number of Jewish messiahs following the death of Herod. Of the many humans presenting themselves as messiahs, the three prominent messiahs crucified by Roman authorities were: Yahuda of Galilee (executed 6 CE: Common Era), Theudas (executed 44 CE), and Benjamin the Egyptian (executed 60 CE).

Due to his good fortune and ability, Herod's grandson, Marcus Julius Agrippa (referred to as Herod Agrippa I in *The New Testament*), was able to pacify the orthodox Jews while governing Herod's former kingdom, from 41 CE, until his death in 44 CE. Following his death, the compromises between the Roman and Jewish authorities fell apart, and the conflicts between the orthodox and the Hellenistic Jews – those faithful to their new socio-pacifist doctrines – had recommenced. This conflicts between them and Jewish population disturbed the Roman occupiers, considering that the Jewish defiance was rapidly increasing in scope and volume.

In this very period of Jewish religious and civil unrest, when:
- Messiahs were sought by Jews...
- Individual Jews presented themselves before Jews as messiahs (and were killed by them for doing so)...
- Rebellious teachers (rabbis), with Hellenistic influences, instigated new ideas for a social change in Jewish society...
- Population comprising of differing ethnicities and cultures, speaking in different languages constantly misinterpreting and misunderstanding one another...

CONVERTING A JEWISH POLITICAL CONCEPT INTO A NEW RELIGION

- Roman authorities displeased with rebellions, and who themselves were feared and hated...

...No-one would have foreseen that the Jewish religious utopian-revolutionary ideology would be turned into a new religion that would influence the whole of human existence for over the next two thousand years. The difference to this Jewish religious ideology, from the one practiced by majority of Jews, was that as part of its nationalism it also had at its basis a utopian concern for social human equality, structured on mutual respect and compassion.

Initially, this newly developing ideology was cautiously presented by patriotic Jewish rabbis (versed in the Greek ideals of democracy and Zoroastrian ideas of a benevolent and forgiving god, for whom all are supposed to be equal) to other rabbis and influential Jews. The intention was to engineer a method for Jewish masses to opposing their Roman occupiers by civil unity, rather than civil disobedience and physical confrontations that were occurring everywhere.

The problem with trying to achieve this was that those whom they tried to convince to embrace socialism, equality, and compassion were only accustomed to understanding personal selfishness, an observance of class system, and suspicion and mistrust of others. This meant that the only way the new ideals could be presented with gravitas of conviction to the Jewish population was to repeat the original idea of Moses presenting his god Yahweh to the Hebrews. But instead of Moses and Yahweh, there would be a new version of Moses and of Yahweh, where Moses would now be John the Baptist and Yahweh would be the son of god, Jesus.

And so, the 'Jesus Myth' came to be, in the form of a 'superman' of that era, whose words depicted the political desires of the Hellenic Jews who invented him, with these aspirations presented in guise of superhuman miracles of wishful thinking, not unlike the superhuman miracles depicted in current Superman comic books.

Until 'Jesus Myth' was written and presented as *The Gospel According To Mark*, the Jewish population continued to await the arrival of a messiah welding a sword not a word. They wanted a defining change in their lives achieved by a superhuman worrier who would take on the Romans on their behalf. They wanted an action man, and not the philosophy devised by some of their elite and the scholarly, who secretly attempted to distribute a concept of united Jewish state achieved by non-aggressive means, by advocating the notions of liberty, equality and fraternity applicable to all Jews, as presented by *The Gospel According To Mark*.

PART 2. DISCLOSING THE SHAM OF HUMAN CONSCIO PERCEPTIONS

So why did *Mark* begin a transformation in many Hellenic Jews and gentiles from following the Jewish religion to that of forming a new religion of Christianity? Because the political ideals presented in *Mark* could easily be applied to any human society of the period, not just the Jews. And furthermore, there was the newly invented awesome concept that gave every human the ability to believe in acquisition of eternal life-after-death. If Jesus could ascend to 'heaven' after death, and have an eternal after-life, then so could everyone else, irrespective of their station in life, just so long as they worshipped Jesus and his Father-the-God. (The notion of The Holy Ghost was devised later by 'Spiritual Males'.)

Being the first of four *Gospels* incorporated into *The Bible* (but placed after *Matthew*, which was written much later), its socio-political meanings depicted in *Mark* are easily recognised, unlike the other three Gospels, which misinterpreted *Mark* more and more, until all its political lessons were distorted beyond recognition.

What can also be clearly recognised in the first and the only original literary work inventing the notion of Jesus, is in that the writers of *Mark* followed the successful old Levi formula of god invention, so as to convert the popularly expectant messiah into a new god, Jesus, who was meant to sell to Jewish population the new political ideology incorporated in a religious disguise.

Fist of all, a duplicate of Moses was needed in order to introduce duplicate of Yahweh, for only a persona of god could have the ability to convince the doubters of the veracity of the newly presented notions.

With *The Gospel According To Mark* being the fists book to introduce the 'Jesus Myth', it stands to reason that the crafty scribes who wrote it replaced Moses with that of John the Baptist.

[Mark 1:1-4] *The beginning of the gospel of Jesus Christ, the Son of God. As it is written in Isaiah the prophet,*
"Behold, I send my messenger before thy face, who shall prepare the way; the voice of one crying in the wilderness:
Prepare the way of the Lord, make his paths straight – "
John the baptizer appeared in the wilderness, preaching a baptism of repentance for the forgiveness of sins...

[Mark 1:6-8] *Now John was clothed with camel's hair, and had a leather girdle around his waste, and ate locusts and wild honey.*
And he preached, saying, "After me comes he who is mightier than I, the thong of whose sandals I am not worthy to stoop down to untie. I have baptized you with water; but he will baptize you with the Holy Spirit."

In other words, John the Baptist is introduced with a reference to a familiar figure to Jews: the prophet Isaiah, so that this John is given instant credibility by a direct association. This John is a new version of Moses. He also begins his presence out of nowhere (wilderness) with no preceding explanations, only to disappear from any further narrative after introducing the new version of Yahweh, in form of Jesus. And just like Moses, the introduction of his new god (in form of Jesus) is based on an assertion that while he is "mighty" (as Moses supposedly was) the one yet to come is greater still, being a 'Son of God', by that signifying that he whom he is to baptize is indisputably, and beyond any doubt a god in his own right. How is that for presenting a convincing assertion?

Because all Jews were certain of the existence of their singular god, Yahweh, the scribes writing *Mark* could not possibly replace him with another similar singular god, so they cleverly devised an addition: a 'Son of God' god.

[Mark 1:10-11] *And when he came up out of the water, immediately he saw the heavens opened and the Spirit descending upon him like a dove; and a voice came from heaven, "Thou art my beloved Son; with thee I am well pleased."*

Jesus, therefore, was presented as a human, who waked amongst other humans; who could be seen and touched; a god who as a human could understand other humans, and by that had an affinity to human needs and wants; a god whose intentions were for the benefit of all humans.

This attribute of humanness was taken up by the writers of *The Gospel According To Matthew*, who placed on him a label of sordid parentage, something that could occur within human society.

[Matthew 1:18-19] *"When his mother Mary had been betrothed to Joseph, before they came together she was found to be with child of the Holy Spirit; and her husband Joseph, being a just man and unwilling to put her to shame, resolved to divorce her quietly."*

Still, this was also a god who proved his godliness by ascending to 'God the Father' after his human death, something that the inventors of Yahweh had not thought of at all. After all, in *The Hebrew Bible* Moses had supposedly died and was buried by unknown persons in an unknown place: he was never given the ability to ascend to god, something that the inventors of Jesus had allocated to every human life.

One of the elements of the 'Jesus Myth' – continued from 'Moses-Yahweh' story – is the role of god. In the 'Moses-Yahweh' tale the god, Yahweh, is basically a supporting actor, while the main emphasis is placed on the Hebrew-Israelite prophets and kings. So it is not surprising that only a small level of importance is given to god in the 'Jesus Myth'. God (who is no longer called Yahweh) remains but a minor stand-in actor with virtually no lines, while Jesus, the man-god, is the major star, with all the emphasis beaming on him in every scene.

Even the disciples are depicted as secondary characters with no depth, used as rather dim-witted supporting cast to their master, and his wise and prophetic speeches.

Just as with Moses, in *The gospel according to Mark* Jesus is presented as someone who lived in the past (so that no living persons could dispute his existence, as there was none), and abruptly introduced him as an adult who makes a sudden appearance from a certain location but without any explanation of his past (his past is added by other 'gospel' writers, some hundred years later).

[Mark 1:9] *In those days Jesus came from Nazareth of Galilee and was baptized by John in the Jordan.*

The setting in which this invented Jesus is placed, refers to some existing localities and towns that would have been known to the authors and to the readers, so as to give the story a feeling of authenticity and realism, but it does not place him in any precise historical year. The reason for that was simple. Most towns of the region and of the period were no bigger than large villages, where everyone knew everyone else. Even in the countryside the peasants knew by sight, if not by name, all those who went by on the roads. To give the 'Jesus Myth' a definite historical timing would have meant that there could be those still alive disputing the claims made by authors of *Mark*. These individuals could have been exactly where Jesus of Nazareth was supposed to have been at a particular hour and day, and by knowing for certain that there was no Jesus there, they would have understood that Jesus was a hoax. Besides, Jesus supposedly always attracted large crowds, so it would be unlikely for anyone to have missed all the commotion if he really was there.

Therefore, by intentionally not indicating precisely when Jesus was supposed to have been either as a child, or a youth, or as a young man or even as a mature man, the 'Jesus Myth' was free to fluctuate in indefinable years, making it impossible for anyone to pin down his presence not just to any real historical events but even to any specific years. Such avoidance made it difficult to dispute his existence. The later versions of the Jesus legend altered this era vacuum slightly, by linking Jesus with the period of Herod, but still avoiding what Jesus was said to have done in what particular years.

CONVERTING A JEWISH POLITICAL CONCEPT INTO A NEW RELIGION

With the resentment of Herod still remaining close to a century after his death, the next generation of 'Jesus Myth' authors added two components to the myth: their dislike for Herod, by having him portrayed as an ogre who desires to kill Jesus at childbirth, and the veneration of Jesus, which added more sentimentality and endearment to the myth. By associating Jesus with a historical figure of Herod, they artificially provided the Jesus legend with historical authenticity.

This, of course, was a favorite method by which all the Hebrew-Israelite-Jewish authors connected their non-existent spiritual heroes to some historical events, so as to give their fakes a veneer of physical and historical reality.

Once the fable of Jesus was presented in writing, many other writers (writers being what they are, especially before the invention of intellectual property laws that forbid plagiarism) could not resist being involved with a new myth: making their own versions of it, derived from their own imagination. By such process of character development, Jesus was gradually altered by new writers from being a messianic Jesus of Nazareth, 'Son of Man' in *Mark*, to that of 'Son of David', and then, 'Son of God'.

Until his myth was first released in a written document as *The Gospel According To Mark*, in the mid of the so-called first century CE, the legend of Jesus was basically unheard of in all of Palestine, or anywhere else for that matter. Of course, tales and rumors of messiahs and their deeds – such as that of Yahuda of Galilee (executed 6 CE) – circulated throughout Palestine and Judea for ages, but the names of those self-proclaimed messiahs were never that of 'Jesus'.

Since then, multitudes of humans have been trying to obtain a physical proof of his existence, without any success. These humans try to rationalise the miraculous abilities attributed to Jesus by his inventors, such as walking on water, feeding the multitudes, and his healing of the diseased and rising of the deceased, against the physical reality that such miracles are physically impossible to achieve. And yet, it requires but a small amount of logic to establish the non-existence of a miracle-making Jesus.

By understanding human (conscio) behaviour, which continues to be repetitively predictable, irrespective of the epoch in which human behaviors have taken place, it is possible to establish with absolute certainty whether particular events had, or had not actually occurred. In order to establish the nonexistence of Jesus, all that is required is to consider the written claims made of his miracle-making ability.

PART 2. DISCLOSING THE SHAM OF HUMAN CONSCIO PERCEPTIONS

Presently on the planet, as in the past, wherever and whenever a so-called 'miracle' occurs, it may seem like half of world's population wants to go there and witness it for themselves, and the other half wants to hear, talk, and read about it. Humans of ancient civilizations made pilgrimages to 'holy' sites – just as humans do now – in order to visit the relics where miracles had supposedly occurred, or visit those who supposedly produce miracles. Though all miracles turn out to be either occasional flukes of natural physical occurrences, or self-delusions, or deceptions, this does not prevent miracles from being noticed and recorded by humans, due to their desire to satisfy their curiosity or to record interesting phenomena.

Current humans (conscios) may not realise this, but humans living in the past were no different in character and behaviour to those of the present. This means that whenever any claims of miracles took place within any past societies, those announcements would cause a sensation amongst the populace, and by such commotion would catch the attention of authorities. These authorities would officially record any unusual events – as these may have been responsible for future unrest – bringing them to the attention of their superiors.

It should be further pointed out that Roman authorities were very thorough in recording events, as vigilance from officials was rewarded with promotion, while negligence was punished. The Roman authorities in Palestine would hardy have allowed for any derelictions of these duties. In fact, they were especially vigilant of events taking place in their provinces where there were ongoing conflicts and rebellions. It is, therefore, highly unlikely that all the Romans on their guard in Palestine would have failed to notice and to make official records, and dispatches, about a human who had the powers of a god or a superhero: having an ability to heal the sick, the deaf, the mute and the blind; revive the dead; cure lepers; walk on water; turn water into wine; and instantly multiply limited quantities of food – and so on, and so forth. The reason that the Romans in Palestine made no mention of Jesus, and had no records on Jesus, was simply because he never existed, and having never existed, could not have made the miracles by which he would have been noticed by the Roman authorities, as well as, the Jewish and gentile population of that region and of that period.

In addition, any human who could publicly brings the dead back to life would become an instant celebrity superhero, with his fame spreading far and wide by word of mouth, with all the scholars, historians, healers, and all national officials from all over the region – including Romans – wanting to meet him. After all, having the acquaintance of someone who heals everyone, and raises the dead, could be very helpful to one's health. Were such a person to have existed, the Roman offices in Palestine, and in Rome, would have been filled to the rafters with official reports and documents on such a miracle-maker. Ordinary Jews and the Jewish priests alike would have instantly worshiped such a superhero. Romans – who loved all kinds of gods – would have especially loved to claim him for their own (as they actually did later with directives from Roman emperor Constantine): a living god producing miracles. After all, how could one not worship such a human who could do what is physically impossible?

In fact, then as now, such a miracle-making superhero would have been constantly mobbed by massive and relentless crowds of the infirm, the crippled, the diseased, the amputees and the dying, both individually and those supported by their relatives. Rich and poor, they would have come from all over the Palestine, if not the Middle East; all of them desperately seeking miraculous healing in the age when there was no health care. Anyone, with even a fraction of healing abilities attributed to Jesus, would have had no privacy at any hour, day or night, because the infirm, the crippled, the diseased, and the dying – then as now – are persistent and demanding: they want their cure, and they want it now!

But then there would also be the demands for cure from the not so ill, but those with cuts, and with broken limbs, those with toothaches, with earaches, with colds, and any other physical problems, such as diarrhea. A pitiful, ongoing, constant, never ending sequence of pleas, requests, and demands would follow anyone capable of providing them with free and instant cures. And were any of them to be cured, then those cured would have been replaced by twice that number the following day, and the next day, and the next day.

Were such a human miracle-maker to have existed, that human would have been a legend in life, not just after death. Such a miracle-making superhero human would have had thousands of monuments erected in his life to honor such a gift. In his life, such a human would have been worshipped as a living god by one and all, with written records in ad nauseam depicting his miracles. During his life, such a miracle-maker would have had his fame spreading out beyond the confines of Palestine all the way to Rome. Such human miracle-maker would have been summoned to Rome, accompanied by thousands upon thousands of the curious and the sick along the way. Such human miracle-maker would have influenced a change in human history during his life, for by his method of healing, miracles would no longer be viewed as miracles but as an every day occurrence – the normal way of preventing diseases, afflictions, and death, with the former dead walking happily amongst the living.

Of course, no such miracle-making event had ever occurred in either Palestine or Judah, or anywhere else on this planet at any period, because human miracle-makers never existed, do not exist, and never ever shall exist, as physical elements and atoms cannot be instructed by miracles to return to where they long departed.

Fiction writing, however, is not restricted to any laws of physics, for fiction is simply a human brain's output of imagination produced on behalf of its demanding conscio. In the same way, all the human conscios who had ever supported religions, and fables and myths, such as those of Jesus, do so because they want to believe, with this need to believe being not restricted by any physical reality. They believe without caring whether or not what they want to believe in is a lie or a physical truth; they believe because they like to believe in hope, and they believe because they want to be right. So much like the beliefs of any addict.

There is just one single brief reference given to Jesus, but not from Roman authorities and not from any Roman or Greek historians. The one single passing mention of Jesus comes in the Book XVIII, of a twenty book work of *The antiquities of the Jews*, completed in 93 CE by a Jewish priest and scholar Joseph ben Matthias, historically known as Flavius Josephus. In *The antiquities of the Jews*, he haphazardly rewrote the fiction of the *Hebrew Bible*, presenting it as a Jewish history, while deleting all the vile prophets and the general Jewish fanaticism, so as to present Judaism to the gentile readers as a sensible religion.

In his single short reference to Jesus, Josephus makes no qualified historical revelation of proof, only acknowledging the existence of Jesus. In doing so, there is fact that needs to be pointed out about Josephus. As a historian he was no historian, just a re-processor of existing material, which he would rehash without any new research, providing no new uncovered material that would add new factual information to historical events.

As a Christian sympathizer, he would have been more that versed on Christianity, considering that copies of *The gospel according to Mark* had been in circulation for over forty years before he produced The antiquities of the Jews. It is also historically acknowledged that a later copyist has tampered with his comment about Jesus. His brief reference to Jesus provides no physical proof of Jesus' physical existence, as he merely expressed a biased, unsubstantiated conjecture of hearsay: a mere indication of approval to an existing myth.

In reality, with exception of some historical figures mentioned in *The New Testament* – so as to provide some authenticity to the fictitious events described in *The New Testament* – all the humans mentioned in *The New Testament*, including Jesus, all his disciples, his family and all his friends, are fictitious characters, including both Peter and Saul (Paul), for outside *The New Testament* they are totally absent from any Jewish, Greek and Roman official records that exist of that period. It is presumed that there are letters of correspondence written between the likes of Peter, Paul and the authorities of their churches. So where are these letters? There are none in existence. Indeed there are various fake documentations in existence, mostly from the Middle Ages (not unlike that of 'Shroud of Turin) but nothing tangible and real.

Of course, there were many 'Christian' churches set up by commercially astute 'bishops' who saw a way of gaining some wealth from a new religion (in a way of all religions), who conducted correspondences the records of which do exist, but they were no former disciples of Jesus.

And for all the basilicas, churches, and cathedrals dedicated to the likes of Peter, Paul, and Mary, apart from hearsay, deceptions and myths, there is no physical proof – whatsoever – to support the physical existence of any of them.

There are a number of ironies to the 'Jesus Myth'. One of those is that the utopian-nationalistic ideology presented by the Hellenistic Jews for communist ideals ('communist', derived from a 'commune' or 'communal'), who wanted to organise a kingdom of the Jews during the Jewish uprisings (between the period of Herod's death and the Jewish revolts of 66-70 CE), did not altogether go unnoticed by the Jews. For it is these exact concepts that were adopted and practiced by every Jewish 'kibbutz' (collective or 'commune') in the new state of Israel, after 1948 CE.

After all, it is not difficult to grasp the meanings of loaves and fishes presented in *Mark*, in guise of miracles performed by Jesus and his disciples in feeding the masses: that being the first in history proclamation of communist ideology, stating that a society that shares equally its bounty of the land and the sea cannot experience hunger, but instead achieves a surplus:

[Mark 8:19-21] *"...When I broke the five loaves for the five thousand, how many baskets full of broken pieces did you take up?" They said to him, "Twelve." "And the seven for the four thousand, how many baskets full of broken pieces did you take up?" And they said to him, "Seven." And he said to them, "Do you not yet understand?"*

To further illustrate the methods by which the conspiratorial Hellenized Jewish revolutionaries communicated their ideas to those whom they wanted to influence, is shown in the following quotation from *The Gospel According To Mark*.

[Mark 4:10-12] *"And when he was alone, those who were about him with the twelve asked him concerning the parables. And he said to them, "To you has been given the secret of the kingdom of God, but for those outside everything is in parables; so that they may indeed see but not perceive, and may indeed hear but not understand; lest they should turn again, and be forgiven."*

To paraphrase the passage: "You, whom I trust as I trust my closest twelve conspirators, already understand our quest to form a Jewish state – a kingdom of Jews. But in seeking supporters for our cause, our message has to be carefully presented, so as to disguise its real intent from the spies and the faltering who are present in the crowds. We have to be careful not to articulate too precisely our intent to some supporters, in case those who are doubtful and wavering in indecision may change their minds of supporting our cause and denounce our anti-establishment intentions to the authorities, in exchange for rewards."

By presenting a restrained but honorable revolutionary as a way of explaining the methods by which a Jewish state could be achieved, confirms that the original use of Jesus symbol was intended for political rather than a religious purpose. The "kingdom of God" was not a place in heaven but an intended Jewish state on Earth, as otherwise there would be no reason to maintain secretive use of parables. After all, Romans may have been tolerant and accepting of all religions, but not of any subversive anti-Roman sentiments.

The argument for a strongly united and determined Jewish state repeats throughout *Mark*:

[Mark 3:24-25] *"If a kingdom is divided against itself, that kingdom cannot stand. And if a house is divided against itself, that house will not be able to stand."*

And:

[Mark 3: 27] *"But no one can enter a strong man's house and plunder his goods..."*

[Mark 4:25] *"For to him who has will more be given; and from him who has not, even what he has will be taken away."*

All these references relate to how best to secure and maintain a Jewish state.

The miraculous healing-ability, attributed to Jesus, was simply another method used by the inventors of Jesus in *Mark* to express their hidden desire for equality of all Jews. For them, all the unclean spirits, lepers, and the diseased, were merely cryptic expressions depicting those Jews who were either unsympathetic to the notion of equality, or those who were lowly humans, shunned and disrespected by the rest of the Jewish society, or those who were ethnic non-Jews, living amongst the Jews.

By having Jesus produce his miraculous work with the ill of Jewish society, the authors of *Mark* were indicating that it is necessary to cure Jewish prejudices and their opposition to the concept of equality, before a Jewish kingdom can be achieved. In 'healing', the intent was to heal the Jewish inequality and societal divisions of the period, and not the physical healing of the bodies. That is why Jesus is not credited with healing broken limbs, or colds or diarrhoea.

Some doctrines of equality presented to Jews by various revolutionary rabbis and Hellenic Jews of the period – as endorsed by the inventors of Jesus – would have seemed very alarming to many Jews. For instance, despite reassuring statements that a Jewish national strength could be achieved from sharing possessions between all the Jews – as symbolized by loaves and fishes – there were other, more radical proposals. The most radical of these (later actually employed by communist nations), was expressed in the following passage, which begins innocently enough, with a wealthy, virtuous young man making an enquiry of Jesus.

[Mark 10: 17] *"Good Teacher, what must I do to inherit eternal life?"*

Jesus replies that he should sell all his possessions, hand out the proceeds to the poor, and then join Jesus in his quest. On hearing this advice the wealthy, virtuous young man departs disappointed, obviously not anxious to be parted from his wealth. This is when the authors of 'Jesus Myth' disclose their true political and social intentions, when they have Jesus make an astonishing declaration to his disciples – the very thought of which supposedly frightens them.

[Mark 10: 23-25] *And Jesus looked around and said to his disciples, "How hard it will be for those who have riches to enter the kingdom of God!" And the disciples were amazed at his words. But Jesus said to them again, "Children, how hard it is for those who trust in riches to enter the kingdom of God! It is easier for a camel to go through the eye of a needle than for a rich man to enter the kingdom of God."*

The reason for the astonishment of the disciples was not because some young Jew wanted a place in heaven without giving up any of his possessions. In his life, such a young Jew would have already given a small fortune in gifts and sacrifices to the priests. What astonished and frightens the disciples was the revelation that the rich Jews could have no place in the state of the Jews, or the kingdom of God (kingdom of the Jews).

With these words the authors are disclosing their proclamation that not only should the Jewish state be based on equality of all Jews through tolerance and respect, but that the equality should apply to all of their possessions, as well. They declare that if the rich choose to belong to the kingdom of the Jews, they would have to give up all of their possessions to the state, or to the kingdom of the Jews.

[Mark 10: 26] *And they were exceedingly astonished, and said to him, "Then who can be saved?"*

Meaning: who can be exempt from giving up all of their possessions, for we, ourselves, are not rich, but we do possess our houses and fishing boats.

[Mark 10: 27] *Jesus looked at them and said, "With men it is impossible, but not with God; for all things are possible with God."*

Meaning: no human can be exempt from sharing communally all of that they own, but afterwards, once the state of Jews is established, the Jewish state will help all; for all things are possible in a communal state.)

[Mark 10: 28-31] *Peter began to say to him, "Lo, we have left everything and followed you. Jesus said, "Truly, I say to you, there is no one who has left house or brothers or sisters or mother or father or children or lands, for my sake and for the gospel, who will not receive a hundredfold now in this time, houses and brothers and sisters and mothers and children and lands, with persecutions, and in the age to come eternal life."*

Note the word 'persecutions', as this is where the authors openly declare that the state would need to resort to persecutions of the rich, in order to force them to share with the state their possessions: rewarding those who willingly contribute to the state with redistributed wealth during their lives, and rewarding them with eternal life after death. This is confirmed with the following words from Jesus:

[Mark 10: 31] *"But many that are first will be last, and the last first."*

Meaning: in the kingdom (state) of the Jews, the current rich and the leaders who regard themselves as elite will no longed be so, whereas, those who were deprived will be advanced, and looked upon with favour. (That is exactly what was to occur with the communist uprising in Russia, thousands of years later.)

It is surprising that *Mark*, with its call for a social revolution amongst the Jews, was permitted to be included in *The New Testament*, even if it was written in the middle of Jewish upheavals that led to the destruction of the Jewish Temple, in 70 CE (Common Era), by the Romans. But obviously being the original source to Jesus legend it could not be ignored. Instead, by censoring and deleting some very socialist-specific passages (a function that was certainly exercised, judging by the brevity of content of *Mark* in comparison to the later versions of *Gospels*), and by placing it after *Matthew*, the notion of the Jewish nationalism was diminished in favour of transforming Jesus into an idol to be worshipped.

There is a vital point of difference between mythologies and religions. While mythical legends are freely accepted to be nothing but human fantasies, the religious fiction is zealously defended to be true. There is, also, a point of similarity between mythologies and religions. Despite being known to be fantasies, the legendary heroes and gods of myths had humans acting as their priests and oracles, taking care of temples built for the mythical gods, and collecting tributes and gifts on the gods' behalf (but basically for themselves). This procedure is very similar to that of religious gods, where humans acting as their priests take care of temples – or churches, or cathedrals – built for them (the priests), and collecting tributes and gifts for their god, or for their order, or for the church (but still basically for themselves). From these points of difference and similarity it is possible to ascertain that whether legends are mythological or theological, there will be humans who are prepared to exploit these fantasies for their own, human, profit. Such was, and is, the case with the legend of Jesus.

Those who came after the inventors of Jesus, had recognized that there were opportunities to exploit monetary revenue and donations from the faithful, the hopeful, the fearful, and those expecting miracles from their new prophet: a son-god. It would be they and not the Jewish priests who would own 'the Jesus brand' and earn tributes. After all, the 'Jesus Myth' had the selling power and could appeal to anyone, not just the Jews.

And as mentioned earlier, what really made the 'Jesus Myth' so appealing to many Hellenic Jews and gentiles of that period were the inventions of ascendency and resurrection – when Jesus would supposedly return to Earth from heaven on 'the judgment day' (appropriated from Zoroaster), when all humans would be judged for their sins and punished accordingly.

Here was the USP (unique selling proposition), as modern marketers would refer to the difference between the Judaic religion with its horrid god Yahweh, and the new Christian religion where god-the-father was kind and welcoming of anyone, and where each worshipper could have eternal life-after-death and be fairly judged for any sins on 'the judgment day'.

These notions were especially appealing to the poor, who could imagine themselves ascending to heaven and then seeing all the greedy and rich being punished on 'the judgment day'.

Still, it was soon discovered by those who made Jesus into their intellectual property, that it is not enough just to own the product. In order to produce revenue, the product requires a worldwide distribution of franchises. The problem with franchises (initially comprising of independent groups of Hellenistic Jews and gentiles forming their own ministries), is that without legal copyright ownership of the product (where the product is the 'Jesus Myth') it is impossible to hold onto the ownership of the product and regulate it, because all who are involved feel that they also own the product and present their own version of it.

Some wanted to be priests from piety, or desired contributions, or desired to share the aura and authority of Jesus, so that they could dispense their self-important influence over others. Some enjoyed the thrill of danger at being rebels for god, as members of a banned religious sect. And others still, wanted to be martyrs, because that would supposedly provide them with a certain entry to heaven: for what is the consequence of danger in comparison to a promise of eternal life and a place by the side of Jesus, right next to God-The-Father? The appeal of such reckless abandon in the name of god had influence some religious zealots to lose their lives back then, just as such practice continues to be used by current religious fundamentalists.

The element that eventually united these sects was an accepted brand name of 'Christians', derived from Greek 'khristos' (anointed one), and the continued writing of literature on the subject of Jesus. There were many Gospels written in the name of non-existent prophets, besides those of *Mark*, *Matthew*, *Luke*, and *John*. It is most probable that there were Gospels written in the names of individual disciples and probably even in the name of Virgin Mary.

Of all the *Gospels*, four were chosen for inclusion into *The New Testament*. The original, *The Gospel According To Mark*, written in crude, plain Greek; then it was *The Gospel According To Matthew*, with a strong Judaic style; followed by *The Gospel According To Luke*, a Hellenistic work of literary style, written outside of Palestine; and finally *The Gospel According to John*, a text of simple Greek.

As there was no notion of Jesus prior to *The Gospel According To Mark*, each new version of *Gospels* had used *Mark* as its source. The amusing feature of the three later *Gospels* is that these works plainly show that while each later version had adapted the information obtained from a previous version, whether through poor translations, poor scholastic abilities, or with full intent, they all misunderstood the original meanings presented in *Mark*, providing instead their own distorted versions of events and meanings.

To provide an example of this, here is a passage from four different *Gospels*. They begin with *Mark*, as *Mark* was the first version that started the legend of Jesus. This particular passage (which is not unique to being used as an example, as other passages could have been used with similar effect), deals with the arrest of Jesus, and clearly shows how each successive adaptation adds more distortion, until, by the third adaptation, the original meaning of *Mark* is obliterated.

[Mark 14:46-51] *And they laid hands on him and seized him. But one of those who stood by drew his sword, and struck the slave of the high priest and cut off his ear. And Jesus said to them, "Have you come out as against a robber, with swords and clubs to capture me? Day after day I was with you in the temple teaching, and you did not seize me. But let the scriptures be fulfilled." And they all forsook him and fled.*

And a young man followed him, with nothing but a linen cloth about his body; and they seized him, but he left the linen cloth and ran away naked.

In this passage, 'slave of the high priest' is a spy in the service of the high priest, while 'and cut off his ear' means: the high priest lost his spy because his spy was struck down, and not that the spy lost his ear. There is also description of a young male in a linen cloth, which is left out by other authors, as it made no sense to them. What that youth represented was either a sabogeat (homosexual) or a male prostitute, who was meant to indicate that such people also had the right to share in the equality of all Jews. "And they all forsook him and fled." This indicates that the crowd felt shameful and left Jesus alone.

Now compare the same passage from *Mark* to that of *Matthew*.

[Matthew 26:50-56] ... *Then they came up and laid hands on Jesus and seized him. And behold, one of those who were with Jesus stretched out his hand and drew his sword, and struck the slave of the high priest, and cut off his ear. Then Jesus said to him, "Put your sword back into its place; for all who take the sword will perish by the sword. Do you think that I cannot appeal to my Father, and he will at once send me more than twelve legions of angels? But how then should the scriptures be fulfilled, that it must be so?" At that hour Jesus said to the crowds, "Have you come out as against a robber, with swords and clubs to capture me? Day after day I sat in the temple teaching,* and you did not seize me. But all this has taken place, that the scriptures of the prophets might be fulfilled." Then all the disciples forsook him and fled.

Now, the same passage from *Luke*.

[Luke 22:49-54] *And when those who were about him saw what would follow, they said, "Lord, shall we strike with the sword?" And one of them struck the slave of the high priest and cut off his right ear. But Jesus said, "No more of this!" And he touched his ear and healed him. Then Jesus said to the chief priests and captains of the temple and the elders, who had come out against him, "Have you come out as against a robber, with swords and clubs? When I was with you day after day in the temple, you did not lay hands on me. But this is your hour, and the power of darkness."*

Then they seized him and led him away...

Finally, the same passage from *John*.

[John 18:4-12] *Then Jesus, knowing all that was to befall him, came forward and said to them, "Whom do you seek?" They answered him, "Jesus of Nazareth." Jesus said to them, "I am he." Judas, who betrayed him, was standing with them. When he said to them, "I am he," they drew back and fell to the ground. Again he asked them, "Whom do you seek?" And they said, "Jesus of Nazareth." Jesus answered," I told you that I am he; so, if you seek me, let these men go." This was to fulfill the word which he had spoken," Of those whom thou gavest me I lost no one." Then Simon Peter, having a sword, drew it and struck the high priest's slave and cut off his right ear. The slave's name was Malchus. Jesus said to Peter, "Put up your sword into its sheath; shall I not drink the cup which the Father has given me?"*

So the band of soldiers and their captain and the officers of the Jews seized Jesus and bound him.

These four samplings illustrate how with each progressive version of the 'Jesus Myth' the scribes of the *Gospels* felt at liberty to alter and embellish their own versions, as they saw fit, even when misunderstanding and misinterpreting the previous versions, as shown most noticeably where the ear of the slave keeps being cut off. By the *John* version – not having understood the figurative meaning of the original – not only does the ear becomes 'right ear' that's lobbed off, but the slave acquires a name of 'Malchus'.

[John 18:10] *The slave's name was Malchus.*

And 'cutting of the ear' is also evolves from being 'someone' to that of 'Simon Peter'. (A supposed fisherman with a sword?)

Such inconsistencies are present throughout the *Gospels*, where each version pretends to present the actual words uttered by Jesus, which are different on each occasion and in each version. (So which Jesus, then, was the original, and telling the truth?) And nowhere does it explain as to who were the scribes who were supposedly recording the words of Jesus in each version, and how could these scribes be present everywhere that Jesus was, and how could they record different words for different versions from the same source; that source being the same Jesus? In comparing the words given to Jesus in each version of the four *Gospels* indicates that even if just one version was the truth then all other versions are lies pretending to be truth. And since *Mark* was the first version of this deceptive fiction, then how can all the priests, clergy and the faithful of today still quote from all versions knowing they are certain lies?

The fact is, with a stroke of their quills, the unknown authors of the Gospels changed or retained what they wanted to, callously pretending that the words they wrote down were actually those of Jesus; writing fabricated lies while being aware that according to the ethics of their religion telling of lies is a sin, as deceit is a sin, for which they would be punished by Jesus when he would return on the Judgment Day. So much for Christian honesty, integrity, and sincerity when one can write religious fiction and lies for profit, knowing full well that what they wrote was fiction and lies, intended to deceive the gullible.

And so, by use of their imagination the later authors of the *Gospels* had altered the original symbolical image of Jesus from a 'Son of man' messiah desiring a Jewish kingdom, to that of 'Son of God', a loving and forgiving redeemer in heaven, who is also, contradictorily, the threatening last judge.

Such penchant for absolute distortion of historical facts, using fantasy, embellishments, and absolute disregard of physical reality are typical of *The New Testament*, just as they were of *The Old Testament*. Of course the same conscio attitude applies to all religions. The fundamental purpose of all those controlling religions is not to convince their faithful through an understanding of physical reality, but to compel them to accept and believe without question all the inaccurate, irrational and false notions they, conscios, had their brains invent for them.

As history proves, all efforts to develop a peaceful opposition to Roman occupation by use of 'Jesus Myth' had failed. Most Jews were adamant that any action had to involve bloody violence, and not pacifist dreams of "turning the other cheek".

After the Jewish rebellion against Rome, in 66 CE, the Roman army under command of Titus retaliated by destroying the Jewish Temple and most of the Jerusalem in 70 CE, with the Jews expelled and forbidden to worship their religion ever again.

Towards 130 CE, the Jews again repopulated the city. Between 132 and 135 CE, they had once more launched an unsuccessful revolt against Rome. On this occasion it was Hadrian who executed the Jews, rebuilding Jerusalem as a Roman city, the remnants of which have lasted into the present.

Meanwhile, the 'Jesus Myth' did not perish with that of Jerusalem. There were many who perceived its value, spreading the myth across the boundaries of Roman Empire, but not in Palestine.

By the end of second century of the Common Era (200 CE) – that is: two hundred years after the supposed birth of Jesus – the Christian 'Spiritual Males' conscios had formulated their religious structures. The Jewish Yahweh (the one god) had his singularity taken away, with the god-the-son given the premier status, living "up in heaven" with the angels.

As all gods require conflict to be relevant, so copying the Zoroastrian model, the Christian myth had been provided with an 'evil' antagonists, comprising of the devil – the anti-Christ – and a bevy of ugly and nasty anti-angels, all of whom live 'down in hell'. The 'Spiritual Males' conscios had their brains also developed structures of religious service procedures, the structures of priestly hierarchy, structures of monasticism, and structures for collecting contributions from the faithful.

There is another service the early Christians would have Jesus-the-idol perform for them: serve as a marketing device like no other before of since.

While he was first invented to promote equality and unselfishness amongst the Jews, with an intention of uniting them by pacifist means into a Jewish kingdom, the Jews and non-Jews who actually did accept the notion of Jesus did so for a totally different reason: to profit financially and influentially from such a divine godhead. This was done by transforming the patriotic/spiritual entity into a 'brand' – a devise to assist sales of goods and services – in whose name they, as so-called Christians, would have the authority to acquire financial profits by instigating that the faithful spend money in his name.

Right from the beginning of the formation of Christianity into an official religion of the Roman Empire, as decreed by Emperor Theodosius I, in 380 AD, the 'Spiritual Males' in charge had banned all their followers from practicing any kind of so-called pagan rites, while actually adopting these very rites into their Christian religious calendar. The celebration of the year's rejuvenation in spring became that of Jesus' birthday, while the winter solstice became the commemoration of Jesus' death and simultaneously a celebration of his ascent to heaven.

To the very present, throughout Christian societies the yearly customs of commemorating the birth and death of Jesus are conducted by means of expectant purchasing and presentation of gifts to relatives and acquaintances. But while these customs may appear to represent the acts of unselfish generosity and charity of individuals, they are in fact the acts of selfish expectation of individuals – where those who give expect to receive even more in return – which only benefits the revenues of the manufacturers, the retailers and the churches.

Apart from the motivation to yearly seasonal sales in most regions of the planet, the notion of Jesus had always been used to sell fraudulent relics to the beguiled, and the foolish faithful. There were the endless offers of fraudulent nails and splinters from the cross, on which Jesus had supposedly been crucified. But if the very idea of the cross and the nails were to be realistically examined, then even in these objects there is evidence to disclose the lie of crucifixion.

Were Jesus to have actually existed and sentenced to death he would not have been crucified on a large cross. There are two reasons for that. If one was to know the region of Palestine at all, the most noticeable feature of its landscape is a lack of large tall trees. That means the timbers for the cross, or crosses, would have cost a small fortune, as all timber was imported at a great cost, especially large beams. Of similar prohibitive cost were all metals, so that hammering bronze or even iron into shape of nails would equate to a great expense. Considering that Romans were frugal, preferring to lavish expense on themselves rather than those condemned to die, they crucified the condemned head down, tied with rope to wooden stakes where wood was plentiful, and never large and long wooden beams. And where timber was scarce, they made alternative arrangements, such as burying the condemned to their necks in sand or soil. Apart from these improbabilities to crucifixion, the concept of a 'Christian cross', as depicted in paintings of crucifixion, had not been invented at that period, considering it represents little practical use. If Romans of that period needed to add a crossbeam to a vertical beam they would place it on top of the beam, with side angles to support the crossbeam. Otherwise, a cross for them represented an equal sided cross. And they most definitely would not use a carpenter to cut into the vertical and horizontal beams so as to fashion a flush-sided cross for some undesirables they chose to execute. Finally, in a land of little wood, were large, long wooden beams to be left unattended, even with bodies attached, there would be those who would steal them overnight, leaving behind the bodies. And finally, put a nail into the myth of 'Jesus Myth' on a cross, had Jesus and his companions to have been crucified on 'Christian crosses', then this would mean that such crosses were in use by Romans in the past, as they would not have invented special cross devises for three undesirables. There is no archaeological evidence to show that such crosses had ever been used by Romans before and during that period.

It is for these reasons, that no Roman authority could, or would, permit an expenditure of a substantial fortune in wood, carpentry, and crafted metal to be wasted on any inconsequential criminals. This would have included the execution of Jesus and his companions.

Then there were, and continue to be, the shrouds, that supposedly were used to cover the dead body of Jesus, prior to his resurrection. The most famous of these remains the Shroud of Turin, where a piece of cloth depicts a washed-out, 'photo negative'-like, two-dimensional image of a naked, long haired, bearded male, modestly covering his genitals with his hands. Shown directly above the top of its head is a representation of the back of a head.

The Shroud of Turin is supposed to represent an image of Jesus' body after death, the image itself supposedly produced by spiritual radiation emitted by his body. This Shroud, in particular, had been used to attest the physical existence of Jesus, despite that the official physical and chemical analyses had dated the article to be less than a thousand years old. This means that the cloth was produced over a thousand years after the presumed existence of Jesus.

Actually, the Shroud needs no chemical tests to prove that it is a fake. All that is required is some logical consideration and a bit of common sense.

Here is the rational: Human bodies are three-dimensional. In covering a lying human body, a sheet does not remain rigid like a board above the body but drapes in folds around the body, from the top downwards to all sides. This means that a sheet covering a lying body radiating some magic radiation onto the covering sheet would not achieve rigid contours of a human body with precise anatomical formation. Once the sheet was spread out, the image on it would be that of a bloated and abstracted distortion of a human body.

This can be very simply proved by covering the face of a lying body with a sheet, then marking all the facial features, including the ears, of the face on top of that sheet. When that sheet is removed from the face and spread out, all the facial features marked on the sheet shall appear as a distorted and bloated caricature of the face. Were the Shroud of Turin authentic representation of any human, the image on the fabric would be that of a wide distortion of a human body, and not that of a two-dimensional image of a slender human body.

Apart from that, if a head of a three-dimensional body is measured from the top of the forehead to the back of the head, the distance of the top of the head is around nine inches or twenty-three centimeters, for an average head size. This would mean that if a sheet covering a human head was tucked in under the head, and if on the same sheet a front-on graphic representation was done of the face, and a front-on graphic representation was done of the back of the head, then the distance between the top of the head of the face, and the top of the head of the back of the head would be of some nine inches, or twenty-three centimeters. They would definitely not be butting one against the other as depicted on the Shroud of Turin. As no human has "A" frame head, then that becomes another physical proof that the Shroud of Turin is but another religious hoax perpetrated upon the ignorant and those blinded by their faith – the emperors without clothes.

But what should be recognised as being despicable, is not the fact that the religious ignoramuses – the current custodians of the Shroud of Turin: the Roman Catholic Church – were fooled by fraudsters, but that the very same religious dupes are continuing their insidious attempts to fool others into accepting the fraud of the Shroud as a genuine article.

Around 313 CE, – more than tree centuries after the supposed birth of Jesus – after having been introduced to Christianity by his mother, Helene, Constantine the Great was the first 'Dominant Male' to attribute to Christianity his military success in civil wars, which he had instigated, and by which he acquired his supreme command over the Roman Empire.

Christian 'Spiritual Males' greatly benefited when Constantine became an emperor in 324 CE. Constantine considered himself to be the successor of the early evangelists, by presenting himself as a thirteenth apostle. He devoted his life and office to legitimising Christianity. As with all other Roman religions and possessions, if Rome could not suppress it then Rome would own it, outright.

His legacy to all his efforts was a transformation of the Roman empire, controlled by a 'Dominant Male' emperor, into a Christian Roman empire, controlled by a 'Dominant Male' emperor and a 'Spiritual Male' pope – the so-called living representative of Jesus Christ, the messiah/idol, on this planet.

Most humans then as now, be they Christians or not, have no real understanding as to why the Roman Empire had converted itself into a Holy Roman Empire. The historical depiction of Christian martyrs dying for their faith on stakes, and in arenas with wild animals, certainly did not sway the Roman opinion to accept Christianity. Nor did the members of the fledgling Christian cult.

So why, then, did the Romans convert to Christianity? It is depicted that supposedly Constantine became a Christian because it inspired him with its spirituality. This is far from the actual truth.

The fact of the matter is that it was Constantine who was the first to see the full potential – the 'big picture', so to speak – not of Christianity of the period, but of the 'Jesus Myth'. The 'Jesus Myth' was a 'big idea' whose 'brand' was undervalued by the faithful of the period. Constantine recognised that 'Jesus Myth' was both a tool and a weapon of imperial proportions; a far more a powerful method of governing all, because Jesus represented a pathway to immortality in heaven, or immortality in hell, with the key element being 'immortality'. And while the concept of immortality was not new – having been invented by humans well before Egyptians – he understood the value of it. Whether Jesus was real or not was of no consequence. What mattered was that those who believed the 'Jesus Myth' believed that he lived and that he died and had risen from the dead and ascended to his god-the-father. And if few did believe in 'Jesus Myth', then all of his Empire could be made to believe it.

What this meant that he, as an emperor, already possessed the power of control over each and every life of his subjects. But that was it. Once his subjects died he no longer had any control over them. But if he had the control of religion that promised life after death, then that would allow him – and all the emperors after him – to control his subjects both in this life on Earth and in life after death. Therefore, any disobedience to the emperor in this life would mean damnation in the next. And to achieve this beyond-life control over all his subjects, all he needed was to own and spread the underrated 'Jesus Myth' and the oppressed Christianity movement.

Furthermore, here was a unique entity for any Roman emperor, with which to unite all the dominions of Rome, so that they would have no ability to rebel and cause uprisings against him and Rome, as the same religion would bind all of them to him: the owner of Christian faith.

This then became the new direction for the Roman Empire: a sinister, serious business of converting all its citizens – and those of its empire – to Christianity. This would be done according to the authority of the Holy Roman Empire and not according to any apostles or disciples, or even Jesus; all of whom would be converted to images exclusively exploited by the authority of the Holy Roman Empire. And even if there was to be a 'Spiritual Male' Bishop of Rome (Pope) to head the Christian religion, he too would be owned by the Emperor, and do his bidding (or be replaced).

It was no accident that the seat of power for the Holy Roman Empire became established in Constantinople, the former ancient Greek city of Byzantium (Istanbul, Turkey). It had to be where the authority of Rome could control the 'Jesus Myth', if they were to own it. And to own it they had to establish themselves in total control of the whole region that produced the origins of the 'Jesus Myth', so as to appear being as one with it. Both Constantine and his successors knew the order of importance, were they to become owners of the 'Jesus Myth': first convert the Middle East and only then enforce conversion on Rome.

This priority was based on a single intent: conversion to Christianity would cause the Romans of Rome to lose many personal freedoms they enjoyed and fought to protect from the beginning of the Latin tribes. For unlike the old Roman Empire, the new Holy Roman Empire (as a Christian organisation), would not allow anyone free thought, or free speech, or free choice of religion: only the Roman Christian thoughts, only the Roman Christian speech, and only the Roman Christian religion would be permitted. All else would be suppressed by destruction, obliteration, and official murder.

And so it came to be, as the 'Dominant Males' of the Holy Roman Empire wanted. With the Holy Roman Empire the lie of the 'Jesus Myth' became not just a religion but a way of life, simply because a 'Dominant Male' Constantine had perceived how society can be totally controlled by a notion of a symbol, or an idol, from their birth to death; totally, completely, and in perpetuity. Controlled in such a way that those controlled end up upholding and defending their own suppression. Not unlike prisoners of no crime but that of stupidity, defending their own incarcerations while praising the warden for their imprisonment.

19

The Muhammad deception

Human history is a proof to the continuous emergence of new 'Dominant Males' whose attitudes towards the 'Dominant Males' already in power are premised on: 'anything you can do I can do better, or larger, or longer, or harsher, or bloodier.' Basically, their desires are to want what others have. It does not mean that in their efforts to realise their ambitions they are immediately confrontational. Going by record of history, however, if given even a slightest opportunity of provocation, sooner or later they end up being confrontational and in physical conflict. For 'Dominant Males' this is unavoidable: their attitude allows for only one kind of solution to virtually any problem – physical violence. Also, throughout human history, many 'Dominant Males' were not content just to be rulers. Like Constantine, some wanted to be simultaneously 'Dominant Males', 'Spiritual Males', and 'War Males'. Muhammad was one of those.

Almost six hundred years after the supposed death of Jesus – at the peak of the Byzantine Christian empire – a forty-year-old Arab, a married minor merchant, born and residing in Mecca (a town in Arabia), began to appropriate elements of Jewish, Christian, and Zoroastrian religions. He was called, Abu al-Qasim Muhammad ibn 'Abd Allah ibn 'Abd al-Muttalil ibn Hashim. He became known as: Muhammad.

Muhammad recognized how influential and powerful the idolised stature of Jesus was, who, as most humans, he believed had actually lived. But unlike the Hellenized Jews who rejected violence when inventing Jesus, Muhammad concluded that any efforts to unite humans into a religious kingdom based on equality of all individuals, under a single god granting eternal life (as is proposed in *The Gospel according to Mark*) was only feasible with the assistance of threat or use of physical violence.

When his brain worked out what he (conscio) wanted to do, he was no longer young. He became a human in a hurry. The shortcut to his solutions did not equate to inventing something new, but simply to appropriate from others that, which already existed. Muhammad simply stole from other cultures and religions any intellectual property that he felt was necessary for his religious structure, claiming those notions as his own. He was also not going to be averse to forcing his religion down the throat of others. For Muhammad, Jesus may have failed in his conquest by word; he, himself, was going to succeed in his conquest by sword.

Before Islam, the Semite Arab tribesmen and merchants believed in inevitability of fate (kismet), but not in supernatural or spiritual powers. They believed that irrespective of an influence of gods – the highest of which was Allah – the same Semite god YHWH as that of Semite Hebrew Yahweh [the phonetic pronunciation of "Allaaah" for Arabs has a similarity to that of "Yahwaaah" for Jews) – all would still result in that which was pre-ordained.

After introduction of Islam (meaning: surrender [to the will of Allah]), the Muslims (meaning: those submitting [to god]), retained the fatalistic notion of 'what will be, shall be', despite that they were expected to dispense with personal opinions of fate, in acceptance of a singular 'will of Allah'.

As a trader in the region where other traders brought with them knowledge of foreign religions from surrounding and distant lands, Muhammad became inspired. Around 610 CE (Common Era) – in a re-enactment of Yahweh revealing himself to Moses – according to Muhammad, the angel Gabriel came to him in visions and pronounced him to be the 'messenger of Allah'. From that moment on Muhammad presented himself as a fifth, the final, and the greatest prophet of god (positioned before those of Abraham, Noah, Moses and Jesus); supposedly receiving verbal messages directly from god.

(Not a bad deal for a middle-aged, second-rate trader, living on support from his wealthy wife.)

During his life Muhammad kept all his revelations unrecorded. After his death in 632 CE, these, so-called messages from god, were written and compiled in 650 CE, as *Qur'an*, or *Koran* (recitations): the sacred scriptures of Islam, obtained from beyond any earthly sources, written on a 'preserved Tablet' (not unlike the fabled 'ten commandments' tablets of Moses – and so where is that 'preserved *Qur'an* Tablet' now?).

Mohammed's first wife's cousin, Abu Talib, a Christian, was very instrumental in 'interpreting' visions that Muhammad was receiving, liking them to the notions of Jewish, Christian and other religious prophets. By such means the initial doctrines of Islam were developed, upon which other laws and instructions were added throughout Muhammad's life (that is: whenever the need for extra laws was necessary to endorse Muhammad's sovereignty).

The earlier god Allah was transformed from having other gods beneath him into becoming a singular god, not unlike that of a Hebrew-Israelite-Jewish god, Yahweh; but unlike Christian god, Allah was allocated no sons and no holy spirits (apart from the occasional use of 'jinn'). Just like Yahweh, Allah was given the credit for all creation by, supposedly, simply commanding: 'Be.' But just to be on the safe side, the Judaic 'six days of work and a Sabbath rest day' are also included in *Qur'an*:

[The Heights (Al-A'raf) 7:53] *"Your Lord Allah, who in six days created the heavens and the earth and then ascended His throne. He throws the veil of night over the day. Swiftly they follow one another."*

Allah, according to Muhammad, is present everywhere, but resides nowhere specifically. Everything in nature that Allah was to have created lives in independent units, yet as a cohesive part of the whole. Everything has its own measure over which it cannot overstep, because of restrictions set by god. God, Allah, reigns supreme in heavens as on earth: all-powerful, independent and self-sufficient. According to the rules set by Muhammad, Allah never speaks directly to any human, but only as an inspiring voice in the head of Muhammad, or through an angel messenger.

As the last and the premier of all prophets, according to Muhammad (and according to his astuteness), he was blessed not to produce or experience any miracles, a requirement placed upon lesser prophets, where: Abraham was saved from fire; Noah saved from rising floodwater; Moses saved from the pharaoh; Jesus had the miracle of divine birth by Virgin Mary, and saved from crucifixion by being taken alive to heaven. Therefore, as the greatest prophet and beyond miracles, Muhammad produced no miracles (which means that he was shrewd enough to know that no one can perform public miracles, and this was his excuse from being exposed as a fraud by those expecting miracles of him).

This, however, did not prevent Muhammad from appropriating from the *Hebrew Bible* concepts of non-reality and applying to them his imagination. For instance, Adam, according to Muhammad, was created from clay at the same moment as his parallel 'jinn' was created of fire. (As an Arab, selling a new religion to Arabs, Muhammad could not rid his theology of jinn, or jinnee, so dear a mythological deity to Arabs, as Loki is to the myths of the north Europeans.) The 'man' – aside from Adam – according to Muhammad, was created from clots of blood (exactly whose blood is not specified):

[The Blood Clots (Al-'Alaq) 96:1] *"Recite in the name of your Lord who created man from clots of blood!"*

In Judo-Christian doctrines, Adam disobeys god and is punished. But in *Qur'an* he is forgiven because man is, supposedly, the noblest of all creations, created to have all nature be subservient to him:

[The Cow (Al-Baqara) 2:36-38] *... But Satan made them fall from Paradise and brought about their banishment. "Go hence," We said, "and may your offspring be enemies to each other." Then Adam received commandments from his Lord, and his Lord relented towards him. He is Forgiving One, the Merciful.*

"Go down hence, all," We said. "When Our guidance is revealed those that accept it shall have nothing to fear or to regret; but those that deny and reject Our revelations shall be the heirs of Hell, and there they shall abide forever."

Having produced an 'original sin', as in the *Hebrew Bible*, in *Qur'an* Adam, as man, is rebellious and full of pride, which is considered to be a 'cardinal sin', for his presumed equality with god.

As in *The New Testament*, the conflicts between the good and evil are included in the doctrines of Islam, so that humans can struggle with their morality and pride, while seeking enlightenment and the true path to god. Satan (Iblis or Shaytan) is the evil, ever-present, former high-ranking angel, who fell from grace for not honouring man:

[The Cow (Al-Baqara) 2:35] *And when We said to the angels: "Prostrate yourselves before Adam," they all did so except Satan, who in his pride refused to and became an unbeliever."*

[Al-Hijr 15:32-37] *"Satan," said Allah, "why do you not prostrate yourself?" He replied," I will not bow to a mortal created of dry clay." "Begone," said Allah, "you are cursed. My curse shall be on you till Judgment-day." "Lord," said Satan, "reprieve me till the Day of Resurrection." He answered, "You are reprieved till the Appointed Day." "Lord," said Satan, "since you have led me astray, I will seduce mankind on earth: I will seduce them all, except those that faithfully serve you."*

According to Muhammad (or rather, according to the various notions appropriated – read that as stolen – from Judaic, Zoroastrian, and Christian religions) Satan's task is to beguile man into temptation, sin, and error, and to inflame man's pride, by that strengthening his disobedience to god. But god is always prepared to forgive truly repentant sinners. Until the Last Day of the world when the hour of the judgment comes, the Satan can legally have his sport with man.

PART 2. DISCLOSING THE SHAM OF HUMAN CONSCIO PERCEPTIONS

On that Last Day, the dead will be resurrected and judged individually by Allah, with the sinners burning for eternity in hellfire, both physically and spiritually, while those saved will enjoy – for a similar eternity – physical bliss and the divine happiness and pleasures. Being an Arab, Muhammad made sure that his kind of sexist heaven would appeal to his male Arab converts, while ignoring women's heavenly rewards, by depriving them of equality with men:

[The Merciful (Al-Rahman) 55:43-77] *...But for those that fear the majesty of their Lord, there are two gardens planted with shady trees. Each is watered by a flowing spring. Each bears every kind of fruit in pairs. They shall recline on couches lined with thick brocade, and within their reach will hang the fruit of both gardens. They shall dwell with bashful virgins whom neither man nor jinnee will have touched before. Virgins as fair as corals and rubies. Shall the rewards of goodness be anything but good? And beside these shall be two other gardens of darkest green. A murmuring fountain shall flow in each. Each planted with fruit trees, palms and pomegranates. In each there shall be virgins pure and fair. Dark-eyed virgins sheltered in their tents whom neither man nor jinnee will have touched before. They shall lean back on green cushions and rich carpets. Which of your Lord's blessings would you deny?*

This bountiful sexual heaven filled with countless virgins was also part of earthly incentive, but once again only for the Muslim males, as emphasized throughout the *Qur'an*:

[The Believers (Al-Mu'minun) 23:1-3] *Blessed are the believers, who are humble in their prayers; who avoid profane talk, and give alms to the destitute; who restrain from carnal desires (except with their wives and slave-girls, for these are lawful to them) and do not transgress through lusting after other women...*

Muhammad introduced the law of polygamy – permitting each male to have up to four wives and numerous female sex slaves – overturning the previous Arab custom of matriarchy, where only the women were allowed to have multiple husbands. He, himself, had circumvented his own law, by having nine wives – one of the relations being incestuous – apart from the countless slave-girls in his harem. To permit Muhammad his extra wives, Allah, of course, came to his aid, through Muhammad himself, with a special dispensation:

[The Confederate Tribes (Al-Ahzab) 33:49-50] *"Prophet, We have made lawful to you the wives to whom you have granted dowries and the slave-girls whom Allah has given you as booty; the daughters of your parental and ma-*

ternal uncles and your paternal and maternal aunts who fled with you; and the other women who gave themselves to you and whom you wished to take in marriage. This privilege is yours alone, being granted to no other believer. We well know the duties We have imposed on the faithful concerning their wives and slave-girls. We grant you this privilege so that none may blame you. Allah is forgiving and merciful."

As to the persona of Jesus, who is mentioned in many sermons of *Qur'an*, Muhammad always stressed that 'son of Mary; is a mortal, doing Allah's work on earth:

[Ornaments of Gold (Al-Zukhruf) 43:57] *"Jesus was no more than a mortal whom We favored and made an example to the Israelites." ...And when Jesus worked his miracles, he said, 'I have to give you wisdom and to make plain to you some of the things about which you differ. Fear Allah and follow me. Allah is my Lord and your Lord: therefore serve Him. That is the right path."*

Whatever was necessary to be appropriated from the *Hebrew Bible* and *The New Testament* was done so. Here is an example with a direct reference of doing so:

[Victory (Al-Fath) 48:30] *"...Thus they are described in the Torah and in the Gospel: they are like the seed which puts forth its shoot and strengthens it, so that it rises stout and firm upon its stalk, delighting the sowers. Through them Allah seeks to enrage the unbelievers. Yet to those of them who will embrace the Faith and do good works He has promised forgiveness and a rich reward."*

Initially Muhammad had a great deal of opposition from local tribal leaders in Mecca, who recognized Muhammad's motive of usurping power by stealth, by setting one leader against another. After the death of his wealthy first wife, Khadijah, in 619 CE – who was some fifteen years his senior – and the death in the same year of his Christian uncle, Abu Talib, Muhammad was forced to leave Mecca, having lost the protection of his clan of Hashim. He failed to find support in other places, returning to Mecca in 620 CE, under protection of another clan. In Mecca he began negotiations with clans from Medina, a town north of Mecca, requesting their permission to emigrate there.

In the summer of 621 CE, twelve men from Medina, on a pilgrimage to the main pagan shrine in Mecca, had supposedly introduced themselves to Muhammad (how much like the twelve disciples of Jesus), and promised to promote him and his teachings in Medina.

In 622 CE, Muhammad left Mecca with 75 Muslims, including two women, all of whom had taken an oath to be warriors for Muhammad. This famous exit to Medina (hijrah), became the traditional starting point of Islamic history and calendar: July 16, 622 CE.

In Medina, Muhammad came across a number of Jewish clans that for generations had lived peacefully amongst the Arab clans, keeping to themselves, while the Arabs fought each other. Muhammad's resentment of this Jewish presence was the impetus that he used as his cause to mediate the warring tribes – gradually instilling his religious leadership.

With an incitement of hostilities between the local population and the Jews, Muhammad also began to encourage the resumption of banditry by his followers. They were sent on razzias (raids) of Meccan merchant caravans passing near Medina, on their way to Syria. Muhammad personally led three such raids in 623 CE, which were failures. In 624 CE, he sent a small group of bandits towards Mecca, which succeeded in attacking a caravan from Yemen. In March 624 CE, while leading another bandit raid with 315 Muslims on a Meccan caravan returning from Syria, the bandit Muhammad failed to capture his quarry but had a military clash with a force of 500 Meccans, who came out in protection of their trade route. Despite a neutral exchange, the elated Muhammad proclaimed himself victorious.

From then on, Muhammad altered his position from being a cautious minor 'Spiritual Male' to that of a zealous and extremist 'Spiritual Male' and a 'War Male'. He organised to have his Medinaen opponents assassinated, for making fun of him. He began to use his position of a spiritual leader to incite the Arab clans against Jews – with whom the Arab clans had previously maintained a peaceful co-existence – calling for a crusade to rout them. Members of Jewish clans who resisted and refused to recognize Muhammad as a prophet were slaughtered; their females and children sold off as slaves. Muhammad also made a change in his own doctrine of worship: no longer would Moslems face Jerusalem for their daily prayers, but would instead face his birthplace of Mecca.

Muhammad had tasted blood and liked it. Victories or defeats meant nothing to him as long as his followers continued to support him with their swords. Meanwhile, he continued to gain alliances through kinship, by taking on more and more wives. His dream was to return to Mecca as a spiritual leader-warrior: a 'Dominant Male, 'Spiritual Male', and a 'War Male' all in one. This he achieved in 630 CE, by entering Mecca with 10,000 Muslim warriors to no resistance, for by then Mecca was long in decline. There was nothing for those living in Mecca to fight for.

With the spreading of Muhammad's dominance, many Arab males began to feel that there were many incentives and benefits in becoming a Muslim. There were the sanctioned-by-Muhammad opportunities to plunder Jews and non-Muslims, a function that endeared Muhammad to many Arabs. Then there was the economic structure of equality, applicable as a support to all males of the same religion, who were prepared to instigate not just a nation of Islam but also a world of Islam, by continued expansion of Islam beyond their borders.

Furthermore, there was the overturning of the previous Arab traditions and laws directing females to be the recipients of any inheritance and possessors of numerous husbands. Now, with directives from Muhammad, only the males could inherit, and possess not only four wives but also female sex slaves. There were also the financial and self-preservation incentives to being a Muslim, for while being required to pay a small Muslim tax, this was much better then paying a crippling non-Muslim tax, which applied to Jews and Christians. It was also much better to embrace Islam than to face the consequences of refusing to do so, which was officially decreed to be death to all other races (apart from Christians and Jews) choosing to remain non-Muslims. On top of all that, there was the promise of immortality in paradise, after death.

Following the completion of the *Qur'an*, the resultant work was presented as an infallible word of god. It comprises of adapted segments from the *Old* and *The New Testaments*, intermingled with notions and instructions relevant to the Arab culture. The writings are in form of sermons, which repeat the same points in many chapters, and constantly refer to the same Judaic and Christian prophets as subject matter, and the "eye-for-an-eye" revenge philosophy, so dear to the Hebrews/Israelites/Jews, and now the Israelis:

[The Cow (Al-Baqara) 2:178] *"Believers, retaliation is decreed for you in bloodshed: a free man for a free man, a slave for a slave, and a female for a female."*

Qur'an is a representation of how Muhammad used the rebukes, expressed in the *Hebrew Bible*, to his own advantage. For instance, the fault-casting prophets of *The Old Testament*, such as Jeremiah, positioned the blame for any of their political, military, and national misfortunes on religious laxity and disobedience, claiming those misfortunes to be punishment from Yahweh. In *Qur'an*, Muhammad uses such Judaic self-blame to chastise the Jews:

[(The Table (Al-Ma'ida) 5:18] *The Jews and the Christians say: "We are the children of Allah and His loved ones." Say: "Why then does He punish you for your sins? Surely you are mortals of His own creation. He forgives whom He will and punishes whom He pleases."*

[The Table (Al-Ma'ida) 5:78] *"Those of the Israelites who disbelieved were cursed by David and Jesus, the son of Mary: they cursed them because they rebelled and committed evil and never restrained one another from doing wrong. Evil were their deed. You see many of them making friends with unbelievers. Evil is that to which their souls direct them. They have incurred the wrath of Allah and shall endure eternal torment. Had they believed in Allah and the Prophet and that which is revealed to him they would not have befriended them. But many of them are evil-doers."*

[Noah (Nuh) 71:26] *And Noah said, "Lord, do not leave a single unbeliever in the land. If you spare them they will mislead Your servants and beget none but sinners and unbelievers. Forgive me, Lord, and forgive my parents and every true believer who seeks refuge in my house. Forgive all faithful, men and women, and hasten the destruction of the wrongdoers."*

But while there are passages in *Qur'an*, which admit that there are pious and honorable Jews and Christians:

[The Cow (Al-Baqara) 2:62] *"Believers, Jews, Christians, Sabaeans – whoever believes in Allah and the Last Day and does what is right – shall be rewarded by their Lord; they have nothing to fear and regret."*

[The Imrans (Al-Imran) 3:14] *"Yet they are not all alike. There are among the People of the Book (Jews and Christians who have books of Testaments) some upright men who recite all through the night the revelations of Allah and worship Him; who believe in Allah and the Last Day; who promote justice and forbid evil and compete with each other in good works. These are righteous men: whatever they do, its reward shall not be denied them. Allah knows the righteous."*

... much too often such humane sentiments are overpowered by prejudicial and hateful remarks, and instructions, such as:

[The Imrans (Al-Imran) 3:85] *"He that chooses a religion other than Islam, it will not be accepted from him, and in the world to come he will be one of the lost."*

[The Table (Al-Ma'ida) 5:51] *"Believers, take neither Jew nor Christian for your friends. They are friends with one another. Whoever of you seeks their friendship shall become one of their number. Allah does not guide the wrongdoers."*

[5:72] *"Unbelievers are those that say: 'Allah is the Messiah, the son of*

Mary (Jesus).' For the Messiah himself said:' Children of Israel, serve Allah, my Lord and your Lord.' He that worships other gods besides Allah shall be forbidden Paradise and shall be cast into the fire of Hell. None shall help the evil-doers."

[Iron (Al-Hadid) 57:26] *"We sent forth Noah and Abraham, and bestowed on their offspring prophethood and the Scriptures. Some were rightly guided, but many were evil-doers. After them We sent other apostles, and after those Jesus the son of Mary. We gave him the Gospel and put compassion and mercy in the hearts of his followers. As to monasticism, they instituted it themselves for We had not empowered it on them, seeking thereby to please Allah; but they did not observe it faithfully. We rewarded only those who were true believers; for many of them were evil-doers."*

With no physical records kept by Muhammad himself, about himself and his beliefs, all of what has been written about Muhammad by his Muslim followers was done centuries after his death. The results are that both he and his eclectically formed religion have been vastly fictionalised. Not unlike the four Gospels on life of fictitious Jesus, there were four notable compilations on life and teachings of Muhammad between 750 and 950 CE, produced by Ibn Ishaq, died 765 CE; Ibn Hisham, died 833 CE; al-Bukhari, died 870 CE; and al-Tabari, died 923 CE. This does not mean that these authors were any less refrained from fiction writing regarding Muhammad to those who wrote the Gospels, even if there was the physical historical truth to Muhammad's existence, unlike that of all the Jewish non-existent prophets and apostles, and the Christian ('son of Mary') Jesus.

Many years after Muhammad's death, Muslim mulas (teachers) and scribes made additions to *Qur'an* and to Islamic religious and social customs. They also gave rise to many Muhammad legends. One such legend concerns Muhammad's ascendancy to heaven – Mi'raj. In this tale, the archangels Gabriel and Michael are ordered by god to invite Muhammad to heaven. But first they purify him in his sleep. They open his body and remove all traces of sin. Then they take him to Jerusalem, from where he is led by Gabriel (Jibril) up a stairway to heaven, proceeding past the seven levels to the throne-room of God. On the way they are met by prophets of *The Old Testament*, including: Adam, Joseph (Yusef), John (Yahya), Moses (Musa), Abraham (Ibrahim), and Aaron (Harun). They also meet Jesus (Isa). Here Moses informs Muhammad that God considers Muhammad as a more important prophet than Moses, and advises that for him, the 50 daily ritual prayers assigned to Muslems be reduced to just five a day. (There were many lazy privileges to being a top prophet.)

Mi'raj was a typical form of one-upmanship, as practiced by all 'Dominant Males', and 'Spiritual Males'. If the Christian writers had Jesus ascend to a god in heaven after his death, then the Islamic writers had Muhammad do the same, but in his life not after death, and he was positioned closer to god than other prophets and apostles – being thus acknowledged by all, including god, that he, Muhammad, is second in rank only to god himself. Compete with that Christians!

Irrespective of any fictitious exaggerations that had been employed to endorse and elevate Muhammad's abilities and actions, as well as, to disguise his shortcomings, he remained a 'Dominant Male', 'Spiritual Male', and 'War Male' to the end of his life. He continued to covet the wealth of other nations, which he raided; he was avaricious for possession of his Islam and all that his Islam possessed; he was intolerant of any opposing views from those around him, and of those Muslims critical of him and his actions; he was hateful of all those who were not Muslims – that is, those who did not belong to Islam, and by that did not belong to him. But while he may have prevented a number of Arab revolts opposing him (to the end of his life) – by that allowing for future Islamic assaults of other kingdoms to be realised, he, nonetheless, could not prevent future religious schisms occurring to Islam, which would lead to ongoing conflicts between Muslims, just like those that constantly occurred, and continue to do so between Christians.

To this day the main religion of Islam is divided into many sects, of which the main two factions are those of Sunni and Shia. The Sunni and Shia are in constant conflict over their interpretation as to which of them are the true representatives of Muhammad.

Shia Muslims regard the fourth caliph (religious ruler), Ali, as the Muhammad's true successor. The Islamic nations that are predominantly Shia are those of Iran and Iraq. Alternatively, Sunni Muslims believe that the first thee caliphs were orthodox and are true representatives of Muhammad. The main Sunni nations are Egypt, Syria, Jordan, Saudi Arabia, Yemen and the United Arab Emirates.

There are many other non-Muslim nations that have large Muslim populations, where both Sunni and Shia used to live with little conflict. For instance, China has more Muslims than Syria, while in Russia more Muslims reside than in Jordan and Libya combined.

While there is a small number of democratically-run Muslim-majority nations, (their number gradually dwindling), most of such nations have historically been controlled (and continue to be) by individual 'Dominant Males' rulers, who, by ruthless means, enforced some levels of harmony between their Shia and Sunni populations, and by that maintained relatively peaceful affairs of the state. When some Islamic states became engaged in wars towards the end of twentieth century, over their religious differences (and territories), the Western and American political and financial interests became involved in trying to influence these Arabic Muslim nations to depose their dictator rulers. In turn, these Arab dictators retaliated by threatening the Western and American interests. When the armed invasions, instigated by Western and American politicians, did achieve the elimination of the unwanted Arab dictators, the repercussion only resulted in escalating political and religious divides and discontent within these Muslim nations. Beginning with the first year of the twenty-first century, many Arabic want-to-be 'Dominant Males' and 'Spiritual Males', in the name of Sunni or Shia Islam, (with or without military support from opposing Western nations) have become responsible not just calamitous carnage, upheavals and armed revolts within their own nations, but have also instigated fears in all countries across the globe by acts of ruthless terrorism.

As far as all the warring and murdering Muslim 'Dominant Males' and 'Spiritual Males' are concerned, there is but one principle on which they all base their religious and political understanding: that they are right and all others are wrong, and in being wrong they must submit to them who are right. No other outcome is to be tolerated or accepted by these Muslim 'Dominant Males' and 'Spiritual Males'.

The overall outcome of such Muslim 'Dominant Male' and 'Spiritual Male' irrational zeal, and the irrational and incompetent European and American involvement in Arabic affairs, has been the tides of Muslim refugees. The endless strife in their homelands have given millions of Muslims a reason to invaded European countries using all the illegal means at their disposal (including the employment of 'people smugglers'), in search of all that they never had back home. But in seeking better lives in secular states, they refuse to relinquish their Shia and Sunni schism.

Because of the dogmas of their religion, which they openly consider to be right and all other religions to be wrong, Muslims do not easily integrate into new cultures, preferring to remain in their own enclaves. They stubbornly maintain the sexism and sexual intolerance of their religion and their 'Dominant Male' culture, with demands for introduction of sharia law into the countries they gradually inhabit, in direct opposition to the religion and culture of their adopted lands. And that inevitably sows the seeds of conflicts and antagonism between the Muslims and the local populations, despite any political efforts to devise peaceful solutions.

As a human species, both Arabs and Hebrews evolved culturally to be Semites: humans speaking Semitic languages; sharing common gods, including Yahweh. Then the Hebrew Semites made a claim on Yahweh, by further appropriating religious notions from Babylonians and Canaanites and then Zoroastrians, allocating these to their god. Later their god was altered by Hellenic Jews, who, together with non-Jews invented a myth of a man-god idol Jesus; consequentially being responsible for the invention of Christianity. Later still, the Judaic and Christian religions – and their respective gods – were once again appropriated; on this occasion by a Semite Arab, Muhammad. Muhammad took the Christian notion of the man-god idol of Jesus, and replaced him with himself for Muslims.

The result of all these appropriations is that these three religions of Judaism, Christianity, and Islam are all based on the same fiction of the Hebrew Bible. And yet, for all the sameness of fiction in these three religions, their 'Spiritual Males' proclaimed – and continue to do so – that only their own religion is the true religion; and only their god is the true god. To emphasize the difference of their religions, these 'Spiritual Males' propagated – and continue to do so – hatred and bias towards the religions of others: which, in fact, are virtually the same, by having the same components of prophets and religious traditions.

The basic reason why 'Spiritual Males' strive to distance their religion from those of their 'Spiritual Male' opponents is to remain relevant to their believers – those whom they convince to share their views. If they could not do this, then they and their religion would become irrelevant to other humans. So, the best way for them to retain their importance to the masses is by engineering conflicts between their religion and those of others. For once the masses became engaged in conflict, they become blindly biased towards their religion, and hostile towards religions of their enemies. Once this happens, the pride of their respective religious zealots propel the masses along the path of destruction where they become too preoccupied with hate and survival to see themselves being manipulated by their 'Spiritual Males'.

The invention of Islam, and its antagonistic and intolerant attitude towards all other religions, had given the Judaic and the Christian 'Spiritual Males' and 'Dominant Males' – with their own intolerant religions – a new foe to hate and to battle.

(Any claims made by Muslims that Islam maintained, and maintains, a tolerant attitude towards foreign religions are avoiding a historical fact that the only method by which Islam spread on this planet was by intentional violence of physical conquest, acquiring converts with threats of death.)

With Islam being a third hateful religion coming into play after Judaism and Christianity, all their 'Spiritual Males', 'Dominant Males', and 'War Males', had engaged into their bloody work of killing each other, which continues to this day. And why would this not be so, considering that each one of these religions is similarly sexist, biased, oppressive, suppressive, and intolerant; all of them similarly loathing and hating the others in equal measures; all of them similarly giving joy of power to all their 'Spiritual Males', 'Dominant Males', and 'War Males' who thrive on human conflict.

Still, it is to the credit of human conscios that although all of their differing racial societies had invented different gods for themselves, not all their 'Spiritual Males' devised religions as savage and selfish as those of the 'hateful trinity' of Judaism, Christianity, and Islam.

20

The Eastern way

In the same periods as human civilisations were developing by the river Nile and the rivers of Mesopotamia, other major river waterways, such as Hindus River in ancient India, and further east, the Yellow River of China, had also served as cradles of early civilisations. Not surprisingly, the human centres in the East were also coming up with gods and religions. But while all religions (without any exceptions) are based on human invention and fabrications, unlike the 'Spiritual Males' of the Middle East (Southwest Asia) and Europe, the South Asian and East Asian inventors of religions had a different approach to spirituality.

On the whole, these religions were more interested in presenting concepts of tolerant acceptance of earthly life, rather than dictating spiritual dogmas on behalf of any specific bullying god, who supposedly tolerates no opposition or criticism. These Eastern religions were not based on principles of: "Commit yourself to me, your one-and-only god, and I will support you in your desire of conquests; rewarding you with a place in heaven," but more of: "Know your way to selflessness before expecting heavenly rewards."

In the years when Cyrus was releasing the Jews from their Babylonian exile, important events were taking place in India and China almost simultaneously. In India, the Buddha was supposedly beginning his physical and mental journeys of development, moving away from Hinduism; while in China, Tao and Confucius were influencing humans of those societies. Their religio-phylosophies produced no causes for humans to fight over, unlike the intolerant religions of the 'hateful trinity', which continue to provoke much conflict and destruction between humans.

The thinking behind the Eastern religions provided no cause to inflame 'Dominant Male' aggressions, instead adding more reasons to abstain from militant violence. If the 'Dominant Males' of Asian fought each other – which they constantly did – it was for the usual reasons: prestige, greed for power, and fear of each other's power – definitely not with any intentions of installing one religion over another.

INDIA: ACCEPTANCE OF ALL GODS

'Hindus' was a name coined around 1830 by the British authorities, to classify the inhabitants of lands surrounding Hindus River in India – later to become part of Pakistan. The religion of these Hindus evolved from Vedism, a religion of ancient Aryan invaders, who comprised of related nomadic tribes from southern Russian steppes and Central Asian plains, who entered the Indian subcontinent in about 1500 BCE (Before Common Era). These Aryans brought with them horses and chariots, Sanskrit language (which is related to the early Iranian), and a sacrificial religion that gradually absorbed some of the indigenous religious elements.

In about 900 BCE, the rites of this religion – relating to human behaviour and worship – were recorded in Rgveda, a collections of Vedist, or Brahman sacred hymns. Those hymns re-defined the various important gods and deities of Hindu religion.

Of these, Rudra, a feared deity associated with mountains and storms, evolved into a prime god Siva (Shiva). Visnu (Vishnu) – initially a minor god of Rgveda – was elevated into one of the most important and popular Hindu divinities. The god Indra remained a favourite with the Aryans. He – in a reflection of early Mesopotamian myths – was attributed with the creation of the Universe after slaying the great dragon, Vrtra.

The Rgveda was produced to justify a class system, which the Aryan invaders implemented over the indigenous population. The 'Hymn of the Person' (Purusasukta), referred to four classes, or castes (varna) of Indian society, with the top three belonging to the Aryans and their descendants. In the top class were the Brahmins (Brahmanas), who claimed superiority over all other social classes, because they were the priests controlling the religion, and supposedly only answerable to the gods – (how alike to the Levi priests of the Israelites). The second class belonged to Rajanyas (later, Ksatriya), the tribal aristocrats and the warriors. The third class comprised of Vaisya, the landholders. The Sudra, the servants, belonged in the fourth class, comprising of the indigenous Indian population. This lowest class was meant to be meekly subservient to the other three higher classes.

While such socio-religious class system of the castes had remained rigidly unfair and biased, the same could not be said of their religion. The Indian Brahmins 'Spiritual Males' kept constantly altering their religion and the functions of their gods. This re-inventing, amalgamating, separating, re-defining and theorizing within the Hindu religion continues into the present. Some variations and changes were regional, while others were national, which were – and are – either accepted, tolerated, or ignored by the various factions of the faithful, but with no malice or ill will.

These various forms of Hinduism have developed some basic doctrines, which most Hindus accept as being fundamental. First, there is the belief in ultimate reality of Brahman as the 'All', which applies to all. This is the uncreated, infinite and eternal entity, and the state of 'being and non-being'. Brahman is the creator, preserver and re-absorber of everything. While impersonal, Brahman can also be perceived as the high god Visnu or Siva, with sublime and charming qualities. This central belief in, and search for, the 'One that is All' has been in essence the Hindu religion's quest for the 'ultimate reality'. As a deity, however, Brahman is not worshipped. Instead, while giving a worshipping preference to either Visnu or Siva, the Hindus consider Brahman, Visnu and Siva, as a Hindu Trinity, representing the 'One with three forms'. However, both Visnu and Siva are meant to be definite gods.

Hindus also accept as natural the concept of transmigration and rebirth: 'karman', whereby after death and a stay in heaven (or hell) a being is reborn; returning as a life form that best represents the good or bad deeds performed during its previous life. Entrapped in the process of rebirths (samsara), a being is supposedly condemned to wander through a restless and unending stream of periodic returns to life, in one form or another. In this way, any earthly process is viewed as cyclic, where all worldly existence is subject to the cycle that has no beginning and in most instances no end. With the concept of 'karman', all fortuitous or unfortunate events are considered to be a rewards or punishments for one's own past deeds.

Such a belief causes very large numbers of Indian in India, who are poor, sick or diseased, or who are born into the low classes of the Hindu caste system, to be shunned and viewed with contempt. But for a fluke of birth these unfortunate humans are considered to be former sinners, who are now condemned to atone for their former wicked lives.

According to Hinduism, the only way to break with 'karman' and reach salvation and eternal peace of Nirvana, is by obtaining (through meditation) a unification of human personality and Brahman; while simultaneously giving up all human attachment to worldly objects and possessions.

Hinduism also incorporates, in principle, all forms of beliefs and worship without judgment or elimination. The Hindu religion is open to reverence of any divine manifestation, whatever that may be, without any imposition, intolerance, or prejudice. In Hindu religion, no ideas are finite or final, including the existence or non-existence of a singular or numerous gods. To Hindus the religious truth is transcendental: it is not conceived in dogmatic terms – (apart from the despicable injustice of the caste system). For many Indians, the coming of the Buddha addressed this injustice (so much like the return of Jesus for the Last Judgment).

All in all, this convoluted religion had provided a freedom for any of its faithful to believe it whatever elements of Hinduism they individually choose to believe, without forcing upon anyone any dogmatic notions, as it is done in many other major religions. This freedom causes no disharmony or malice within the ranks of its faithful, allowing them to live in peace with their religion, instead of striving to prove that their religion is right and all other religions are wrong; and for that needing to be destroyed together with their population of followers. So any turmoil and violence that flares up in Hindu nations is caused by political and social dissatisfactions and not religious differences in Hinduism.

THE BUDDHA: GIVING IT ALL AWAY

It is presumed that around 534 BCE, Siddhartha Gautama, a prince, was born in Lumbini (a border region between Nepal and India). From early childhood he was supposedly prophesied to become either a powerful monarch or the Buddha: the 'enlightened one'. At the age of 29, he left his palaces, his wife, and his son, to pursue his quest of personal spiritual enlightenment. He supposedly wanted to find the absolute truth, salvation, and Nirvana. (The indescribable state of Nirvana is supposedly achieved when the 'karman' and the succession of lives and rebirths of 'samsara' have been overcome.) So begins the myth of Gautama, the Buddha.

Siddhartha Gautama was possibly not a total fiction, but neither was Gautama the legend he became. In early life he was a troubled, unhappy individual, who had sought to find a solution to his mental state, and found it not in spiritualism but in realism of physical existence. Although, understandably for that period, there was no knowledge of absolute physical reality, those who came up with the notions attributed to the Buddha came close to understanding the existence of the physical process of change, and of the existence of conscio.

The reason for why Gautama must have been a historical figure is because unlike the 'Jesus Myth' there was no one at that period to promote him or gain from his deeds. He was all by himself doing, supposedly, what he wanted to do with no populist agenda.

All that he achieved for himself, and taught to others, had to have been noted, remembered and passed on by word of mouth, from one generation to the next for many centuries. This would have undoubtedly resulted in substantial distortions and additions between his actual words and those that were finally written down. After all, the deeds and words attributed to him in various scriptures had been written hundreds of years after his death, by many priests who included their own perceptions, considering that the person described as being Gautama had never written a word about himself or of his notions (just like Jesus and Muhammad, who became the symbolic idols of major religions).

Whatever his real life may have been, in his legend he just had to have been a prince. The reason for that is because human conscios do not respect, uphold, or trust anyone who is just an ordinary person. For conscios, if they are expected to respect, uphold and trust anyone, then that anyone had better be someone high-ranking and of 'noble blood': at least a prince or a king, no less.

This is because humans (conscios) know that they are untrustworthy liars and cheats, and therefore would not trust someone similar to themselves. Instead they erroneously presume that those of noble blood, of enormous wealth, or of high rank are more honourable (which they seldom are). Nonetheless, from such admiration of royalty, most human conscios are more likely to believe a word of a dishonest prince than that of an honest pauper.

Were Gautama not to have been presented as a prince, no one would have cared to accept the teachings of a human with a humble, or a common family background – even were those teachings to describe how to acquire Nirvana. Certainly not the nobility and the rich of that culture, for they all looked down at those considered beneath them. Not even beggars and the destitute would accept the teachings of an ordinary person. Only by adorning the legendary figure of Gautama with an impressive title of 'royalty' does the legend acquire a grandeur that disallowed an offhand dismissal. It also signifies that if such pursuit is good enough for a price then it is worthy of anyone else. And if a noble, rich, and powerful man discards earthly titles and possessions to acquire sublime spiritual reward of nirvana, then how simpler and easier it must be to achieve this by all those having no titles or wealth in the first place.

The fact remains that despite an intensive search to locate the origins of Gautama, no such physical proof had ever been uncovered. Obviously there never was a royal family of Gautama. Nearly all, if not most of what was written about Gautama – if that was his actual family name – contains no undisputable proof. He may have been seeking teachers to instruct him in the way of the truth; who, apparently, proved unsatisfactory. He may have tried to achieve enlightenment by strict regimes and starvation, only to recognize the fact that dying of malnutrition does not lead to salvation. He may have reasoned how and what was important to a body and a 'mind' (a brain and its conscio), to achieve harmony between them. Whatever he may have done it is all a conjecture, based only on ancient texts by those who never met him by hundreds of years.

What can be said of the Gautama legend is that it includes gods, and as gods do not and could not physically exist, that alone renders his story with distinct fabrications.

For example, when he supposedly resolved not to get up without attaining the enlightenment he was seeking, at that moment he was approached by Mara, an evil god of temptation (just like the supposed temptation of Jesus by the devil before his own quest, over 500 hundred years later), who, with his demonic servants, reigned the world of passion.

Of course, in his legend Gautama overcame all offers of temptation from Mara, and in the next twelve hours of meditation acquired all the knowledge. In the first four hours, from 6 pm to 10 pm, he acquired the knowledge of his former existences. From 10 pm to 2 am, he attained the 'superhuman divine eye', which allowed him to see deaths and rebirths of beings. In the last four hours, Gautama discovered the 'Four Noble Truths': 1: man's existence is full of 'dukkha': conflict, dissatisfaction, sorrow, and suffering; 2: 'dukkha' is caused by man's selfish desire; 3: man has a choice to be liberated by attaining Nirvana (Pali Nabbana); 4: the way of achieving this is by following the Noble Eightfold Path: right view, right thought, right speech, right action, right lifestyle, right endeavor, right mindfulness, and, right concentration.

But while these doctrines came to be written by religious scribes centuries after his death, they do indicate the rational of a human – or more likely, a composite rational of a group of humans – who were able to (thanks to their brains) perceive the reality of human need in physical existence. It is by such realisations that whoever is represented by Gautama, he or they deserve the title of the Buddha. For when surrounded by self-centred 'selfish fear', causing greed for power and possessions (as continued to be practiced by humans to the present day) he supposedly was able to recognise that his perception of 'self' or 'I' could be more beneficial to him if directed in pursuit of the conscious selflessness.

This doctrines present, quite correctly, 'anatta', a view that the body, feelings, consciousness, and thinking, are not the self (because conscio

PART 2. DISCLOSING THE SHAM OF HUMAN CONSCIO PERCEPTIONS

does not exist without its brain to maintain it), or the soul, because of 'anicca', pronounced 'anateca', means that: all is impermanent, being affected by change. With these notions Gautama came close to understanding why humans are human. What he failed to realise (for such realisation is impossible without the knowledge of physical reality) is that reincarnation does not, and cannot physically exist despite what humans may choose to believe. All that is physical cannot be redone after it's been undone. That which had changed cannot be physically be remade in to what it was in the past because all the physical particles and atoms cannot reform themselves after they disperse as energy.

And yet, only by having a conscio voluntarily disassociate itself from its selfish fears and selfish desires – while accepting reduced levels of selfish needs of the body – can that human brain allow its conscio to becomes consciously free of human wants and desires. Only then, when life is no longer restricted by such human conditioning can 'Nirvana' (happiness in physical existence, and not in some spiritual sense) be provided by its brain.

At the age of 35, in 528 BCE (some nine years after Cyrus released the Jews from their Babylonian bondage), Gautama was said to have attained Enlightenment and an Awakening, becoming the Buddha and a teacher.

Whatever the distortions and fiction that may have been added onto his existence, were he to have existed as an individual, then it would have been probable that he was a self-disciplined human who understood and tolerated human weaknesses in others, treating them all with affection and respect.

As a 'Spiritual Male', the Buddha was supposed to have been a social reformer, who condemned the Indian Hindu caste system and proclaimed human equality of all castes. This proclamation remained unheeded by majority of Indian population (until this was achieved through actions undertaken by a unique Indian politician, Mahatma Gandhi, in the middle twentieth century CE).

According to his legend, Gautama had also expressed that economic welfare for the underprivileged improves moral development of all, while any attempts to suppress crime through punishment was futile, because crime can only be reduced if the economic conditions of the poor were improved. Such philosophy of equality, anti-violence, and tolerance appealed to many Hindus, especially those from the lower caste classes, the poor, and the destitute.

Gautama, the Buddha, is attributed with miracles (as is the case with most 'Spiritual Males' after their deaths). However, it is written that in life the Buddha forbade his disciples to perform miracles, suggesting that the tricks of miracle making be left to the Hindu 'munis' – a fraternity of wizards, tricksters, and shamans. Instead he reasoned that to convey his truth so that it would be understood and accepted would be a miracle in itself.

Not surprisingly, irrespective of Gautama's knowledge and teachings, his entity had been converted by humans into what they understand and appreciate: an idol, whose statues are displayed to be worshipped, in exchange for collection of gifts by the Buddhist priests.

CHINA: GODS FOR THE MASSES; PHILOSOPHY FOR INTELLECTUALS AND EMPERORS

The Chinese 'Dominant Males' were always clear on why they fought each other: to have whatever the other one had, never for some religious advantage. The basic religion was based on what one's own ancestors presumably thought of you. This was accepted as a premise for superstitions, not to be taken seriously, but neither to be ignored.

Apart from such ancestor worship, from late sixth century BCE (Before Common Era), two schools of spiritualism and philosophy were being firmly established in ancient China. They were the Tao and Confucian schools of thought. Together with the later introduction of Buddhism, they had influenced Chinese thinking and development in every cultural sphere: from politics and the social systems, to religion, arts, and sciences.

The WAY of TAO

While the founder of Confucianism, Confucius, is a historical figure, known to have physically existed, the founder of Tao is as legendary and fictitious as that of Moses and other prophets and apostles of the *Old* and *The New Testaments*.

Lao-tzu ('Master Lao') is the name by which the nameless author of Tao-te Ching – one of the most famous books of ancient China – is usually known. He is considered as the first philosopher of the Taoist school; his family name was supposedly Li, and his first name Erh. Factually though, his presumed employment in the royal court of Chou dynasty; his meeting with the young Confucius; his travels to the west; his writing of the book of thoughts at the request of the guardian of the pass to the West, Yin His; and finally, his disappearance (just like that of Moses), were all fabrications.

PART 2. DISCLOSING THE SHAM OF HUMAN CONSCIO PERCEPTIONS

Even the 'Book of 5000 characters', Tao-te Ching – traditionally attributed to him – could not have been written by one writer, as it contains proverbs from the era of Confucius to those of much later period. The contents, as a whole, date to around 300 BCE, but many proverbs had been in existence long before then.

Actually, there is no great mystery to the work of Tao-te Ching, for it is a small compilation of publicly circulating proverbs amongst scholars, produced by various Chinese schools of thought (philosophy), in form of astute – but unsought – advice to an emperor. The intention of giving advice to a monarch was to show that scholars actually had the ability to influence a head of state, and consequently, the ruling classes of nobility as well.

A wide distribution of these scripts had allowed Tao-te Ching to be considered as one of the most sacred scriptures of the Taoist religion. This is despite that the original does not exist, and the contents having been innumerably altered, reinterpreted, and miss-transcribed in the last two thousand years – and that is just in China alone. Regardless of Lao-tzu never existing, this did not prevent the 'legendary sage' from becoming deified by the Chinese, as Lao-Chun (Lord Lao), and worshipped as a great saint and savior of mankind.

Tao, or the 'way', refers to a code of doctrines and behaviour. Taoism is a ritual worship of the 'way' as indicated in the book, Tao-te Ching, paraphrased as: 'Classic Methods of Applying Influence', and the later related books, Chuang-tzu and Lieh-tzu.

The primary intent of Tao was to impact this form of thinking onto the Chinese 'Dominant Male' rulers, and 'Dominant Male' emperors, by advising them to adopt an attitude of a sage: a wise person. It was envisioned that in maintaining a reserved and uninvolved relationship with their subordinates and their subjects they would benefit with a more efficient method of retaining control.

[39-41] *The best rulers are hardly known by their subjects;*
The next best are loved and praised;
The next are feared;
The next, despised, of whom liberties are taken:
For they are contemptuous of their people,
And their people become disgusted with them.
When the best rulers achieve their purpose
Their subjects claim the achievement as having
   occurred 'naturally'.

And another example:

[131-133] *Do not control people with laws,*
Violence or surveillance,
But control them with inaction.
For:
The more morals and taboos there are,
The more intolerance afflicts people;
The more factions there are,
The more weapons divide people;
The more laws and taxes there are,
The more theft corrupts people.
Take no action, and the people nurture each other;
Make no laws, and the people deal fairly with each other;
Own no interest, and the people cooperate with each other;
Express no desire, and the people harmonize with
each other.

Tao also advised the rulers to keep their subjects in ignorant contentment, assuring that such state leads to social harmony:

[8-10] *Not praising the worthy prevents contention,*
Not esteeming the valuable prevents theft,
Not displaying the beautiful prevents desire.
In this manner the sage governs people:
Emptying their minds,
Filling their bellies,
Weakening their ambitions,
And strengthening their bones.
If people lack knowledge and desire
Then they cannot act;
If no action is taken
Harmony remains.

In the form of wise advice to the emperors and rulers, the 'way' of Taoism views all, and comments on all, impartially and impersonally, fully convinced that human developments are not advancements but regressions. Civilization is considered as a degradation of the natural order, where the ideals of life can only be represented by the return to original purity of pre-civilization.

[43] *If we could abolish knowledge and wisdom*
Then people would profit immeasurably;
If we could abolish duty and justice

> *Then harmonious relationships would form;*
> *If we could abolish artifice and profit*
> *Then waste and theft would disappear.*
> *Yet, remedies used treat only the symptoms,*
> *And therefore are inadequate.*
> *People need personal remedies:*
> *Reveal your naked self and embrace your original nature;*
> *Bind your self-interest and control your ambition;*
> *Forget your habits and simplify your affairs.*

For the individuals, life and death are presented as continuation of eternal transformation from non-being to being, then back to non-being, with the wisdom of life being present in acceptance of conformity to the natural universe, where progress is not enforced upon it. In this manner, it is thought that to achieve the 'way' was not to take any action, because by avoiding unnecessary action, the 'way' would result in an appropriate and just outcome.

According to Tao, there is no true achievement without 'wu-wei'. 'Wu-wei' was not a representation of inaction but a subtle, unperceptive activity of no action. By such methods of inactivity, Tao considers that when any actions are instigated by power of a monarch, or a magician, then the universal force of the cosmos or drastic human relationships and striving would come into effect.

In which case, Tao implies, that results of such extreme qualities will inevitably revert to the very opposite of those qualities. This is the Yin and Yang (the 'dark side' and the 'sunny side' [of a hill]), which always reverts to what they were previously; as in: white having been black, as black was white. This would mean, for instance, that an intentional quest for power – with all its drastic and overwhelming actions and efforts – would result in futility, due to the very complexity, inability of control, and the unknown outcomes of the actions themselves.

> [91] *In its rising (of the way) there is no light,*
> *In its falling there is no darkness,*
> *A continuous thread beyond description,*
> *Lining what cannot occur;*
> *Its form formless,*
> *Its image nothing,*
> *Its name silence;*
> *Follow it, it has no back,*
> *Meet it, it has no face.*

A Tao puritan, therefore, should live a life of an innocent child (where innocence is not stupidity) untarnished by knowledge and unrestricted by morality, both of which are imposed by society.

For Tao any such influence should be no more than, 'a gentle breeze through the reeds'. While on surface the flux of change was accepted as inevitable, it was considered that within itself there remains an unchanging unity of the 'permanent Tao', that which is nameless.

Tao advises self-suppression of thought and reasoning as an assured pathway leading to the self-realization and to universal wisdom, with an accompanied advice for maintenance of a conscious rejection of greed, and less selfish attitude to personal actions and possessions.

[194-196] *Truthful words are not beautiful;*
Beautiful words are not truthful.
Good words are not persuasive;
Persuasive words are not good.
He who knows has no wide learning;
He who has wide learning does not know.
The sage does not hoard.
Having bestowed all he has on others,
He has yet more;
Having given all he has to others,
He is richer still.
The way of heaven benefits and does not harm;
The way of the sage is bountiful and does not contend.

Immortality for Tao was a serious matter. Long life and the vital life force, or life energy, were considered inseparable from saintliness. A body at birth was presumed to have been allocated a certain quantity of various vital life energies, derived from Heaven and Earth. Tao was believed to be the process and the technique of retaining longevity of life. It was seen as vital to preserve these life forces, preventing them from being needlessly squandered, as an absence of them was to result in death. Because this vital energy and spirituality could not be distinguished apart, for in Taoism old age, in itself, becomes sainthood. By following the 'way' and preserving personal life force energy a Taoist sage would become a saint, because he had been able to cultivate himself through a long existence.

Due to such reasoning there are many categories of immortals in Tao. Of these, the high saints ascended to heaven in the broad daylight, while lesser immortals would do so only after their death. The highest level was awarded to the fictitious Lao-tzu. Not unlike the *Dead Sea Scrolls*, a scroll found in a walled-up desert library of Tun-huang, the *Book of the Transformations of Lao Tzu* (*Lao-tzu Pien-hua Ching*) depicts Lao-tzu as a cosmic, ever-present, all-capable god; the origin of life.

While the philosophy of Tao was aimed at individual thinkers and scholars, its principle notions were incorporated into a religion, comprising of good and evil gods, as well as eternally living sages and spirits of human ancestors.

After gradual penetration of Buddhism into China, Tao also became controlled by structures of organized hierarchies of hereditary priesthoods, operating temples dedicated to the veneration of popular Chinese gods and deities. But because Chinese 'Dominant Male' rulers and emperors constantly elevated or lowered the status of various gods, and disliked Tao priests and monks for being a burden on the state for not working for the state and paying taxes, Taoism never became ingrained as that of, for instance, Hinduism in India.

The relationship of Tao: the philosophy, and Tao: the religion, had been accepted as a union of mind and spirituality; and yet, remained independent of one-other.

Whether consciously or not, over two thousand years the subtleties of Tao had an enormous influence upon all levels of Chinese societies, to the extent that both the ruling emperors and their empire had become isolationist, as suggested by the Tao proverbs.

> [50a-50c] *Therefore the Sage embraces the One, and is a model for the empire.*
> *He does not show himself, and so is conspicuous;*
> *He does not consider himself right, and so is illustrious;*
> *He does not brag, and so has merit;*
> *He does not boast, and so endures.*
> *It is because he does not contend,*
> *That no one in the empire is in position to contend him.*
>
> [104-105a] *When the Way prevails in the empire,*
> *Fleet-footed horses are relegated to ploughing the fields;*
> *When the way does not prevail in the empire,*
> *War-horses breed on the border.*
> *There is no crime greater than having too many desires;*
> *There is no disaster greater than not being content;*
> *There is no misfortune greater than being covetous,*
> *Hence in being content, one will always have enough.*

The profound influence of Tao had shaped the national character of the Chinese and their national attitude towards their neighbouring and distant nations, up to the early twentieth century CE. Unfortunately for the Chinese, the humans of other nations were taught the very opposite to Tao by their religions. These religions encouraged their worshippers to desire all that China had, even while China had been willing to remain isolated, and to dismiss anything that other nations had to offer.

In its current state, the Chinese political leadership had altered its attitude towards Taoism. From 1970's it dismissed isolationism and any political or military reluctance to advance its causes in the world arena. As to the current notions of religious influences, while the overall populace may still observe ancestor-worship customs, their primary gods are now those of 'luck' and 'money'.

Confucius: Humanity of all for all

Unlike Lao-tzu, Confucius, whose family name was K'ung, and who became known as K'ung-fu-tzu, meaning: Master K'ung (later Latinized to 'Confucius') had certainly existed, living in China between 551 and 479 BCE. Like other saints, however, his life and reputation had been elevated by the authors who recorded his life after his death. That is only to be expected of humans who feel that a teacher – if worth learning from – needed to have had a high rank earlier in life. For that reason, just like Buddha – who supposedly gave it all up after being born a prince – Confucius had also been awarded a high birthright from being born of a noble family, but in his case an impoverished one.

While scraping for his meager existence from an early age, Confucius still managed to educate himself. He personally witnessed the results of injustice and oppression practiced by various 'Dominant Male' feudal lords, forcing their own and their emperor's subjects to be over-taxed, starved, and subjected to forced labor. This compelled Confucius to attempt providing some relief to the sufferings of the general Chinese population. He reasoned that the solution was in having the 'Dominant Male' rulers re-evaluate the purpose of their power.

As far as Confucius was concerned, their power was not intended for their pleasure but to provide happiness to their subjects. He, therefore, advocated reductions in taxation, legal justification for severe punishment, and avoidance of needless military conflicts between aristocrats.

Although his ideas did not obtain for him a government post he was seeking, they did attract to him young disciples. He began educating his students by talking to them as individuals or in small groups, rather than presenting authoritarian, dogmatic lectures. His teaching method consisted mostly of asking questions, to which he expected his disciples to find appropriate answers for themselves.

"If, when I point out one corner of the subject,
The student cannot work out the other three for himself,
I do not go on."

He, however, did not consider himself to be always right, or to be even understood. The core of his teachings was that the development of any human has to be based on ethics in all dealings between humans; sincerity in all communications with everyone; and education to develop abilities and strengthen the character, so as to have that human understand all that which surrounds him, without having him wallow in ignorance by merely depending on presumptions. He felt that ability was not dependent on a birthright, but rather that education gave opportunity for the capable to develop themselves for their own and their society's benefit.

Confucius regarded the process of government as a craft requiring an application of ethics on all levels, including cultural. He had his disciples study history, poetry, music and procedural rites, as well as schooling them in the theory and practice of human relationship, and how to conduct themselves in a wide variety of situations.

These methods of education proved to be profoundly revolutionary for that age. By declaring the right and duty of every individual to make personal basic decisions (choices), he was challenging the premise of authoritarianism, prevalent in government operations of that period. By having the government services accept as employees even those of the poorest and the humblest-of-birth students, he became influential in breaking the monopoly of the aristocrats.

Until then, aristocrats were the only ones who could afford the education necessary for entry into the government service. Confucius believed that anyone possessing virtue and ability could properly govern, while those lacking these qualities had any rights to power. Certainly not the aristocrats, who presumed themselves to be descended of divine ancestors, and who, they proclaimed, gave them the right to rule. As far as Confucius was concerned, even the hereditary rulers should confide all administrative duties to ministers selected on their ability and virtue, who then should do their best for the state; running it as a cooperative enterprise.

Confucius considered war as evil, yet, if unavoidable, then to be conducted vigorously, with the main potential for success in battle being the clear understanding by the army as to the cause for which they were fighting, and the justice of it.

The central element in his philosophy was humanity:
"Virtue is to love men;
And wisdom is to understand them."

"The truly virtuous man,
Desiring to establish himself,
Seeks to establish others;
Desiring success for himself,
He strives to help others succeed."

Confucius broadened the concept of the family – which was always important in the Chinese culture – to that of family of mankind, endorsing one of his disciple's expressions:
"Within four seasons,
All men are brothers."

It is ironic that Confucius, who never believed in, nor worshipped any gods during his life, was made into a major deity after his death. While Confucius was to have praised as virtuous those who stood in awe of heaven, he considered most of the religions of his day as sheer superstitions, condemning many of their practices.

Chinese Buddhism: The Way of the Zen Buddha

The third religion that influenced Chinese thought and the national religious consciousness was Buddhism. Unlike Tibet, where Buddhism was embraced with fervour by a population living in a cultural vacuum, Buddhism was not readily accepted in China.

In China, Buddhism met Confucian tradition of rationalism; the speculative forms of Taoism; the ancient culture of ancestor worship; and the authoritarianism of the state. The Chinese state did not appreciate the Buddhist concept of monastic organizations. The general populace however found the notion of spiritual salvation appealing, while the concept of gradual attainment of enlightenment through meditative purification was simply either ignored or dismissed.

By the second century CE, Buddhism in China had sufficiently altered from its Indian origins to be accepted as another extension of Taoism. The concept of 'samsara' (reincarnation) and the state of salvation (Nirvana) merged with the Taoist ideal of 'non-activity' and 'avoidance of direct action'. The result of such assimilation was a new form of Taoist mysticism: Ch'an (Zen) Buddhism, or: Chinese Buddhism.

In its development, Zen Buddhism influenced evolving movements of Confucianism, allowing Confucianism to acquire an added dimension, and to delve beyond its concerns for the state and society.

In assessing the Eastern and Asian religious philosophies of Buddhism, Taoism, and Confucianism, a conclusion can be reached that the three teachers – the semi-legendary Buddha, the legendary Lao-tzu, and historical Confucius – had produced a calming effect on very large populations of India and China, for a very long period. Confucianism produced a moral and political system that fashioned society and the Chinese empire; Taoism represented more personal and metaphysical human preoccupations; and Buddhism presented fundamental physical concepts, like the suppression of the individual ego and the rejection of the illusory nature of the physical world.

These three philosophies had effectively influenced the 'Dominant Male' rulers of those regions to practice humanity and humility on behalf of their subjects, while the subjects were taught to accept and enjoy their lot on Earth, and, if so desired, attempt to attain peace and freedom beyond death.

In essence, these religions had cocooned India, and especially China, from having any desire, or need, of making any sustained extensive overseas explorations, and of conducting any dialogue with other foreign powers. This state of isolated self-sufficiency remained until the Muslims, then the European Christians, came rudely marching in.

Some humans (conscios) may see this as a failing of those Buddhist, Taoist, and Confucian cultures based on non-aggressive notions, in preventing India and China becoming militarily avaricious, like the European Christians, and the Arab Muslims, with their domineering and intolerant attitudes towards other nations' methods of worship and trade. Those who think this way give little thought to what kind of bloodshed and destruction the countries of the East and West have been spared, thanks to the legacies

THE EASTERN WAY

of the three kind teachers. Should the 'Dominant Males' rulers and emperors of China and India had religions similar to those of Judaism, Christianity or Islam to guide them, then in all probability they too would have behaved as hateful, domineering bullies, ready to conquer all lands before them – like the murderous butcher Genghis Khan – just because they were there for the taking.

In contrast to the Mongols, India and China had the advantage of developed systems of governing the most heavily populated regions on Earth, and being then the richest. Therefore, had the nations of India or China rolled out their vast armies in conquest of foreign lands, behaving with ruthless imperialistic frenzy – as been done so, or attempted, by nearly all other nations on Earth – it is most probable that the current borders of this planet, imposed by the empire-driven Christian and Moslem 'Dominant Males', would have been totally different for centuries to come. And so would have been the racial makeup of humans in all those regions.

This is not to say that India and China had no 'Dominant Male' rulers, who were bullies and butchers. They did. But on the whole, under the influence of these three religious philosophies of Buddhism, Taoism, and Confucianism, most of those 'Dominant Males' were content to retain what they had, without the greed to possess the world; nor to thrust their religions onto all humans, as Christianity and Islam constantly did, and continue to do. These religious imperialist invaders of Asian countries, especially the Christians, thrusting forth their religions of intolerance, contemptuously considering the large nations of India and China as weak, and therefore, open to exploitation and pillage of their wealth.

Having been taught to experience the harsh lessons of military aggression and imperial exploitation by their former European 'Dominant Male' masters, the current large Asian nations have put aside the gentle teachings of the Buddha, Tao, and Confucius. These have been replaced with their 'Dominant Male' military and industrial ambitions based not on religious but political ideologies, involving selfish aspirations to achieve imperial expansion by means of nuclear weapons, economic market dominations, domestic and international political manipulations, and claims of foreign land ownership through supposed past ancestral links, as in Tibet and Kashmir and the South China sea.

The problem for all those in future, is that these two large Earth regions – currently populated by over forty percent of all humans on this planet – are anxious to experience the wanton greed of the 'western world'; something they were forbidden to do in their past by their cultures and religions that discouraged selfishness and greed. By having replaced the philosophy of 'Eastern' selflessness for the 'Western' ideology of desire, the current Asian nations had become so very much like all other 'Dominant Male' imperialists of the past.

And yet, while these south eastern and eastern Asian 'Dominant Male' political and business rulers replicate the behaviors of their former conquerors and imperial masters, they still remain devoid of the religious dogmas of the western nations, which proclaimed their usual desires of wanting to dominate 'the world' in the name of their god, because supposedly their god had sent them specifically on such missions. It is by having no such divisive cultural and religious falsehoods – requiring justification of 'sin' – that leads the current Asian politics, financial dealings and business to ruthlessness. That is because for them all human actions, no matter how selfish, are permissible and forgivable, just as long as the intent is to give advantage to the 'family'; where for these human cultures the 'family' has relevance beyond paternal groups, signifying: business, political, military, and gangster structures.

By discarding their knowledge of selflessness – as taught by their past gentle teachers – the current enormous Asian populations are primed to fulfill the desires of their current 'Dominant Males', which are singular: to dominate their former masters – their former Western rulers – at the expense of any resultant discomfort and suffering that may be brought upon their own Eastern populations.

21

The way of 'Dominant Male'

Before Constantine the Great came up with his inspired 'eureka' moment, no 'Dominant Male' had considered that a nation, or an empire united by one religion would be good for a Leader (or an Emperor) to rule that nation (or an empire). Not even the Jews, who were presented with that notion by the *Gospel* of *Mark*.

From the earliest epochs, 'Dominant Males' fought each other with banners of their political allegiance and those identifying their regiment, but not that of their religion. Religion played no significance in their conflicts, because every soldier had differing belief to that of his comrade, while many worshipped simultaneously multitude of gods and spirits.

It was Constantine who thought to change all that. He would have all of his armies and populations believe in one particular religion. His successors succeeded in doing just that, where there was not just one god, but a trinity: God the Father, God the Son, and the Holy Spirit.

The reason that Constantine's 'Dominant Male' successors continued with this plan is because it is very helpful, if not indispensible tool for ruling. The advantages were not only in having the population united in one religion, but for the 'Dominant Male' rulers as well.

To illustrate this the 'Dominant Male' rulers who succeeded Constantine obtained the following advantages to their ability to rule, at having become Christian:

• They acquired a dispensation of righteousness for all of their deeds, and they were now proclaimed to rule, in perpetuity, in the name of a righteous god.

• They reinforced their control over their subjects by wielding not just the physical power over them, but a spiritual leverage, as well.

• They expected more local compliance and loyalty, from having an affinity of faith with their subjects.

• They gained greater security for their domains by making alliances with other kingdoms ruled by 'Dominant Males' of the same faith.

• They had justification to defend the notion of their god in anyway they saw fit, in the process invading, attacking, pillaging, and killing humans liv-

ing in other regions of the planet, especially if the residents of those lands were practicing a different religion to their own.
- And, just in case there was a heaven, no matter what wickedness they were guilty of, they had the spiritual security of a promise to an afterlife – for where else would a ruler go, but to heaven.

By adopting religions, the 'Dominant Males' have become used to ordering their subordinates to go into armed conflicts and to their deaths by evoking their gods, so as to justify their actions (from which only they, the 'Dominant Males', actually benefit, as they remain alive while those whom they order to die do so). And for all those who should perish, well, there is the promise of their ascendancy to god in heaven, and the eternal life: be that a Christian afterlife, or an Islamic afterlife.

There is another use for evoking gods, as far as the 'Dominant Males' are concerned: an absolution to killing. Despite the fear, guilt, and conscience with which human brains try to influence their conscios, in order to have them be kind to all other human brains, many conscios retain a weakness for killing. They like to kill not just to alter a physically threatening situation, but because killing makes them feel superior to those they have killed, kill, or want to kill. By killing, conscios feel that their 'superior will' is being done. Besides, there is the override negotiator component to every conscio, which conscios use to justify to their brain the reason for killing, and by that hoping to avoid the consequences of experiencing the pain of mental agitation and anguish, issued by their brain's conscience.

In their futile efforts to alleviate their conscience, some try to transfer the blame for their actions onto their victims, whom they have killed. Others seek to alleviate their guilt by transferring the blame onto those in authority, who had given them orders to kill.

'Dominant Males' had long understood this blame-game, and have learned to use it to their advantage. In their efforts to prevent any hesitations from their subordinates to kill when ordered, aside from using threats, the 'Dominant Males' have learned to issue wavers for killing.

Apart from the 'War Male' officers, who are trained to like killing other humans (by overriding their conscience), most other humans introduced to killing other humans fear the pain of their conscience, and from that have a fear of killing. So, in order to make them feel relieved and more accepting of killing other humans and to become more enthusiastic participants in such bloody deeds, the 'Dominant Males' – with the aid of their

'Spiritual Males' – have devised absolutions of the crime of killing enemies, in a form of forgiveness issued from their god.

By rendering the 'sin' of murder to be nullified by god for the duration of the physical conflict, the human combatants embroiled in such conflicts become answerable to two authorities: that of the state (that being the authority of the 'Dominant Males'), and the god (whose authority is the 'Spiritual Male'). With god, presumably, giving his permission to kill on behalf of the 'Dominant Males' (the state), supported in this by the 'Spiritual Males' (representing god), these two colluding entities can demand killing from anyone with no possible refusals. Any disobedience of an order to kill becomes indefensible, because both the state (represented by 'Dominant Males'), and their god (represented by 'Spiritual Males'), demand killing of an enemy; whereas in any civil court the intent to kill equates to murder.

With such entrapment, most humans embroiled in processes of military murder usually suffer the consequences of their bloody deeds, for murder of other human brains may be forgiven by the 'Dominant Males' and their 'Spiritual Males' – in the name of their fictitious gods – but never by human brains involved in doing the killing.

Currently, the after-effects (that often occur many years after the events) for the so-called 'war combatants' – or more accurately: 'War Males' – are described as: 'panic attacks' and 'post-traumatic stress disorders', where former or still active 'War Males' (and these days, 'War Females') suffer ongoing symptoms of depression, uncontrollable crying, spasms or stiffening of the body, racing heart-rate, difficulty of breathing, constant head aches, and reoccurring nightmares. These afflictions can affect humans (conscios) who consider themselves healthy and mentally capable of controlling their fears and bloody surroundings.

But what these 'War Males' – and all the psychologists who prepare them for conflict and then treat them afterwards – do not understand is that the brains of the 'War Males' do not appreciate what they do: they, the 'War Male' conscios, blatantly prepare to kill, and then do so, other human brains. This their brains cannot forget nor forgive. And although these hurt and disappointed brains do not begin to punish their conscios with fears, guilt, and pain immediately – because they, themselves, have to experience what they present to their conscios – sooner or later they decide that they simply cannot allow their conscios to go unpunished.

The only way that some human brains may excuse their conscio's participation in killing of other brains, is when that conscio has no alternative but to defend itself and consequently its brain. But even then, having survived, that brain will mourn those brains who had died, which is reflected in the feeling of at least sadness if not depression and despair.

There are many ways that humans had been – and still are – duped into joining warring schemes instigated by their 'Dominant Males' politicians and 'Spiritual Males' priests. Apart from a financial incentive or the patriotic appeals promoting hate and fear, the other method is the call-to-arms in the name of god. This hook often works for 'Dominant Males' because those who join up are inevitably too obtuse to rationalise that they had never, ever, received an adequate explanation of what 'in the name of god' means. After all, had anyone, ever witness a god publically evicting an order: "You shall go to war on my behalf!" Never. Or, had they ever received a direct physical letter from god directing them to fight, as they do with a conscription notification? Never! Furthermore, had there ever been but one officer of any armed forces – of whatever rank – who had experienced god placing into his hand an executive order directing him to kill his enemies? Never!

So if there had never, ever, been an occasion when god – any god – had physically given instructions for humans (conscios) before witnesses, to butcher each other, what makes them presume that those evoking god are telling the truth? Conscio acceptance of lies, their ignorance and a refusal to recognise sham: that is what makes them accepting. This is all the reasons that dupes need to dutifully follow their 'Dominant Males' who dupe them with notions of non-existent gods, because those gods, supposedly, command the dupes to expose their bodies and their brains to danger.

(It is also amusing to consider how dupes fail to reason that if their god was so powerful and mighty, then why would soldiers be even needed, when their god could simply dispense with the enemy using his godly super powers?)

'Dominant Males' had always existed, then as now. Humans (conscios) of today, who claim to pledge their careers in politics to better the lives of their voters, and to serve faithfully and honestly their constituency, their society and the nation, are in the league of 'Dominant Males' no different to those of the past.

On the whole, if looked at without delusions or sentimentality, any human (conscio) that enters politics is doing this for one purpose only: to rule, to command, and to enforce their opinions and decisions onto others under their control.

Of course, in most nations where the so-called democracies present a liberal outlook, the voters have a power of choice to elect or not to elect their representative politician, and by that have some power over their politicians. This power of an individual voter has been cleverly circumvented by the invention of political parties, where irrespective of individual politicians being discarder by their voters, it is the majority of winning politicians of a whole political party that rules the county, the state, or the nation.

We (conscios) understand and like lies (using them ourselves daily) so that we can instantly recognise political lies. Despite this, there is little what individual voters can do about this, as concentrating on our own lives is more important to that of political lies. So unless media steps in to divulge political lies for what they are, causing indignation and requests for retribution from the voters, many political lies continue to be used as political tool to belittle others and to avoid disclosure of truth.

That is why those who do rule, often do so with disregard (and often with impunity) of their voters, knowing full well that while in power the laws they introduce shall remain for a long period, and therefore, shall have a long-lasting impression of their personal will having been done, irrespective of any future consequences.

Having party systems, democratic politicians in power to rule, often make choices not just for the benefit of their nation but for personal gain of influence over others, be that even other politicians; for obtainment donations for re-elections, and even for personal gain of wealth.

They continue to dictate their 'wants' and 'desires' onto others with the constant referrals to god, just as those in the past. Their repetitive expressions of, "God bless (followed by the name of their nation)", 'Thank God', "God be with you", "God knows", and so forth, are not expressions of their faith. Instead they are defensive acts of self-protection, signifying that god may intervene favourably or not, and by that alter the outcome of their reasoning and the implementation of their decisions, as if it is something that is out of their mere mortal control.

So how then the politicians of today differ to the 'Dominant Male' rulers of the past? Basically they do not. They are still the 'Dominant Males' irrespective of their gender. The only difference is that now when media cries out about a failed or disreputable politician "falling down on his/her sword", they do so figuratively, and not literally as in the past.

PART 2. DISCLOSING THE SHAM OF HUMAN CONSCIO PERCEPTIONS

And what of those whom "Dominant Males" rule? Not much has changed there either. We (conscios) continue to believe their half-truths, lies and deceptions, with hope that whatever eventuates our personal outcome is favourable, disregarding all others. We humans (conscios) resolve to be foolish now, as we had been thousands of years ago.

Meanwhile, while "Dominant Males" rule by evoking god for their own political benefit, the 'Spiritual Males' fulfill a different objective. Their aims comprise strictly of obtaining financial wealth from their faithful.

22

The God business

If only there was a heaven for humans. How happy they would be, one would think. But if admission to heaven required a lack of sin, then heaven would have been an absolutely desolate place, for even god would have been evicted for lying and supporting evil causes. For proof of this consider that each religion of the 'hateful trinity' claims their god to be a singular male, who had been quoted by each religious writings, in which he is stating different sentiments for each religion. This would make god a liar, as differing versions cannot all be right, and that would involve lies from god in some of the religions. Furthermore: all warring sides in a military conflict claim to have god on their side. Therefore, as each side regards their enemies as being 'evil', this would mean that god ends up aiding 'evil' causes of one of opposing sides, especially that of a losing side.

Sadly for humans (conscios), there is no heaven, irrespective of what the 'Dominant Males', and their servants, the 'Spiritual Males', may care to claim. Whether the humans on this planet agree with this physical truth, or not, this fact shall remain: there is no one super-being who possess the ability to instantly produce from nothing any form of physical matter, whether as objects or life forms, including himself. There is no super-being that can destroy or cause any forms of physical matter to instantly cease to exist, or vanish, be that objects or life forms, including himself. And there is no super-being who has the ability to change any occurrence in the physical process of change, in space. Such super-beings, or gods, simply do not exist, never did exist, and never, ever, shall exist.

There is no living, consciousness-possessing Christian God; no living, consciousness-possessing Moslem God; and no living, consciousness-possessing Jewish God. There are no living, consciousness-possessing gods of any kind, anywhere on this planet or throughout the Universe, or anywhere in Eternity. There is no heaven, no hell, and there is no purgatory. There is no living, consciousness-possessing devil, by any other name or title; there are no living, consciousness-possessing angels. There is only the physical process of change, which is physical change occurring to all that is physical, everywhere, simultaneously, with all of them formed of inunens (to be explained in the following part of the book).

What gods there are, are those in the imaginations of humans. Human conscios had their brains invent gods, as they had their brains invent everything to do with their human activities and cultures. Everything invented, formed, built, and constructed for humans, has been done by human brains. Languages: by human brains; arts: by human brains; and religions: also by human brains!

The invention of gods and divinity had been an important device for 'Dominant Males', and their servants the 'Spiritual Males', by which to retain control of other humans at all levels of social structures of all societies. Every new 'Dominant Male' and 'Spiritual Male' that came along, appropriated and then altered the structures of rule and worship that existed before, convincing those who followed them that they, as leaders, are right and all those before them were wrong. This is why Christianity – as any other religion – is fractured, having constant schisms, rebellions, and separations.

Taking Christianity for an example: the early Christians could not agree on the structure of their religion, and so broke up into Catholicism and Orthodoxies. Then came 'Spiritual Males', likes Martin Luther. His anti-Papal religious doctrines, arising from his business and ecclesiastical disagreements with the Roman Catholic Church, resulted in formation of many Protestant Christian denominations.

One such version was devised on behalf of a 'Dominant Male' monarch, King Henry VIII, king of England. Henry desired: his 'Spiritual Males' fulfilled. And so, a new Church of England was devised to assist Henry to legally wed, and divorce (and behead) a number of wives in succession; an activity his previous (Catholic) religion disallowed.

It was inevitable that all such religious divisions resulted in bloody and violent conflicts between Christians, with hatreds and loathing on all sides. Each faction proclaiming god to be on their side, and that only their version of Christianity was most favoured by god.

No wonder that the 'Dominant Males' and their 'Spiritual Males', representing the three Yahweh-derived religions of hate, have been responsible for so much human misery and turmoil. The legacy of these religions continues to this day, with the blood of innocent children, and the ignorant and hateful women and men being shed in the name of their god, all over this planet.

To prevent such religious turmoil from continuing into future, humans will need to finally understand why these religious conflicts, both internal and external, actually occur.

The real reason for religious conflicts is this: while being presented as disagreements on some points of ideological religious perception, the actual disagreements erupt from the inability of 'Dominant Males', and, or, 'Spiritual Males', from agreeing as to who shall control a particular, lucrative business of religion.

From the moment the first religions came into practice, they had one purpose only: to bring income to those who operate them. From then on, all religions that have ever been invented and practiced by humans, had, and have, but one fundamentally common characteristic: they are not free of cost to the faithful. All religions expect payments to their god, or gods, from their worshippers; whether these gifts and contributions comprise of food, goods or money; and whether these payments are disguised as sacrifices, contributions, gifts, legacies, dispensations or donations. They are all part of the process of wealth-accumulation conducted by 'Spiritual Males' of all religious denominations and institutions. Most religions, and their 'Spiritual Males', encourage their worshippers to contribute as much as they can, on an understanding that the more they give the better are their chances of attaining 'heaven', and the closer they would be placed to their god.

Of course, this is not the only revenue-making activity for religions. Most religions were always a big business. They made a great deal of money from merchandising of religions artifacts – be they jewelry, books, calendars, statuettes, paintings, and icons – based on visual representations of their corporate possessions: those being their saints and their gods, presented as idols. Most religious organisations had and have enormous investment portfolios in real estate, from which they derive lease and rental incomes. Most have investment portfolios in banking, financial institutions and share markets. Most religions operate private schools and universities, through which they obtain the next generations of the faithful while also acquiring substantial incomes from the education fees. Many religions operate large private manufacturing and agricultural businesses, under corporate business names, from which they derive large profits. Most religious institutions pay no taxes on their assets and incomes to their states, saving themselves huge amounts of money. They pay meager wages to their priests – their sales representatives – expecting most of them to make their own living from alms, charitable contributions, and payments for their services performed on behalf of the religious individuals within human communities. No religions pass up an opportunity to make more money by extracting payments for their places of worship when used for rituals of marking births, weddings, and deaths of their faithful. Some religions operate charitable institutions, but even those organizations are not allowed to operate at a loss, and are used to enhance their religion's public relations image.

Because religions are a very big multi-national businesses, just like all other multi-national corporations selling their goods and services to the public, religions do all they can to ensure that their right to operate their businesses remain unchallenged by competitors or critics. For that reason they spend a great deal of energy and money to produce and maintain their influence over societies in which they operate.

To this end, they conduct public awareness activities to display their 'good' charitable deeds and their 'caring' attitudes. They subtly instill their oppressive and repressive notions of obedience, disguised as 'righteous' forms of social behaviour onto their faithful and supporters; once this is done, they use the numbers of their indoctrinated worshippers as leverage of influence over politicians. They organize and support political parties that represent, or endorse, their prejudicial views. They own, organise, and operate media outlets that publish and broadcast only their views on social and political issues, as well as, financially support publishing and broadcasting opinion makers: those who indorse and agree with their religion's biased political and social views.

The difference that exists between the business of religion and those providing other goods and services, it that while the businesses of goods and services sell tangible goods and services, the businesses of religion sell only intangible sham.

Most religions consider that with sufficient prayers and belief, their god can answer these prayers with a miracle, granting those wishes. Whether it be a cure that is needed for a damaged or a diseased body, a reversal of financial misfortunes, or millions of other selfish human wants and desires, each and every religion is prepared to present their god as the only way to achieve the desired outcomes. As far as religions are concerned, all human burdens and expectations can be met with sufficient prayers to their god, financial contributions to the church, and faith in the 'Spiritual Males'. By such means, 'hope' – which is a human conscio's natural expectation of the impossible, or at least improbable to eventuate – has been seized and claimed by religions to be their exclusively property, which is supposedly managed by their god, for he, and only he is supposedly able to give or not to give, according to his choice. So physical change is what god wants it to be. "Too bad for reality," chuckle the 'Spiritual Males', despite that the reality of life never, ever, diverts from physical reality, not even for them.

What religions must like about themselves, is that despite being big corporate business they take no responsibility for failure. Unlike other businesses of goods and services that have to provide guarantees and warranties for what they sell, religions do not. If, by chance, that which was prayed for eventuates even remotely resembling the selfish request, god is given credit; and in all instances where prayers do not eventuate, god is never blamed; for that is 'the will of god'. What a successful business formula! One in a billion prayers eventuate: religion wins; most prayers do not: religion wins.

Then there is the religion-controlled monopoly on issuing immortality to humans. Whatever the chance outcome to the life of a body, religions are ready to offer a better alternative after the end of life. According to religions, simply pray, pay, and the gods in heavens will welcome those who do, to live, presumably, forever in peace and comfort. Whatever the diseases or mishaps that put an end to life, religions have taught humans (conscios) to expect a salvation in the afterlife, without ever having provided guarantees and tangible proof of this, simply because heaven is their intangible product that brings them their tangible monetary revenues.

For all the readers who may disagree with this view of those who operate religions, there are two simple questions:

What would happen to all religions in the world if the 'Spiritual Males' made their religions free of any cost, or contributions, to their worshippers?

And what would happen to religions if any payment, in exchange for worship, were to be considered as an affront to their religion and to their god?

The answer is that they would all go out of business. After all, that is exactly what is occurring to all monastic institutions across the world. Without financial support and income religions would have no incentive to exist. Those who are involved with religions have to live as all others. To do so they need money. Without any money they would need to find other means of employment to obtain money. After all, the consideration of performing religious rites for no payment would have little appeal to 'Spiritual Males'.

Fortunately for religions and their faithful, as well as, their 'Spiritual Males, and the 'Dominant Males' who support religions for their own benefit, this is not about to happen, just yet. That is because the controlled populations enjoy being controlled. After all, they have been shoved, prodded and pushed by 'Dominant Male' conscios to such an extent – and for so long – that they consider themselves to be lost without guidance made of restrictions. (This is despite that most of them naively consider the notion of 'freedom' being imperative to their social existence.)

But while the religious masses are fearful of losing their religions, the 'Dominant Males' and the 'Spiritual Males' who orchestrate the illusion of religious wellbeing, are themselves fearful of losing their religious 'customers'. For who would then provide them with the lives that they have come to not just to appreciate and enjoy, but also to expect?

The result of these combined insecurities force the 'Dominant Males, the 'Spiritual Males' and their believers to cling to each other, forever fearful of being alone.

However, if humans continue to cling – in their insecurity – to 'Spiritual Males' and to their religions, they should at least acknowledge one undeniable truth: like it or not, religions they follow require changes.

The scriptures of Judaic, Christian, and Islamic religions retain teachings that are not compliant with even the minimum notions of humanity. Therefore, all religious writings containing any references to: sexism or sexual gender discrimination; prejudice; racism; vilification of alternate religions or non-religions, must be revised or deleted altogether – without any exceptions or reservations.

Furthermore, all religions, without exception, should embrace the concept of 'freedom of choice'. This could be an open declaration, included into the religious texts of all religious teachings, advising and acknowledging, that:

• Every human, whether a participant of their religion or not, has the right of choice to worship, or not.

• No followers of any religion have the right to force onto others their religion.

• No worshippers of any religions have the right to scold or criticize any humans who think, or worship differently to them.

- No human is to be denied the liberty to believe – physically unhindered, unmolested, and unharmed – in anything in the Universe, whether existing physically, or existing only as a figment of human imagination.

What is often misunderstood by those who profess to be religious, is that it is inconsequential whether one worships or not, but whether in doing so – or not – this improves the quality of their life's journey, without impeding anyone else' life journey. If any human (conscio) requires the support of any religion, then they should use it. If, instead, they have no dependency on religion, then so be it.

The only thing that matters to a brain of a believer in god and a brain of a non-believer is that they, human brains, do not believe but know that it is only they – the brains – who support the conscios of both a theist and an atheist, as no other entity in physical existence can do this.

23

Thinking aloud

Once upon Earth existence there were no domesticated animals. Referred to now as 'wild animals', these animals had no man-made restrictions of movement. Their only constraint of roaming anywhere they could was that of Earth's seasons. The herds of herbivore grass-eaters followed the seasonal plant growth as their food source, while the carnivore predators followed the herbivore herds as their source of food.

In the same manner to the early wild animals, once upon Earth existence there were no domesticated humans. Early human groups and tribes had no borders to content with, spreading out in all directions on Earth, in similar pursuit of animal herds to that of the carnivore predators.

Since those early days human conscios have learnt to subdivide most of Earth's surface, and to subjugate each other within the borders of those subdivisions.

But such restrictions have not deprived their imaginations from pretending that they, conscios, remain entitled to forms of freedoms their distant ancestors may have enjoyed. They continue to refer to "their basic freedoms" of living, speaking and thinking as they themselves would have it, without any restrictions or prevention from any opposition.

This, of course, is nonsense, as human conscios gave up the rights to these, so-called, freedoms long ago.

Once individuals accepted the concept of 'Dominant Maleism', where a single figurehead was given the right to rule all others by mutual decree rather than a personal physical domination, the subordinate individuals had abandoned their right to do what they would want to do without the permission of their 'Dominant Male'. They would also have to agree to any opinions their 'Dominant Male' chose to express.

As in many societies the opinion of a single 'Dominant Male' became superseded by opinions of other notable individuals, such as 'Spiritual Males' offering their religious notions, this had further eroded the freedom of individual humans. By accepting the religious notions of 'Spiritual Males', humans in all societies became indoctrinated to think what they were taught to think by the opinions of 'Spiritual Males' and 'Dominant Males'.

Even as in many societies the 'Dominant Males' have evolved from a ruling individual to that of a ruling government structures, their laws, rules and instructions (together with the religious laws, rules and instructions) – with threats of punishment for disobedience – continue to influence all individuals to do exactly what they had been taught to do, what to think, and what to say.

What this means is that by being disciplined in what they do, think and say, the population masses of all societies on Earth are conditioned to obey the opinions and directives of their political, business and religious masters.

The problem with this circumstance is not in just what the political, business and religious rulers desire but in the method by which they want to accomplish these desires. What every government, business and religion had always wanted was more financial revenue from more tax payers, more profits from more customers, and more donations and contributions from more faithful. All this equates to an endless desire for population growth. And who is there to facilitate such desire? The business of science.

Indeed, it has always been science that attempted to improve human lives. But now science has become big business – in competition with itself – to provide chemicals and means to sustain the span of human lives, including providing methods to constantly feed more people.

In the process, this planet has been rapidly exploited of all its natural resources and land space, while its land, water and air had being polluted beyond recovery.

Now, without any particular concern, humanity is faced with the most threatening event to its existence: Earth's climate change. After all, as far as human conscios are concerned, what is an inconvenience of temperature rise by few degrees and some destructive weather patterns in comparison to their individual pursuit of presumed freedom, especially that of earning and spending money – exactly what all the governments, business and religions expect them to. It is so much easier to expect human life to go on unchanged, instead of assessing this calamitous threat as a beginning of an end to their existence.

Currently the population of this little planet is approaching eight billion human lives. Note the number: EIGHT BILLION INDIVIDUAL LIVES. With each one of these eight billion conscios – who have no idea they are conscios – wanting all that they possibly can gain in their life, out of life. The poor wanting to become wealthy, the wealthy wanting to become rich, the rich wanting to become even richer. Every single person wanting more, more, more!

This desire for more of everything by a growing population means that a growing demand is placed on business and the business of science to provide more of everything. And this continues to have a direct effect on increase in global warming. The more humans there are the larger is their overall need of survival, the more the business-of-business activity injects more methane gasses and carbons into the atmosphere and water, increasing the global temperature and altering global heat behaviour on Earth.

There are many humans who recognizing that there is a direct connection between capitalism and global warming. What these concerned individuals fail to recognise is that the cause is deeper than any financial system of trade: it is the population that is the problem. If anything were to be practically done about the problem of global warming would require global population growth management and change to human conscio attitude to consumerism: the very elements that governments, business and religions would loath to alter.

So how would such a change be possibly introduced, where most individuals would nominate their own welfare as being far more important to any climate change?

The only way this could be done is by teaching human conscios who they are and where they are: on a small planet – their only place of existence.

Only by knowing these physical facts can they become more willing participants in management of their numbers, with reduction of their wants, including that of their 'freedoms'.

The probability of this occurring is very low: between doubtful and none.

PART 3

DISCLOSURE OF EVERYTHING

Physical reality had always surrounded human existence. A tree had always been a tree and nothing but a tree, even if that tree represented a body-brain life form. A body of water would always be just water and nothing but water. A mountain ridge would be a shape of a land mass and that is all.

Yet, such state of physical reality had never been sufficient for human conscios. For them a tree had to have a mischievous tree spirit that could use the twigs and branches as spindly arms and fingers. Any body of water had to have some kind of a manipulative fish-like deity or god, while a mountain was home to either a spirit, such as an eagle spirit, or a home to some mighty mountain god who slumbered between waking with rumblings and terrible earth shudders.

Irrespective of physical reality that could always be easily seen and examined by one and all, conscios had preferred to apply imagination to add complexity to existence. In doing so, they could, and did form their opinions on everything according to their own perception of reality.

Gradually some human individuals began to examine the surrounding physical existence with more diligence, recognizing attributes of physical reality. This practice became the premise to collection of knowledge, currently referred to as science.

However, the formation of science did not dispel or even reduce the existing human opinions. Human conscios continue to embrace their imagination-derived notions that disregard physical reality in preference to their self-concocted beliefs of all that which have no physical existence.

But how could such acceptance of invented nonsense become so prevalent as to displace common sense, logic and reason?

Because their self-devised notions are ingrained in their consciousness to such extent that they became accepted as being a physical reality – needing no further re-examination from supposedly being beyond any doubt.

Such blind acceptance of beliefs applies to all human perceptions, including their understanding of science.

For that reason, Part 3 of the book provides the basics of all physical existence, beginning with what the smallest particles look like, and explaining exactly how and why these shapes form everything in physical existence, and how and why they behave as they do. Only by such thorough examinations can it become possible to expose all the errors that have become fixed not just in scientific acceptance but also in human cultural behaviour.

All this, of course, you may judge for yourself after gaining the knowledge of all physical existence as it is, and not as human-derived thinking would have it. And it is only by understanding physical reality that you may become capable of separating scientific truth from current science fiction.

24

A brief look at all physical existence

THE PAST WITH NO BEGINNING

Before all else, it is necessary to explain the physical definition of a word 'space'. Apart from all the definitions in dictionaries 'space' is one of two physical states: space can be empty, as in nothingness, or it can be solid space as in somethingness. The difference between the two is that empty space can be traversed and moved through by anything, while the solid space cannot be penetrated by anything at all.

There is another term needing an explanation: 'physical'. The word 'physical' signifies not just an existing body or an object, but that a body or object can be contacted (as in touched) either directly on its external surface or contacted directly at a distance, when a contact is made directly with a body's or object's gravity or electro-magnetic coils. This means that 'physical' applies to all that exists and can be contacted, for only contacts can provide physical change. This also means that a body or an object cannot produce or experience physical change without experiencing physical contacts.

Now then, to begin with, in the beginning of everything there was no beginning. Instead, there always was, is, and forever shall be an absolute 'nothingness' (empty space, or vacuum space), together with a physical 'somethingness' (solid space).

The nothingness is the vacuum space of Eternity. It is Eternity because it is eternal, having always existed as a vacuum space of nothingness. It is also infinite vacuum space, without a beginning or an end in absolutely all directions. Unlike the current scientific presumption that vacuum space is only part of our expanding Universe, the nothingness of the vacuum space of Eternity extends endlessly in all directions beyond our Universe, without any 'upright' position, something that humans are used to on Earth.

Despite that the Eternity is a nothingness it actually exists. Were it not to exist then no physical existence would be possible. There would be absolutely no physical existence of anything, anywhere, including that of vacuum space of nothingness itself. This, of course, is a physical impossibility, for nothingness of vacuum space is already a vacuum space of nothingness. While the vacuum space of nothingness can be considered as something that is in existence, that existence comprises of nothingness at all.

By existing, the nothingness of Eternity makes it possible for all that physically exists (within it, so to speak) to do so at three-dimensional distances from each other – with nothing but nothingness of vacuum space separating them.

Contrary to the current scientific opinions, the vacuum space of Eternity neither expands nor contracts; it is made of no energy or matter; it produces no energy or matter; it absorbs no energy or matter; it restricts no energy or matter. As nothingness, the vacuum space of Eternity is a constant at being nothing. As such, it cannot ever be transformed, deformed, bent, stretched or changed, because there is actually nothing in a nothingness to transform, deform, bend, stretch or change in any way.

The space of Eternity is heatless, and as such is at absolute cold – thousands of degrees colder than freezing point of water. While it may appear black to those who are sighted, it actually has no colour at all, as it consists of nothing.

Despite being a nothingness, the vacuum space of Eternity has an important attribute: it exerts an influence of attraction on all that physically exists, not because it possesses any attraction but because the attraction derives from that and those being attracted.

All these characteristics of nothingness make the vacuum space of Eternity the first of the three constants of all physical existence.

The second entity of all physical existence is the 'somethingness' (solid space). This somethingness has a physical presence, which fills the nothingness of the vacuum space of Eternity. This physical 'solid space' – in contrast to the vacuum space of Eternity – comprises of minute individual units. The reason that they may be described as being 'solid space' is because these physical individual units are indivisible, impenetrable and eternal. In being so minute, they are in a form that is hard to envisage as visible mass, and only begin to resemble visible mass once they unite in unimaginably large numbers to form atomic structures.

These individual units of somethingness of 'solid space' can be called 'energy', and from that, the individual units of energy derive a name of 'inunen' for singular and 'inunens' for plural, as an abbreviation of individual unit [of] energy. The reason for such name is because the first law of thermodynamics states: "Energy can neither be created nor destroyed; it can only be converted from one state to another." If energy is supposed to be "neither created nor destroyed" then this description applies precisely to all inunens (individual units of energy) for they are neither created nor destroyed but converted from one state to another. Or rather, in their active existence they go from one phase into the next, in the process doing work and producing physical change. This physical change can be expressed as being the Physical Process of Change, as this represents all the simultaneous inunen physical activity everywhere, where this results in physical change occurring to everything, everywhere.

Individual units of energy (inunens), in any of their changing phases, exist in incalculable numbers, filling with their presence the vacuum space of Eternity. Without their eternal existence there would be no physical existence of any object or body anywhere.

Therefore, the presence of inunens (individual units of energy, that is: individual units of heat) is the second constant of all physical existence.

As the vacuum space of Eternity provides no physical obstructions or impediments to functions between any individual units of energy, their physical process of change takes place – irrespective of any distance from each other – in the 'present-moment'. Nowhere in vacuum space of Eternity does some physical function occur slower or faster than the rest. Whether the functions in themselves are slow or fast, they are all concurrent, always happening in line with the instant of the present-moment.

This means, for instance, that some particular change occurring on a star in a far distant galaxy is doing so in the same moment as any changes taking place on Earth, irrespective of any distance in between.

This also means that a distant star may have ceased to exist before Earth was even formed, but its light is still is moving towards Earth, giving an impression on Earth that the star still exists. So when astronomers observe that approaching light with telescopes, what they are doing is viewing that light at a shortened distance (thanks to their telescope's lens) in the same instant they are experiencing on Earth, and not peering into the past, as is often incorrectly claimed.

This simultaneity of all physical actions taking place everywhere in the same instant is the third constant of physical existence.

HOW AND WHY INUNENS LOOK AS THEY DO

Ask anyone in science what the smallest particle looks like and how it behaves and the response will be either that of deferral, or dismissal, or an admission that they do not know.

PART 3. DISCLOSURE OF EVERYTHING

Current physics – for all it tends to take credit for – cannot provide a visual representation of what the fundamental particles look like. Instead, it is an accepted practice for them to use terms such as 'matter', 'particles' or 'quark' subatomic particles, when referring to that which currently has no visual interpretation or variables of behaviour.

Yet only by understanding their external structural appearance, and from that their internal function and their overall behaviour, can it be possible to grasp the full consequence of their shapes that are responsible for all that is physical existence.

So that is where we shall start: with descriptions and visuals of 'inunen' (individual unit of energy) physical appearances and what is physically derived by such structures.

The primary aspect of each and every singular, individual unit of energy (inunen) is that it is a singular unit of heat. A singular inunen heat level is so low it is almost that of the absolute zero cold of the space of Eternity. Being such a low level of heat is due to its minute – almost non-existent – size.

Despite being a low-level unit of heat, an individual unit of energy (inunen) is a body. This body comprises of internal and external movement of heat, and as such undergoes changes to that body in a one-directional cycle of four distinctly different phases.

Each of the four phases produces a variance to its body, and this body change results in a change of behaviour. These four phases of every cycle are listed as the following:

Phase 1. In this phase an inunen has a compressed shape. As a free-moving, individual unit, it is free to unite with similarly shaped inunens, and for that reason can be classified as being in a 'uniting energy' inunen phase.

Phase 2. In this phase an inunen retains its compressed shape, but because it is in a close proximity of other similar inunens – those in front and behind it – it is unable to move about freely. In this phase a compressed inunen is classified as being a 'united energy' inunen.

Phase 3. In this phase an inunen takes on a more spherical shape by having expanded its body, and having vacated its position from its surroundings. In this phase it is a 'dispersing energy' inunen.

Phase 4. In this phase an inunen is a 'dispersed energy' inunen, having a fully spherical shape.

Before explaining exactly what occurs to an inunen within each of these four phases, here is a graphic interpretation of a compressed and an expanded inunen body. These images are only meant to give an indication of the overall body shapes that are maintained by waves of internal heat movement with no intention of suggesting that the shapes have hard, smooth surfaces.

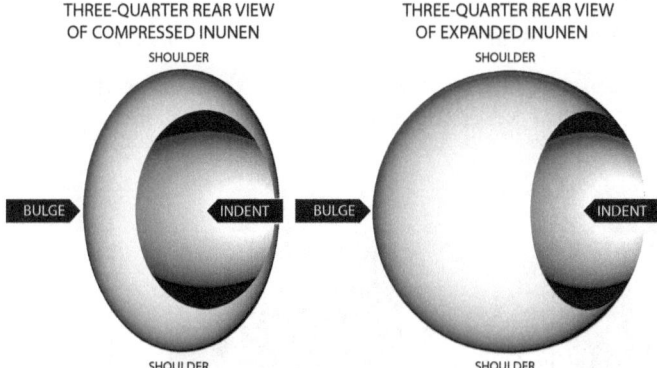

When compressed, an inunen has a surface shape similar to a saucer with rolled over edges: a round flat disk with a bulge (convex) on one flat side and an indent (concave) at the other flat side, with the outer edges slightly rolled over inwards on the indented (concave) side.

In its decompressed, or expanded 'dispersing energy' phase, the physical shape of an inunen resembles a sphere with a large indent at one side. Basically, whether expanded or compressed, the overall shape of an inunen is that of a flattened or a rotund spherical object, with a bulge at one side of its body and a large indent at the opposite side.

The following graphic shows what a compressed and an expanded inunens would look like in cross-section.

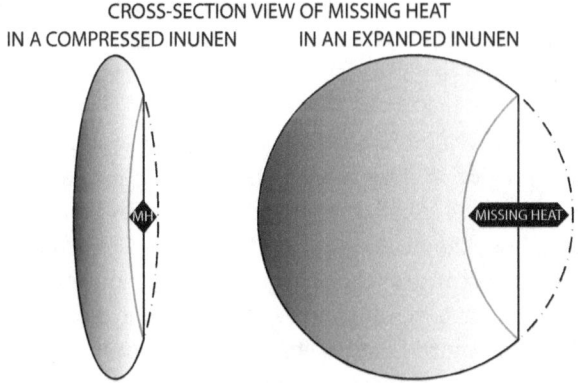

As all inunens are heat, it can be seen that a large chunk of heat is missing on one side, which can be classified as being indents.

PART 3. DISCLOSURE OF EVERYTHING

An inunen body shape comprises of a structure where there is a closed three-dimensional circuit of internal movement, which cannot be broken, as shown below. However, please note that the left, bulging sides of the inunens, is now represented as 'A', and the right, indented cavity sides, as 'B'.

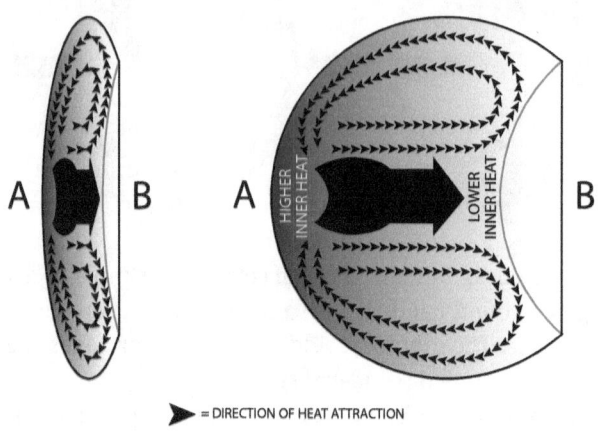

= DIRECTION OF HEAT ATTRACTION

This diagram gives an indication of the internal, cyclical movement of heat, which is the core movement of any compressed and expanded inunen. It is this closed circuit of heat-movement structure that separates it – as a unit of solid space – from the surrounding vacuum space.

What can also be observed, is that at the region 'B', there is less internal heat than at the internal 'A' region, which causes the internal movement of heat to go from where there is a lot of heat at 'A', to where there is little heat, at 'B', so as to fill that region, or internal space, with itself in an effort to become balanced. But when the heat from 'A' reaches 'B', the 'B' region becomes hotter, so that at the internal region 'B' the heat spreads out around itself and moves back towards the region 'A'. This internal heat movement, which is the structure of an inunen body, is called: **internal directions of heat attraction**.

But why could not the heat, upon reaching region 'B', simply turn around and head directly back towards the 'A' region, without any looping?

The reason for that is because lower heat cannot move directly against higher heat. So when more heat reaches region 'B', that region is still being influenced to move forward by the higher heat at its rear. Unable to go fully forward or directly back, the internal heat at region 'B' has no alternative but to spread out three-dimensionally, and return in loops to the rear of the central polarity of the core, from where it had come from, only to repeat the whole process again and again.

In summary, it can be said that internally, the higher heat at 'A' region is attracted to lower heat at 'B' region, having a direction of attraction where its higher heat is directed towards its lower heat.

This internal direction of heat attraction can be expressed as:

(H]>>h(

This represents that: at the internal concave '(' the higher heat 'H' is attracted '>' to the lower heat 'h' at the internal convex '(' but repels ']' the lower heat 'h', while the lower heat 'h' attracts '>' to itself the higher heat 'H'.

It is for this reason when two flames approach head-on they spread out and fly backwards where they meet, without being able to penetrate each other. The only way that the direction of heat attraction can move forward is by moving around itself, towards the lower heat at the rear of the higher heat region, in three-dimensional loops.

This internal heat movement is responsible for another function: an **external direction of heat attraction**. The difference between the internal direction of heat attraction and the external direction of heat attraction is that while the internal direction of heat attraction is actually physically produces a heat movement in form of an impenetrable three-dimensional circuit, the external direction of heat attraction is merely a display of intent by the internal direction of heat attraction. This display of intent is called: 'boundary of influence'. As an external extension of the internal direction of heat attraction, the 'boundary of influence' of one inunen allows that inunen to recognise a nearby presence of another inunens by 'feeling' the 'boundary of influence' of that other inunen with its own **'boundary of influence'**.

The following graphic shows a representation of an external direction of heat attraction ('boundary of influence') and the internal direction of heat attraction of a compressed and an expanded inunen.

PART 3. DISCLOSURE OF EVERYTHING

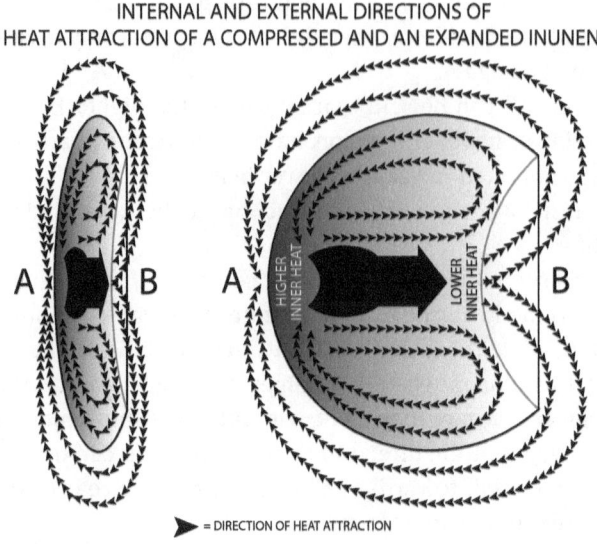

INTERNAL AND EXTERNAL DIRECTIONS OF
HEAT ATTRACTION OF A COMPRESSED AND AN EXPANDED INUNEN

= DIRECTION OF HEAT ATTRACTION

While compressed 'uniting energy' inunens remain alone in vacuum space, their external directions of heat attraction remain inactive. But once a compressed inunen comes close another inunen, both of their external directions of heat attraction ('boundaries of influence') immediately become activated, with the external direction of heat attraction flaring out towards each other, with their 'boundaries of influence' coming into contact with one another, while the main bodies of internal heat of inunens remain apart. Because these 'boundaries of influence' flare out further at the sides while being less prominent at the indent and the bulge facings of inunens, these stronger side attractions can also produce stronger repulsions. For that reason, the compressed inunens prefer to unite with each other at their 'A' and 'B' facing-sides of their bodies, where the external directions of heat attraction are not as pronounced.

What can be seen from the graphic above, is that the external direction of heat attraction – once it is activated – moving in the same direction as the internal direction of heat attraction, is derived from being a projection of internal heat movement. This means that the three-dimensional 'boundary of influence' loops of external direction of heat attraction also function as: (H]>>h(

Where: at each and every internal concave '(' the higher heat 'H' is attracted '>' to the lower heat 'h' at the internal convex '(' but repels ']' the lower heat 'h', while the lower heat 'h' attracts '>' to itself the higher heat 'H'.

This equation for the external direction of heat attraction is exactly the same as the one for the internal direction of heat attraction, where both the internal and external directions of heat attraction move in the same direction. What is unique about the external direction of heat attraction ('boundary of influence') is that it connects the internal direction of heat attraction to its surface direction of heat attraction, by that producing internal and external closed circuits of heat movements that give an inunen its structure and shape.

So what is the external 'boundary of influence' all about?

To explain this, imagine being a sightless body, floating in a pitch-black environment of the vacuum space of Eternity. As a sightless compressed inunen, you cannot hear or see another inunen floating towards you. If that other sightless inunen collides with you, directly touching your body, you both shall be required to alter your body shape and disperse. To prevent this from happening needlessly, inunens, which are structures of heat, emit an external on-and-off pulse of a physical extension of its internal heat direction.

This external extension becomes an external projection of the heat directions, which extends some distance into the vacuum space away from the inunen internal body structure, by that providing a boundary within which an inunen has a controlling influence. This pulse of external 'boundary of influence' functions something like a feeler, which, upon getting a feeling that it is in contact with the feeling pulse of another sightless inunen, becomes 'switched on' permanently. Once the two external directions of heat attraction lock onto each other, their respective 'boundary of influence' (their external directions of heat attraction) allows them to decide what to do with each other, without actually touching one another physically. Depending on their facings to one another, they can either unite or to repel each other, without their individual internal core directions of heat attraction actually physically touching.

This can take place because the underside (concave) regions of the external direction of heat attraction (around the inunen shoulders) direct all external heat inwards, towards the central polarity of the core, as shown in the following diagram.

PART 3. DISCLOSURE OF EVERYTHING

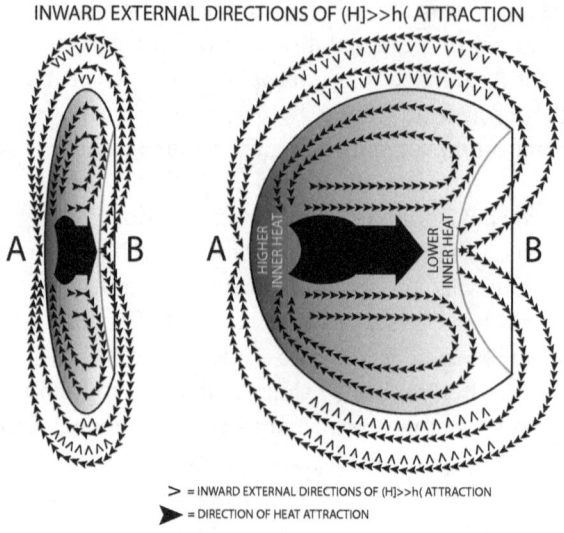

INWARD EXTERNAL DIRECTIONS OF (H]>>h(ATTRACTION

\> = INWARD EXTERNAL DIRECTIONS OF (H]>>h(ATTRACTION
▶ = DIRECTION OF HEAT ATTRACTION

By such process of attraction compressed 'uniting energy' inunens can bring other inunens closer to them, and more importantly, the external direction of heat attraction ('boundary of influence') feature makes it physically possible for the 'uniting energy' inunens to unite. This is illustrated on the following diagram.

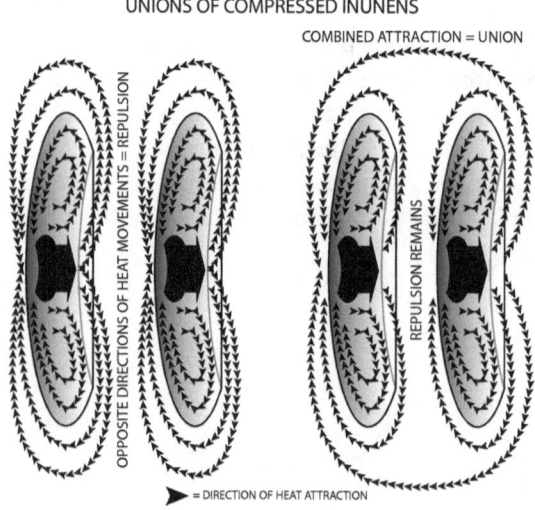

UNIONS OF COMPRESSED INUNENS

▶ = DIRECTION OF HEAT ATTRACTION

A union is produced by aligning their central polarities, and merging their more distant external direction of heat attraction coils into a mutual external direction of heat attraction – or rather a mutual 'boundary of influence' – so that two (or more) inunens can become attached to one another.

However, it also has to be noted that despite forming a bond with each other, some of their external directions of heat attraction coils cause a conflict, as they move in opposite directions. This produces a repulsion between them, which allows them to become united while still remaining apart. The component of remaining apart in any union is important for 'uniting' and 'united energy' inunens. That is because no sooner were they to physically touch one another than they would be required to instantly change their body shape of a new phase, and disperse away from each other and any other 'uniting' and 'united energy' inunens that they had been united with. So it is vital for grouped inunens not to actually physically touch each other.

What physically occurs with unions of 'uniting energy' inunens is the very same process that takes place between the planets of our Solar System and the Sun, except on a much larger scale. Their individual boundaries of influence' (external directions of heat attraction) make it possible for attraction between them, with the simultaneous repulsions preventing them from merging directly with each other and the Sun.

This means that despite their miniscule size, individual inunens produce their individual 'boundary of influence', which can be described as being 'gravitation', for it produces attraction between individual inunens without them coming into immediate contact with each other.

Now that the basic internal and external directions of heat attraction have been explained, it is possible to substitute the equation of (H]>>h(, where 'H' stands for higher heat and 'h' for lower heat, with those of 'positive' and 'negative' values, as shown by the following diagram.

PART 3. DISCLOSURE OF EVERYTHING

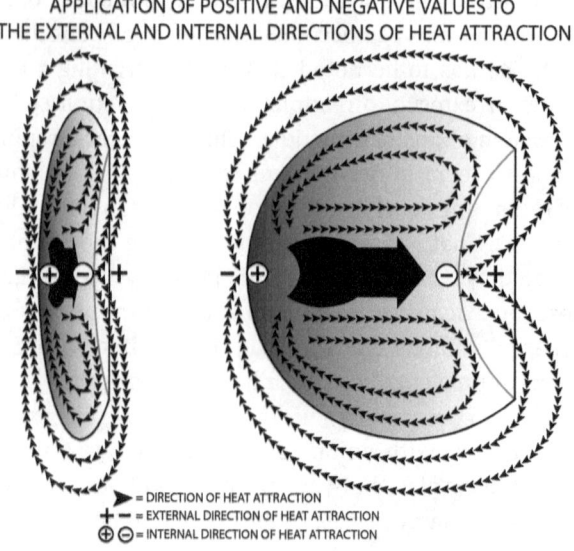

APPLICATION OF POSITIVE AND NEGATIVE VALUES TO
THE EXTERNAL AND INTERNAL DIRECTIONS OF HEAT ATTRACTION

▶ = DIRECTION OF HEAT ATTRACTION
+ − = EXTERNAL DIRECTION OF HEAT ATTRACTION
⊕ ⊖ = INTERNAL DIRECTION OF HEAT ATTRACTION

By substituting 'H' for '+' and 'h' for '–' the statement of:
(H]>>h(can be replaced with: (+]>>-(

Where: while a higher heat '+' repels ']' any oncoming heat, it is attracted '>' to lower heat '–' which, as low heat '–' attracts '>' higher heat '+' to itself.

From this, the internal and internal inunen directions of heat attraction can be expressed as being: >-(+]>>-(+]>

Where: the internal direction of heat attraction produces a responsive external direction of heat attraction, and where they both produce closed loops (circuits) from + to –, because the higher heat '+' will never permit the lower heat '–' to approach it directly head-on.

This equation may seem simplistic but, in fact, it explains the reasons for polarities in any physical object and body, and why attractions and repulsions occur between polarities of these individual objects and bodies, be they represented by magnets, electro-magnets, electric currents, and gravitations of suns and planets.

Furthermore, this equation actually dispels one of currently accepted scientific notions that presumes it is possible to have a totally 'positive' charge (such as an all positive proton) and a totally 'negative' charge, (such as an all negative electron). The fact is, no matter what an object or a body may be, and no matter how large or small they are, they are all simultaneously both positive '+' and negative '–' within the same object or body.

There are no physical objects or bodies that are either all-positive or all-negative. Beginning with each and every inunen, everything that they physically form ends up replicating their behaviour. Everything has positive and negative regions with a central polarity direction, with these functions resulting from internal and external direction of heat attraction.

There is yet another feature to an inunen structure: their internal space carries an influence of attraction.

As the cross-section a compressed and an expanded inunens show in a following graphic, there is a variance to their internal solid space.

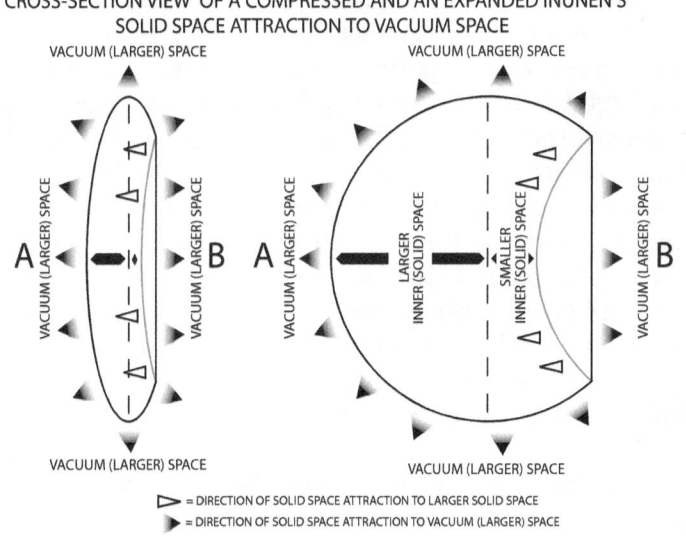

It can be seen from this diagram that internally the concave 'A' side of compressed and expanded inunens is much larger than that of its right, internal convex 'B' side. And if the area of the left, concave, 'A' side is larger, then so would be its three-dimensional internal mass. What these solid space differences indicate is that the smaller solid space at the right, internal convex 'B' side of an inunen, is attracted to the larger internal space at the left, internal concave 'A' side, while larger solid space attracts the smaller solid space. This attraction of smaller solid space to the larger solid space results in the heat circulation within the solid space of inunens moving from the smaller space to that of larger space. This means that once the internal movement of the central polarity of heat reaches the region of low heat it also reaches the region of smaller internal solid space. With the smaller solid space wanting to move towards the larger solid space, and the heat being unable to move directly back towards the direction of the central polarity, the heat in the smaller solid space moves outwards and inwards, by that following the direction of smaller space towards the larger space, while circumventing the direction of the central polarity of high heat.

Internally, therefore, the attraction of smaller solid space to larger solid space can be expressed as:

(S<<s(

Where: at the internal concave '(' a larger internal space 'S' attracts '<' a smaller internal space 's' at the internal convex '(' and the smaller internal space 's' is being attracted '<' to the larger internal space 'S'.

This attraction of a small space to a larger space applies not just to internal space but also to external vacuum space. And with the vacuum space of Eternity being the largest space there is, all internal spaces and all levels of heat are equally attracted to Eternity.

This can be illustrated by having a container of water suddenly being exposed to vacuum space. When this happens the water boils and decreases, before remnant of it freeze. This means that the 'uniting' and 'united energy' inunens forming the atoms of Hydrogen and Oxygen are permitted to be attracted to the larger-space-beyond, which causes them to move inwards into their combined mass, only to produce physical contacts with each other. This, in turn, causes many of them to change into the 'dispersing energy' inunens and to disperse directly into the vacuum space of Eternity.

But how can this be so? Was it not explained that the vacuum space is an absolute nothingness? How can such nothingness attract a physical body of an inunen? Besides, what is a solid space anyway, and how can it have attractions within itself? And besides that, what is all this 'larger-space-beyond' attraction?

To respond to these questions, let us first consider the attraction of vacuum space. As stated earlier, yes, the vacuum space of Eternity is a physical state of nothingness. It is a physical presence that is different to inunens, but simultaneously the same as the inunens, in that they all physically exist. From this, the vacuum space of Eternity is real for inunens, the miniscule bits of heat, which are in pursuit of wanting to be one with the Eternity. This they can never achieve from being three-dimensional circuits of heat. But this does not prevent them from seeking a direct path from '+ ' to '–': where they are the ultimate multiple positive '+' and the nothingness of Eternity is the ultimate singular negative '–'. However, to achieve this is harder than it may seem, because no matter where inunens are – and no matter that the vacuum space of Eternity is infinite in all directions – vacuum space for inunens is always larger beyond them. This causes inunens to want to be beyond where they are.

So what does this mean?

Earlier, it had been shown that internally, within an inunen, a larger space 'S' attracts smaller space 's': (S<<s(and a smaller space 's' is attracted to a larger space 'S', all this expressed as: S<<s. But how can it be shown that inunens are attracted to the largest space, considering that an endless vacuum space of Eternity equally surrounds everything? It is possible to do this by examining a spherical object, surrounded by vacuum space of Eternity.

It is possible to presume that from any point on the surface of a sphere the vacuum space would be equal in all outward directions. But that would be incorrect. If a point is taken anywhere upon the surface of the sphere, the outward space for that point is actually larger on the opposite side of the sphere than it is on its own side of the sphere. To make this explanation more graphic, presume that a drawing compass is used to draw two circles, one circle exactly inside the other, as show on the diagram below.

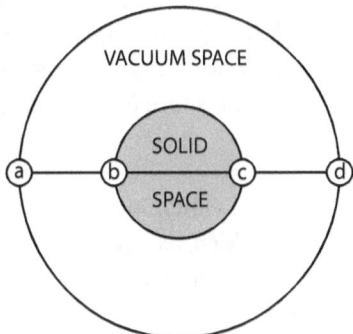

The inner circle would represent a solid space of an object, and the outer circle would represent equidistant vacuum space surrounding that solid object. A line would now be drawn through the centre of the two circles, and the intersecting points labeled from left to right as: 'a', 'b', 'c' and 'd'.

Taking the point 'c', at the right-side of the surface of the solid object, measure the distance between it and point 'd'. Then, taking the same point 'c', measure the distance between it and point 'a'. By doing these measurements, it becomes easily apparent that the distance from point 'c' to 'a' – the point in vacuum space on the opposite side of the object – will always be longer (more distant) than the space directly over or behind a point on the surface of a physical object. The combined distance of 'a' to 'b', plus 'b' to 'c', being longer than the distance from 'c' to 'd', proves that the space is larger on the opposite side of the surface of any physical object or body.

What this signifies is that not only does each point of an inunen wants to be directly beyond the other side of its own inunen, but every inunen wants to be directly beyond all the inunens that are in front of it. It is this attachment to larger-space-beyond that keeps the inunen structure together, preventing the internal and external directions of heat attraction from moving apart too far, as well as, allowing for formation of all 'combined mass' in physical existence.

This attraction is exhibited not just by inunens, but by all that inunens fundamentally form, including humans. For instance: it is presumed that it was human curiosity that drove, and still drives humans to explore and populate every vacant region of this planet. In fact, it was their attraction of larger-space-beyond them that caused them to seek that which is beyond where they were.

Furthermore, it should not be surprising that when humans first get a chance to observe their surroundings from a great height – say from a cliff top or a roof of a tall building – they often consciously experience a thought, or a feeling, of wanting to jump into that void of space before them. They, as small spaces of 'combined mass', are drawn (attracted) to large spaces before them. It is also not surprising their attraction to larger-space-beyond is also attracting them to explore the vacuum space beyond this planet filled by other planets and moons.

As to the internal solid space of individual inunens, this can be explained as being an unequal density of heat, which, as a pulse in pursuit of its lower self, maintains a cycle. It could be crudely compared to a dog chasing its tail, except that this dog and its tail are motionless, and it is the intention that moves. It is this still structure of heat with a moving intent – the internal direction of heat attraction, always moving forward in a closed circuits to where it is not – is that which cannot be penetrated, and as such is the only solid space in physical existence.

This solid space is what the active inunens replicate in forming everything, be it: atoms, living bodies, or structures surrounding living bodies. In each case it is a function of internal space to be provided with some kind of a barrier, or a boundary, or a membrane, to separating itself from the external space.

Take a house, for instance. The inside of a house is an internal space, which prevents the undesirable influences of the outside – the external space – from penetrating. This includes the varying, uncomfortable heat levels. When the external space is cold, the internal space of a house is heated by means of various heating appliances. And when the external space is hot the internal space of a house is cooled, by cooling appliances. Such internal space, within an external space, exists in everything physical, everywhere.

THE FOUR PHYSICAL PHASES IN AN INUNEN CYCLE

Throughout their eternal existence, every active inunen (individual unit of energy) undertakes an ongoing cycle of alternate, temporary physical phases. A phase may last just a few a seconds or billions upon billions of Earth years. There are four phases to every inunens cycle. These four phases are: 'uniting', 'united', 'dispersing', and 'dispersed'.

The first two inunen phases of being 'uniting energy' and 'united energy' produce all the physical unions that result in existence of all physical objects and bodies. The other two inunen phases of being 'dispersing energy' and 'dispersed energy' result in the dispersals of all physical objects and bodies. Each cycle of change – comprising of the four phases – proceeds from physical unions between inunens to that of their dispersals from each other, before unions begin once again, but at a totally different location and incalculable billons of years later.

These four physical phases occur as a progression: from that of being a 'uniting energy' inunen, then a 'united energy' inunen, to that of 'dispersing energy' inunen and finally 'dispersed energy ' inunen.

When multitudes of inunens happen to be in one location of vacuum space, these inunens can experience the progressions swiftly or gradually, or not at all, depending of any physical contacts that may occur between them all.

Phase 1: being a 'uniting energy' inunen

In its first phase – that of 'uniting energy' – an inunen (individual unit of energy) find itself activated and ready for action. Physically, a 'uniting energy' inunen has a shape of a compressed disk with a bulge on one side of its round flattened shape and an indent on the other, with the side edges – or the shoulders – slightly rolled over the indent. Such a 'uniting energy' inunen is actively involved in attracting and being attracted to other 'uniting energy' inunens.

When 'uniting energy' inunens unite, they can only do so indent to a bulge. This combination allows their internal directions of heat attraction to move in

PART 3. DISCLOSURE OF EVERYTHING

the same direction, while their individual external direction of heat attraction keeps them both united and, simultaneously, slightly apart, by being repelled from each other. This repulsion prevents them from physically touching one another.

A union between two compressed 'uniting energy' inunens can be shown as:

>-(+]>>-(+]> >-(+]>>-(+]>

Here it can be seen that at the junction of two inunens coming together indent to a bulge (shown in bold and larger text)...

>-(+]>>-(+]> >-(+]>>-(+]>

...the higher heat at the external surface of the indent, '(+]>', together with the lower heat at the surface of the bulge, '>-(', produce the same direction of heat attraction as that which takes place internally of every inunen: (+]>>-(.

This means that the internal and external directions of heat attractions between 'uniting energy' inunens can allow them to be joined in a united circuit while avoiding a physical contact with one another.

Every individual unit of energy (inunen) in vacuum space of Eternity has no defined program for any activity. A 'uniting energy' inunen may remain alone in vacuum space for billions of Earth years, or be attracted or attract other 'uniting energy' inunens within seconds. Incalculable numbers of such spiralling and spinning inunens blindly attempt to become connected (united) to each other. These attempts are often chaotic, and not always instantly successful, causing physical contacts that lead these 'uniting energy' inunens to disperse as 'dispersing energy' inunens, with this 'dispersing energy' seen and felt as light and heat.

But gradually, in all the disorder and chaos of spinning, tumbling and spiraling in vacuum space, inunens begin to form 'strings of attraction'. These unions of compressed 'uniting energy' inunens – united indent-to-bulge by their external fields of directions of heat attraction – resemble strings, or worms, jerking and flopping about in vacuum space, surrounded by incalculable similar inunen strings of different length in clouds of yet unattached inunens.

At this point, when the individual 'uniting energy' inunens unite with other individual 'uniting energy' inunens, they become 'united energy' inunens.

Upon reaching a particular string length, these 'strings of attraction' close into loops, which then twist themselves into spirals. These spirals are called, 'ununen'; short for: **un**ited **un**its of **en**ergy (pronounced: u-nunen, or as plural, u-nunens). The 'uniting energy' inunens that manage to join in a process of forming a ununen, undertake their first memory impression of what kind

A BRIEF LOOK AT ALL PHYSICAL EXISTENCE

of structure that they are part of. These memories form their identity, which separates them from other, still singular, inunens.

As ununens continue to move in space, their individual attraction to a larger-space-beyond causes them to combine their external direction of heat attraction ('boundaries of influence') with those of other ununen string spirals, by that gradually forming short sausage-like structures with a bend, which replicates the bulge and an indent of an inunen shape. These short sausages are called 'grunun', short for: **gr**ouped **un**its of **un**unens. As grununs, the 'uniting energy' inunens making up the physical structures of these sausages, obtain their second memory impression of the individual physical structure they are part of.

In turn, these sausage-like grunun structures pile together to form rough spherical shapes. It is unavoidable for such structures to have gaps of vacuum space between the grouped grununs.

Each of such individual porous structures is called a 'solid-body' (otherwise known as 'proton'). These unions produce a third memory impression on all the individual 'uniting energy' inunens of that structure.

While these 'solid body; structures may also be known as protons, unlike the current understanding of protons, they are not all positive, but continue to retain a positive and negative polarities. Otherwise they could not be able to form atoms.

As the incalculable numbers of 'uniting energy' inunens manage to unite into ununens, then into grununs, and these into 'solid bodies' (protons) rough-surfaced spheres, their individual 'boundaries of influence' (external directions of heat attraction, that is: gravitation coils) also become combined into a united 'boundary of influence', which has an exponentially stronger and further reach to that of any singular inunen.

At the same period as all this is going on, multitudes of other 'uniting energy' inunens remain unattached (from being unsuccessful in uniting indent-to-bulge or bulge-to-indent with other 'uniting energy' inunens). Clouds of them continue to surround the spheres of 'solid-bodies' (protons). But as the 'solid-bodies' (protons) begin to establish a more permanent extension of their 'boundaries of influence', this causes the clouds of unattached 'uniting energy' inunens (surrounding each 'solid bodies' [protons]) to be repelled and moved back away from the 'solid bodies' (protons).

While remaining at a distance from 'solid bodies' (protons) the clouds of still unattached single 'uniting energy' inunens begin to form their own unions.

These unions take form of a mesh, or a net, that begins to surround each 'solid body' (proton) as a shell. The unattached 'uniting energy' inunens achieve this by forming their own combined field of external direction of heat attraction ('boundary of influence': gravitation coils), which becomes connected to the 'boundary of influence' emitted by 'solid bodies' (protons). As soon as the individual 'uniting energy' inunens combine into a singular shell, surrounding a 'solid body (proton) these shells can be described as being 'energy bodies', and currently known as 'electrons'. The formations of 'solid bodies' (protons) surrounded by 'energy bodies' (electrons) are called 'united bodies', currently known as 'atoms'.

As 'united energy' inunens, they can no longer unite individually with other 'uniting energy' inunens but only disengage.

At becoming 'united bodies' (atoms), this does not prevent the 'solid bodies' (protons), with their 'energy bodies' (electrons) structures, from establishing partnerships with other 'united bodies' (atoms); in this way merging with other similar structures to make more complex atoms and molecules.

Not all 'united bodies' (atoms) succeed in having a 'solid body' (proton) whose 'energy body (electron) shell remains at a large distance from it. In many instances a forming 'solid body' (proton) is incapable of preventing the surrounding 'energy body' (electron) shell from collapsing onto it, covering its surface with 'united energy' inunens. Such a 'solid-body' (proton), covered by its 'energy-body' (electron), is called a 'combined body', currently known as 'neutron'.

Unlike 'solid bodies' (protons) that have rough surface formed of grunun sausages, the surface of 'combined bodies' (neutrons) are much smoother, from being covered by the 'uniting energy' inunens that are part of the collapsed 'energy-body' (electron) shell. This surface covering of 'energy body' (electron) shell makes the 'combined body' (neutron) a bit larger in size to that of a 'solid body' (proton).

If not for the 'combined bodies' (neutrons) there would be no possibility for 'united bodies' (atoms) – comprising of a single 'solid body' (proton) surrounded by a single energy body' (electron) – with a name of Hydrogen (H), to merge into more complex structures.

Phase 2: being a 'united energy' inunen
and
Phase 3: being a 'dispersing energy' inunen

To repeats previous explanation, the phase two of inunen physical behaviour is when individuals units of energy (inunens) – as 'uniting energy' of phase one – began to unite with one another, so as to form ununens (united units of energy, made of spirals of closed loops of inunen strings), then grunens (grouped units of ununens), then solid bodies' (protons), 'energy bodies' (electron shells), and 'combined bodies' (neutrons). With each additional inunen formation – from ununen (united unit of energy) to that of 'united bodies' (atoms), each 'united energy' inunen retains a memory of the structure its part of, and a distance between itself and all other 'united energy' inunens surrounding it.

Although phase two is absolutely different for inunens to that of phase three, these two phases are inevitably linked to one another. This is because all unions taking place between 'united energy' inunens requires some kind of involvement of 'dispersing energy' inunens. This always results in some form of conversion of 'united energy' inunens to that of 'dispersing energy' inunens.

Before this can be done, there is another element applicable to all inunens that needs an explanation, so that both the phase two and phase three can make sense. This particular element is that of intelligence.

While currently this is unknown, the primary function of intelligence is to provide physical change. Change, therefore, is unavoidable as it is the result of intelligence. And how does intelligence reveal itself? It does so in knowing what has to be done as a reaction to an action. This is the function of inunen intelligence, from which all intelligence is derived, including that of humans.

This revelation may immediately provoke a rejection from humans (conscios) presuming that: A:– Intelligence belongs to life forms with brains, making human intelligence supreme, and that: B:– Change has nothing to do with intelligence but with physical actions that may occur at random and accidentally for all sorts of reasons.

Wrong on both counts.

The reason for such out-of-hand dismissal is because change – all physical change, without exceptions – is derived from physical contacts between inunens, who know what to do when this happens.

When a 'uniting energy' or 'united energy' inunen is physically contacted, or physically touched or struck at any point of its body by another 'uniting energy' inunen, or 'united energy' inunen, or 'dispersing energy' inunen, three functions take place.
- One: a physical contact comprises of a transfer of a physical message passed from the contacting inunen to the contacted inunen. For instance, a physical contact may specify: "Sorry, my fault, please take this no further", or: "Thump, take this, and take it all the way as far as it can go!"

While such observation may seem silly, in reality the contacts between inunens are very much like that, except they are more detailed.
- Two: the information of the contact is broadcast to other surrounding inunens.
- Three: the repercussions resulting from the initial contact are never premeditated or preordained.

To be more precise, the following list provides the physical changes that inunens in various phases will experience from coming into physical contact with other inunens:
- When a compressed 'uniting' or a 'united energy' inunens come into contact with another compressed 'uniting energy' or a 'united energy' inunen, they both change into expanded 'dispersing energy' inunens. The physical expansion of their bodies produces the instant of 'effort' or 'work being done', causing their movement of dispersal.
- When a compressed 'uniting energy' or a 'united energy' inunen comes in contact with an expanded 'dispersing energy' inunen, at the instant of the contact the compressed 'uniting' or 'united energy' inunen changes to an expanded 'dispersing energy' inunen – the instant when it physically expands its compressed body, and by that pushes away from its neighbors, producing 'effort' or 'work' of dispersal – while the already expanded 'dispersing energy' inunen changes its phase to that of 'dispersed energy' inunen, expanding its remaining indent to become a full sphere.
- When two expanded 'dispersing energy' inunens contact each other they both remain unaffected, because as 'dispersing energy' inunens they cannot pass their 'dispersal code' (explanation of this is following) to another inunen already in a 'dispersing' phase. An example of this is when two beams of light meet head on they do not cancel out each other.

In producing all these physical changes to their bodies – which is the very cause of all physical change – it is the inunen intelligence that allows them to understand what they need to do, and when to do it.

All this occurs in the following way: When an expanded 'dispersing energy' inunen contacts a compressed 'uniting' or a 'united energy' inunen, at very instant of the contact the contacting 'dispersing energy' inunen passes over its instructions – in the form of a 'dispersal code' – to the contacted 'uniting' or a 'united energy' inunen; in turn, upon receiving the 'dispersal code' from a 'dispersing energy' inunen a contacted 'uniting' or 'united energy' inunen formulates its response in a form of its own 'dispersing code', and passes that response to surrounding 'uniting' or 'united energy' inunens, then converts its compressed body to an expanded body – producing 'effort' or 'work' in the process – and with that 'effort' departing (dispersing) away from where it had previously been. To achieve all this takes but a fraction of a second.

Now just consider the physical and intelligence abilities that allow for such transfers of information and instruction occurring from physical contacts between inunens, by using you as a substitute for a 'united energy' inunen:

A physical contact from a 'dispersing energy' inunen hits you, and in that instant you receive from that 'dispersing energy inunens its 'dispersal code'. You calculate your response to that 'dispersal code' with a 'dispersal code' of your own, let all the surrounding 'united energy' inunens know of your intention in regards to your newly-formulated 'dispersal code', then you inflate your compressed body, and with that expansion push away from the remaining 'united energy inunens and fly off into space, all in a fraction of a second.

Think you can achieve all that with your own intelligence? Then consider that it is this very process, undertaken by every 'uniting' and 'united energy' inunen, which is the reason why to every action there is an unpredictable or predictable reaction. Unpredictable, when the same physical object, such as a roulette ball, constantly produces unpredictable outcomes when halting on a roulette wheel. Predictable, when a specific chemical produces the same behaviour, over and over, when combined with another specific chemical.

In order to examine the 'dispersal code' in more detail, here is a glimpse into the complexity of this process, with the following explanation presenting the whole procedure.

- The 'dispersal code' includes information about the contacting 'dispersing energy' inunen – its credentials, so to speak – listing its ununen structure, its grunun structure, its 'solid body' (proton) structure [or 'combined body' (neutron) structure], as well as, the whole 'united body' (atomic structure). All this information is part of 'dispersal code', and can physically be seen as either 'emission spectrum' or 'absorption spectrum'.

The other component of 'dispersal code' is the 'transfer of intent' information to be forwarded by the contact onto its surrounding inunens and those beyond.

In return, upon receiving the 'dispersing code' from a contacting 'dispersing energy' inunen:

A: The contacted 'uniting' or 'united energy' inunen assesses the contacting 'dispersing energy' inunen's credentials; as to what structure it is representative of.

B: It then assesses the contacting 'dispersing energy' inunen's suggestion of speed and strength of dispersal for itself, and the 'transfer of intent' information to be passed onto the directly surrounding 'uniting' or 'united energy' inunens and those beyond.

- Taking all these factors into consideration, the contacted 'uniting' or 'united energy' inunen calculates its own responsive 'dispersal code' package, including the 'transfer of intent' information, which it retains for itself upon changing to a 'dispersing energy' inunen, so as to present it to another 'uniting' or 'united energy' inunen it may contact in future, be that future of one second duration or incalculable billions of Earth years. Simultaneously, it calculates its own rate-of-speed-of-dispersal.

In producing its responsive 'dispersal code' – within which there is its 'transfer of intent' – the just-changed-at-that-moment expanded-bodied 'dispersing energy' inunen (from 'uniting' or 'united energy' compressed-bodied inunen) is very aware of the information it had received from the contacting 'dispersing energy' inunen', but does not have to follow it implicitly. Instead it can make its own assessment and formulate it in its own 'transfer of intent', which may request a faster rate-of-speed-of-dispersal, or slower, or the same, or none. In this way the responsive 'dispersal code' never results in a foregone conclusion, but is that of an unpredictable outcome, even in instances where the overall outcome of a chemical reaction is predictable.

It is the intelligence of all inunens that allows for all that physically takes place to do so. Were it not for the inunen intelligence there would be no physical activity of any kind, anywhere in the whole of Eternity. And without physical activity there would be no processes of unification and dispersals, from which all that physically exists comes together as objects and bodies, then disperses, allowing for some other temporary object and body formations.

It must be understood, however, that the inunen intelligence is not an intelligence of reason or logic, as humans understand intelligence. Instead it's an intelligence of physical function that behaves according to precise rules of inunen direction of space attraction and the internal and external directions of heat attraction. Inunen intelligence does not take inunens, themselves, into any consideration. All that they do is to function as they know how, where the outcomes of those functions have but one consequence: to participate in the cycle of inunen existence, which is eternal and mindless.

Despite that, there is a very vital significance to inunen intelligence that needs to be pointed out: What the intelligence of the 'dispersal code' with its 'transfer of intent' upholds is the physical limitations of inunen physical ability.

To explain this simply requires anyone on Earth to look around and recognise that at no period of Earth's existence, and anywhere on Earth, had an event ever taken place that could be described as being unreal, supernatural, or an unnatural phenomenon. No time warps or multiple dimensions had ever occurred on Earth, twisting objects and bodies into distorted, stretched shapes, and causing them to appear or disappear, or to fly off into space followed by half of this planet.

The reason that no such events had ever happened is because they physically cannot happen. And that is because inunens have limitations to what they can do upon enacting their 'transfer of intent' of their 'dispersal code'. As physical entities, once 'uniting' and 'united energy' inunens convert to the third phase of their cycle, by expanding their shape to that of 'dispersing energy' inunen (and producing an action of so-called 'effort' or 'work being done') they can physically depart from their environment as light or heat, with that being the limitation of that function.

That, therefore, is all they can do. They cannot devise any distortions, or any deformed formations, or bending of any surrounding vacuum space, or stretching of any 'time', or sudden appearances or disappearances out of nowhere. As is the case with every inunen phase, all that inunens can do is to provide constant limited functions as applicable to each and every of their phases, with these in no way being able to physically exceed those physical limitations, so as to produce any kind of unreality.

That means no amount of human imagination can alter the fact that there is no unreality, no unnatural occurrences, and no abnormal activity, because it is physically impossible for any inunens, in any of their phases to do so.

Mind you, what the inunen intelligence can achieve, when prompted to change simultaneously in incalculable numbers from 'uniting' or 'united energy' inunens to that of 'dispersing energy' inunens, can be viewed as an explosion of a firework, or that of a nuclear weapon, or a dispersing supernova.

It is the very limitations of the 'dispersal code' and 'transfer of intent' of inunen intelligence that are responsibly for all actions being physical actions and not

PART 3. DISCLOSURE OF EVERYTHING

those of unnatural and unrealistic feats. For instance, transfers of 'dispersing energy', such as light, may seemingly passes through substances like glass, where glass is made up of atoms – which comprise of grunens and ununens, and those formed by the 'uniting' and 'united energy' inunens – but that is not so.

According the current understanding of physics, light – as a so-called photon of energy – comes up to an atom from one side, is absorbed by that atom, and then released by that same atom, as an equal amount of photon energy, from the other side. By such process of absorption and release of light energy (in a form of a so-called photon) the atoms (forming the glass) presumably allow the incoming light to pass and exit on the other side. This process is currently known as: "propagation of a photon through a medium". The important factor in this interpretation is that the initial photon of light being absorbed is not the same photon of light that is released.

Where this current "propagation of a photon through a medium" understanding diverts into error is in the currently wrong presumption that energy is transformable with no ill effects, as if energy comes into an atom, energy comes out – no stress to the atom – no loss and no harm done. After all, by current understanding of energy, energy equals mass. Therefore, according to the $E=mc2$ equation, energy presumably can go from being energy to that of mass, with that mass being able to change back to energy: back and forth, from one to another, with no harm or loss done to energy or to mass.

What this equation fails to grasp is that energy, while not having being created nor destroyed but only changed from one form to another **can do this but once in each phase, and then never be able to revert to what it was in its previous phase, in the same physical cycle of existence.**

This means that while energy ('uniting energy' inunens) do form mass (comprising of 'united bodies' [atoms], which themselves are formed of 'uniting' and 'united energy' inunens) they, as 'united energy' inunens, cannot revert to being 'uniting energy' inunens in the same cycle, just as any 'uniting' and 'united energy' inunens, having converted to 'dispersing energy' inunens (from having been contacted by 'a dispersing' energy' inunen), can never revert to being 'uniting' or 'united energy' inunens in the same cycle.

Energy ('uniting and united energy' inunens) can form mass once, and that mass can change to 'dispersing energy' once, in the process changing that mass to 'dispersing energy', so it can never ever be that same mass again in the same inunen cycle. If this were not so, then all atoms would exist unchanged forever, instead of being sooner or later dispersed by the 'dispersing energy' inunen contacts.

A BRIEF LOOK AT ALL PHYSICAL EXISTENCE

Therefore, what happens when 'dispersing energy' inunens of light contact one side of glass pane, is that they cause a chain reaction within atoms ('united bodies') forming that glass, with this chain reaction finally emitting reactive 'dispersing energy' inunens of light from the other side of the glass pane.

To explain this by a very simple experiment, let us consider a number of solid glass balls, or marbles, as shown in a diagram below, placed in contact with each other in a straight row, on a hard surface. (The reason for marbles to be made of glass is because any damage to glass marbles becomes immediately and clearly apparent.) In this formation a glass marble 'B' is placed first in the row; a marble 'D' is placed at the rear of the row, behind marble 'C', while another marble 'A' is placed a little distance in front of the marble 'B'.

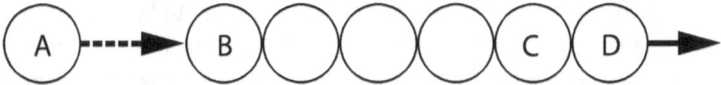

If the back of the glass marble 'A' is softly struck by a tip of a billiard cue, the marble 'A' directly strikes marble 'B', then while the row of glass marbles between 'B' and 'C' will remain stationary, marble 'D', at the head of the row, will be propelled forward with almost the same velocity with which 'A' struck 'B'.

The interesting feature of this experiment is that at the point of contact between 'A' and 'B', both glass marbles sustain some damage: damage that can never be reversed or undone. The same kind of damage occurs between the glass marbles 'C' and 'D', while all other marbles between 'B' and 'C' remain intact.

While, currently, energy is thought to comprise of 'potential' energy, 'kinetic' energy, 'mechanical' energy, and so on, with all of these energies being transferable back and forth, what the above experiment shows that within all physical actions there is damage done (change of conversion of 'uniting' and 'united energy' inunens – forming atoms of the glass marbles – to that of 'dispersing energy' inunens), even if that damage (change) is so minute that it is not be easily seen. This damage (change) is actually the 'effort' or 'work' produced when 'uniting' and 'united energy' inunens expand their compressed bodies at the instant of converting to 'dispersing energy'.

What, therefore, can be learned from this glass marbles experiment is that when a 'dispersing energy' inunen (represented by the glass marble 'A') contacts a 'united energy' inunen (represented by 'B'), both 'A' and 'B' experience a physical change, (as depicted by the sustained damage to the 'A' and 'B' surfaces). As 'A' passes its 'dispersal code' containing its 'transfer of influence' onto 'B', 'B' passes its own 'dispersal code' and 'transfer of influence' onto the next glass marble, specifying what had occurred to it, and what should be done about it. The next glass marble receives this information and passes this on to the next one, and so on, until the glass marble 'C' passes its response to 'D'. On getting the 'dispersal code' and the 'transfer of influence' from 'C', 'D' has no-one else to pass on its 'dispersal code'. It therefore, accepts the message that came all the way from the first contact between 'A' and 'B'. It then expands its body so as to push away from 'C', (this being 'effort', which produces damage between 'C' and 'D'), in that instant passing its responsive 'dispersal code' back to 'C' but also takes the same 'dispersal code' package (with the 'transfer of influence' instructions) with itself, so as to pass this onto the next compressed 'uniting' or 'united energy' inunen it comes across. In this way the information from 'A' is transferred through solid space to 'D', which is accepted by 'D', altered by 'D' and is taken away by 'D' to be passed onto some other inunen. Meanwhile, 'C' having received a responsive 'dispersal code' from 'D' – as 'D' departed – can decide what it will do with that information. It can either do nothing, or it can send its responsive 'dispersal code' package back towards 'B', which may cause further damage to 'B' and to itself from such reverberation.

What all this indicates is that light does not actually pass through glass. Instead the 'dispersing energy' of light causes the 'uniting energy' inunens (that form the atoms and molecules of glass) to receive their 'dispersing codes' and 'transfers of intent' and act upon them by causing the 'united energy' inunens of atoms on the other side of the glass to convert to 'dispersing energy' inunens and disperse as light. This procedure makes it seem as though light is passing through a pane of glass, but actually it does not do so directly.

Also, the 'dispersing energy' inunens of light inevitably cause deterioration to the atoms and molecules that form glass. That is why glass eventually loses its transparency and becomes brittle. This actually applies to everything that light comes in contact with. The 'dispersing energy' inunens of light cause an eventual change to all it contacts.

Furthermore, it is for the same reason that the white light refracts when passing through a glass prism, when the various 'dispersing energy' inunens cause the uniting energy' inunens to form a spread of basic red, green, and blue (RGB) visible colours, as the responsive 'dispersal code' from each 'uniting' and 'united energy' inunen of the glass, changing the initial 'dispersing code' information received from the 'dispersing energy' inunens of 'white' light.

Were light to comprise of photons that could be accepted and released by atoms in a manner of in-out transaction, with no consequences or detriment produced by the responsive 'dispersal code', then the "propagation of photons through a medium" would allow light to pass though anything, including heavy metals. It is only due to the ability of all 'uniting' and 'united energy' inunens to understand the information of 'dispersal code' (with its 'transfer of intent'), and their ability to formulate a responsive 'dispersal code' that provides physical limitations to inunen duration of existence in their 'uniting' and 'united energy' phase. Without the restrictions imposed by the 'dispersal code' there would have been nothing to prevent inunens from remaining unchanged forever.

Here are two more simple examples of what takes place when 'dispersing energy' inunens contact 'uniting' and 'united energy' inunens, causing these contacted 'uniting' and 'united energy' inunens to convert to 'dispersing energy' inunens.

When, for example, the sharp edge of a knife's blade comes into a physical contact with a loaf of bread it is thought that by the effort of cutting the blade forces the bread to part into slices. Or, that if a naked flame comes in contact with a piece of paper, it is considered to be burnt by the flame.

Let us first take the example of a knife and a loaf. What actually physically occurs is that the overall 'dispersing energy' (of the body doing the work of cutting) influences a physical contact between the 'uniting' and 'united energy' inunens of the atoms at the tip of the metal blade and those on the surface of the atoms making up the loaf of bread. This results in the 'uniting' and the 'united energy' inunens of both, the blade and the bread, to convert to 'dispersing energy' and disperse, leaving behind an empty space for the next contact between the 'uniting' and 'united energy' inunens of the blade and the bread. In the process of having the 'uniting' and 'united energy' inunens change to 'dispersing energy' and depart, whole atoms become eliminated, both from the knife's edge and the bread loaf. While more inunens – and by that, atoms – may disperse from the loaf than the blade, bread knives also become blunt, which means that a blade also lose inunens – and by that, atoms – to dispersion.

Should anyone point out that the process of cutting the loaf involves friction, they would be correct. But friction is simply the moment of the physical contact between separate 'uniting' and 'united energy' inunens being influenced by 'dispersing energy', when they, in changing from compressed shape of 'uniting' and 'united energy' to the expanded shape of 'dispersing energy', increase the physical size (shape) of their bodies as they disperse.

Now for the second example: that of the flame and the paper. In this situation the flame is already a 'dispersing energy' being converted from 'uniting' and 'united energy' inunens forming the atoms of oxygen and other fuel, such as wood, oil, or gas. Upon a contact with the 'uniting' and 'united energy' inunens of the paper, the 'dispersing energy' inunens of the flame will cause the 'uniting' and 'united energy' inunens of the paper to change to 'dispersing energy' inunens and disperse into space as radiating heat and light, while some of the paper's 'uniting' and 'united energy' inunens forming atoms of paper will change their atomic structure, and so remain 'uniting and 'united energy' inunens of carbon.

What these simple examples show is that once the 'dispersing energy' inunens influence any combined mass of an object, or a body, in a prolonged and significant way, the remaining structure of that combined mass shall be altered forever.

Clay or ceramic jugs, for instance, are hardened when 'dispersing energy' of the Sun, or fossil fuels dispersed in a kiln, are used to disperse some of the 'uniting' and 'united energy' inunens of water in clay, forming a hardened chemical structure. If that jug is broken (also caused by 'dispersing energy') it can never again be a whole unit of its former self, no matter how well repaired. The regions where it was broken are those where the 'uniting' and 'united energy' inunens had turned to 'dispersing energy' and dispersed, leaving behind nothing but vacuum space, which, on Earth, is instantly filled by Oxygen (O) or other chemicals. And despite any repairs, those dispersed atomic particles of ceramic clay can never again be returned and reunited.

Similarly, once cut or fractured, a sheet of metal can never again regain its former structure in the region where the 'uniting' and 'united energy' inunens have been instructed by the contacting 'dispersing energy' inunens to change to 'dispersing energy' inunens and disperse.

Having explained the basics of the 'uniting', 'united ' and 'dispersing energy' inunen structures and behaviors, it is now possible to approach a further explanation of 'dispersing energy' inunens: that of light and heat.

LIGHT AND HEAT

The third phase of an inunen cycle is the most visual of all. This is when 'dispersing energy' inunens have expanded their bodies, and by that produced 'effort' or 'work', have departed from other 'uniting' and 'united' energy' inunens.

In this, the third phase of their one-directional cycle, they are moving in space as unattached and independent inunens, primed and ready to present their individual 'dispersal codes' to any 'uniting' and 'united energy' inunen they may come in contact with. It is in this phase that inunens – as individual units of heat – can actually display their radiating quality of heat and light, some of which can be visualized by the spectrum of human eyesight.

The one irreversible characteristic of all 'dispersing energy' inunens of heat and light is that they are a 'spent force', which can never be converted back to being 'uniting' and 'united energy' inunens responsible for forming so-called mass. Even if a 'dispersing energy' inunen wanted to become once again part of a structure it had left, this would be impossible to accomplish: its expanded body would not fit in with the compressed bodies of the 'uniting' and 'united energy' inunens.

Such impossibility is assured by the 'dispersing energy' inunen intelligence. Having accepted the 'dispersal code' from a contacting 'dispersing energy' inunen, and made the responsive 'transfer of intent' calculations – which are followed by the chosen intensity of 'effort of work' of dispersing away from the structure they were part of – the intelligence of 'dispersing energy' inunens retains awareness of two functions: the need to pass on their 'dispersal codes', if and when necessary, and in the interim to be unattached and free of any obligations, but that of direct movement, gained from the initial inertia of dispersal.

It is the state of being detached singularities in motion that gives the 'dispersing energy' inunens their unique variance to that of inunens in their other phases. 'Dispersing energy' inunens cannot but be in motion, and as such, are responsible for long-distance change of re-distributing inunens in vacuum space of Eternity.

While the 'dispersing energy' inunens may be referred to as being 'detached singularities in motion', it must be understood that they actually produce dispersals in incalculable numbers. Otherwise, being so small, they would not be seen or felt as 'dispersing energy' in the form of radiating heat or light.

Despite that 'dispersing energy' inunens are everywhere as part of this Universe, their presence cannot be confirmed unless they have a direct physical impact on other inunens, and pass on their individual 'dispersal codes'. It is the transfer of their individual 'dispersal codes' that facilitate the functions of all 'dispersing energy' inunens into producing physical change in all they physically contact.

This means that a fire cooking meat does not actually burns the meat: it is the 'dispersing energy' inunens, appearing as flames, that pass on their 'dispersal codes' to the 'uniting' and 'united energy' inunens of the meat, giving them suggestions as to what rate of change they should adopt to also convert to 'dispersing energy' inunens, with this being the process of cooking.

Similarly, it is the 'dispersal codes' from 'dispersing energy' inunens that make it possible for a human eye to 'see' light.

While, for the sake of simplification, referring to 'dispersing energy' can infer dispersing heat and light, this should not mean that heat and light are the only specific values of 'dispersing energy'. Because of the unlimited scope of 'dispersal code' available to each and every 'dispersing energy' inunen, these can cause a large variety of responses. These include the value of speed and force with which a 'uniting' or 'united energy' inunen converts its body into an expanded 'dispersing energy' inunen, and the chosen value of speed of dispersal with which a 'dispersing energy' inunen moves into space. All these values equate to how 'dispersing energy' inunens are categorized, apart from them all being simultaneously both heat and light.

Considering that inunens (individual units of energy) are individual units of heat, the fact that they are heat does not alter their state of being heat when altering their physical body shapes in their cycle of existence. So when compressed-bodied 'uniting' and 'united energy' inunens expand their bodies upon converting to that of 'dispersing energy' inunens, their bodies alter their shape but not the state of being heat. Therefore, everything physical remains heat, even when heat is either active heat or dormant (inactive) heat. (This is explained in the last chapter.)

With everything physical being heat, this means that light is also heat. But because light is so essential to human lives – with good reason – it has acquired a status of being something special: an entity in its own right. Whereas, in fact, just as everything physical is heat, so it is all light. The difference lies in the limited range of our human eyesight to perceive the full span of movements made by 'dispersing energy' inunens.

The current span of the so-called electromagnetic spectrum is calculated on the movements of 'dispersing energy' inunens according to the frequency and length of their side-view wave-shaped trajectory. This spectrum is scaled from the highest frequency with the shortest wavelength to the lowest frequency with longest wavelength.

The dispersing movements with highest frequency and shortest wavelength are called gamma rays, which represent 'dispersal codes' with request for very fast dispersing speed and wide 'transfer of intent' from any contacted 'uniting' and united energy' inunens. This activity is associated with high-level radiation.

From gamma rays, the gradually decreasing frequency and lengthening wavelength is allocated to x-rays, then ultraviolet rays, white light, infrared rays, microwaves and finally the radio waves. The while light is also divided into segments categorized as visible spectrum, or light spectrum, where the range consists of a blend from purple to green, to bands of yellow and orange, and ending in red.

What the electromagnetic spectrum allocations of 'dispersing energy' inunen movements fail to indicate – simply because current science has no knowledge of inunens – is that all these movements are the result of past efforts of work completed by heat particles (that is: mass), which is why they are in motion, on their way with instructions for further efforts of work to be done by other heat particles in different phase of their cycle, and not some vacuous "energy (electromagnetic) waves". Even if current scientific acceptance has granted light to comprise not just of electromagnetic wave but also of a so-called photon, they miss the implication of light: which is: particles of heat that had already been responsible for physical change, on their way to instigate further physical change. That is the meaning and function of all moving heat (light).

Furthermore, by accepting a flat-image view of moving waves – the kind that are displayed by computers on their monitor screens – they have failed to recognise a simple fact that everything physical in existence is three-dimensional, with this formation having a direct influence on all movements produced by three-dimensional existence.

What this means is that the forward movement of all 'dispersing energy' inunens is also three-dimensional. **This three-dimensional forward movement can be expressed as being in a shape of a spiral, or a coil, or a helix.**

Despite this information being unknown by current science, this phenomenon is simple to explain, using the accepted configuration of an electric wave and a magnetic wave flowing perpendicular, along the centre of each other. If the ongoing crests of the two perpendicular waves is plotted – from that of electric wave crest to magnetic wave crest then back to electric wave crest, and so on – then that plotting reveals a point on crests that moves forward as a spiral. The spiral, or helix, forward motion applies to all 'dispersing energy' inunen movements, whether it visible as light or not.

The proof of this helix movement is found in the way that sunlight travels by application of 'polarizing' filters, such as those used for sunglasses. What such polarizing filters do is to convert overall light into 'flat strips of light', which can run parallel to each other, after having been transferred by 'transfer of intent' through gaps in specially produced molecules. Such polarization of light means that light does not comprise of flat waves in the first place, but that of three-dimensional spiraling movement.

There is also the so-called the quantum "double slit experiment", in which light, or electrons are passed through two vertically parallel slits in a panel to form patterns onto a second panel, positioned behind and parallel to the first. To this day those conducting this experiment are confused by the behaviour of the resultant patterns because they continue to think that light and electrons move forward with a flat, two-dimensional trajectory. It never occurred for them to consider for everything to move three-dimensionally, which it does. That is why the spiraling particles passing through the slits produce patterns of vertical stripes outside the facing of the slits.

There is another vital factor to why the same 'dispersing energy' inunens behave differently. This factor applies to the reason for their varying movements, which current science has no idea of.

To explain this requires a confirmation of some notions that are uniformly accepted by science.

Firstly: as assessed by the "electromagnetic spectrum", many forward movements of 'dispersing energy' inunens have different frequency and wavelength. This means that a 'dispersing energy' near the range of gamma rays moves forward in tiny rotations with small advancing distance between spirals, while a forward movement at the range of radio waves produces enormous rotations, with enormous distances between spirals.

What this also means is that if the same distance covered in space was measured between a gamma ray, with its tiny spirals, and that of a radio wave with its long large wave, then that would show a gamma ray overall movement – if straitened out – being much, much longer in overall distance travelled to that of a radio wave.

Secondly: the speed of light is accepted to be a constant of 186,000 miles per second. As all 'dispersing energy' movements, represented by electromagnetic spectrum, are those of heat – and therefore of light – then that would make all waves of electromagnetic spectrum travel at the speed of light.

In which case a disparity of inconsistency arises: how can a gamma ray wave and a radio wave – equally comprising of 'dispersing energy' inunens – both travel at the same speed of light when gamma ray wave needs to travel an overall longer distance to match that of radio wave? How could it be that if a gamma ray wave and a radio wave were to travel an exactly the same distance in vacuum space, beginning from the same spot, then (despite both of them travelling at the same speed of light) it would result in radio wave reaching its destination long before gamma ray would?

The reason for such occurrence is due to what facing 'dispersing energy' inunens adopt when they travel (disperse).

Everything that 'uniting' or 'united energy' inunens experience in converting to 'dispersing energy' inunens is consistent, with one exception: depending on their 'dispersal code' decision, they have to take appropriate resolution to reflect that 'dispersal code', and this comprises of the facing with which they move forward in their dispersal. The alternatives are: with their indent forward; with their rear bulge forward; or with their sides forward.

Were all 'dispersing energy' inunens to move forward in their 'dispersing energy' phase in the same, then all 'dispersing energy' activity would have more-or-less the same value. This would basically mean the same frequency and wavelength for all dispersals. But because 'dispersing energy' inunens take into consideration where they are dispersing from, and what they shall pass on with their 'dispersal code' as a reflection of this, their choice of forward facings makes it possible for all kinds of dispersal values.

Primarily it can be considered that an indent-forward facing represents the mid-range of 'dispersing energy' dispersing value, where the frequency and the wavelength of the three-dimensional forward movement represents the white light region of the electromagnetic spectrum. This indent-forward facing can be classified as being 'light', because apart from being visible to eyesight, its physical impact is moderate, and its speed is supposedly the constant of speed of light.

The 'dispersing energy' inunens that disperse with higher frequency and shorter wavelength, in the region of gamma rays on the electromagnetic spectrum, do so their bulge forward, and can be classified as being 'fast heat' or 'heat'. As 'heat', their bulge-forward facing represents the reverse internal direction of heat attraction. This means that when making a physical contact their impact is counter to that of indent-forward facing of 'light', making their 'dispersal code' more forceful and determined, and their 'transfer of intent' inviting a more resolute and further-reaching action from the 'uniting' and 'united energy' inunens they physically encounter.

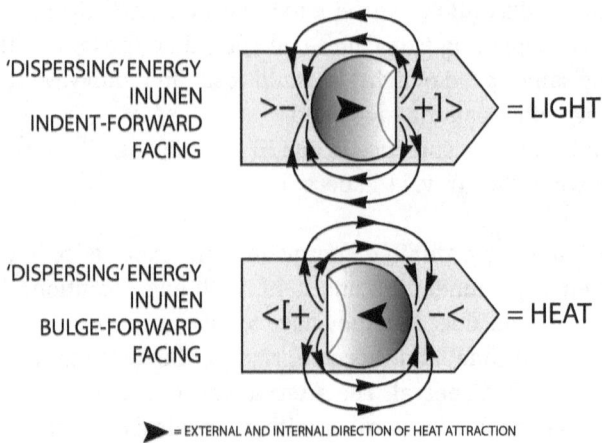

▶ = EXTERNAL AND INTERNAL DIRECTION OF HEAT ATTRACTION

Their bulge-forward movement may be slower and they do not project themselves as far as visible light does, but unlike the indent-forward movement of 'light' the impact from 'heat' is more pronounced, and can be quickly felt and seen in the destructive dispersals.

At the opposite side to gamma rays on the electromagnetic spectrum, are the radio waves that are produced by 'dispersing energy' inunens that move forward in large three-dimensional waves with their sides – or their shoulders – forward. Because of their size, these 'dispersing energy' inunens require less duration of travel to those of 'light' or 'heat'.

Side-forward facings can be made to disperse as closed loops, or circuit formations. This occurs when a straight path of 'energy bodies' (electrons) – moving back-and-forth along a broadcasting antenna as 'uniting energy' inunens – are compelled to change to 'dispersing energy' inunens. This they do on mass in pulses, joined together by their external direction of heat attraction; dispersing into air or vacuum space as three-dimensional closed loops reminiscent of grunun bent 'sausage' shapes, as radio and TV signals.

Just as there is a variety to 'dispersing energy' inunen forward-moving facings, these three-dimensional waves move forward with different spins. Some spirals move forward left-to-right, while others move right-to-left. These differences influence different waves from combining into a double helix, which can and does occur as 'tandem' forward movement, or as 'tandem radiation'.

As to the reason for the helix, or spiral, forward 'dispersing energy' inunen movements – rather than a flat, two-dimensional wave – apart from the earlier explanation, the 'dispersing energy' inunens are also influenced by the larger-space-beyond attraction.

For a 'dispersing energy' inunen, a larger-space-beyond becomes relevant both at the sides and in front. This causes them to move not just forward but also to the side, so that once having reaching a sideway point in space, a sideway point in space on the opposite side becomes the larger-space-beyond. By chasing a never attainable larger-space-beyond at their sides results in 'dispersing energy' inunens having a forward movement of a spiraling three-dimensional wave.

Phase 4: being a 'dispersed energy' inunen

Here is a question: "What happens to a beam of light when it is blocked by an obstacle, such as a wall?" The answer is that a beam of light does not simply vanish, or stop existing at its point of contact with an obstacle. The 'dispersing energy' inunens of light are merely converted, at the point of contact with that obstacle, from 'dispersing energy' phase to that of a passive 'dispersed energy' phase, while the contacted 'uniting' and 'united energy' inunens that form the atoms of the wall are convert to 'dispersing energy' phase.

After the conversion of 'dispersing energy' into 'dispersed energy', the 'dispersed energy' inunens produce a transformation from partial spherical 'dispersing energy' inunens to that of completely spherical 'dispersed energy' inunens.

On becoming a 'dispersed energy' inunen, this inunen stops being a spiraling light or heat moving forward in one direction. Its small remaining indent inverts by pushing out, so that the whole shape becomes a sphere – a ball. Its internal direction of heat attraction becomes a central core with no polarity, where the internal higher heat becomes prevalent everywhere just beneath the surface, pointing towards the core. This means that its external field of direction of heat attraction (boundary of influence) ceases to function. With no external direction of heat attraction, a 'dispersed energy' inunen can no longer have an ability to repel. The only ability it retains is to attract, that is, have other inunens be attracted to it because of its inward direction of heat attraction. This allows 'dispersed energy' inunens to become directly attached to another 'uniting' or 'dispersed energy' inunen, because there is only the inward direction of heat attraction to hold them together, as any physical contact between them cannot produce any further expansions to their bodies, and from that, no further 'effort' or 'work' can be achieved.

PART 3. DISCLOSURE OF EVERYTHING

CROSS-SECTION VIEW OF A
COMPRESSED 'UNITED ENERGY' AND AN EXPANDED 'DISPERSING ENERGY' INUNENS
BEING ATTRACTED TO A 'DISPERSED ENERGY' INUNENS
BY THEIR OWN EXTERNAL DIRECTION OF HEAT ATTRACTION

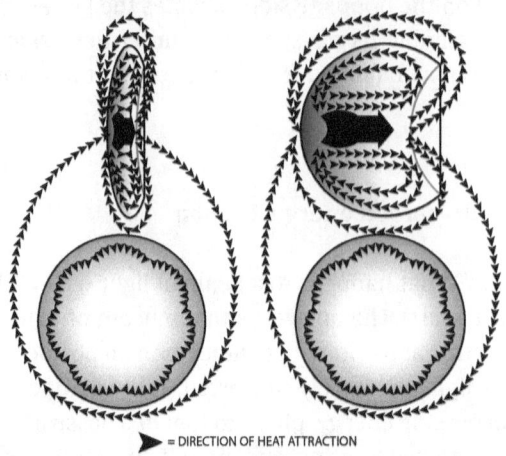

▶ = DIRECTION OF HEAT ATTRACTION

By not being able to produce any change from a direct physical contact with an individual 'uniting', 'united', and 'dispersing energy' inunens means that the individual 'dispersed energy' inunens can be embraced by their host 'uniting', 'united', and a 'dispersing energy' inunens and carried along. But should, for instance, two 'dispersing energy' inunens carrying a 'dispersed energy' inunen each contact any 'uniting' or 'united energy' inunens, then the two 'dispersing energy' inunens change to 'dispersed energy' inunens (and by already being in close proximity to the 'dispersed energy' inunens they were carrying) the four 'dispersed energy' inunens become united, because of their individual inward direction of heat attraction. This kind of attraction between 'dispersed energy' inunens and all other inunens produces faster unions between 'dispersed energy' inunens than if they only merged between themselves, as in being ununens and grunens forming atoms (solid bodies of protons).

Furthermore, should the 'dispersed energy' inunens come across each other while being carried by their host, the attraction between the 'dispersed energy' inunens can alter the movement of their host to that of their own. This can result in matter and light being directed by 'dispersed energy' inunens to follow their path of attraction to other 'dispersed energy' inunens. The underscore of this is that light, as 'dispersing energy', can be influenced to 'bend': not that light actually bends, but by being carried by 'dispersing energy' inunens, the overall attractions between 'dispersed energy' inunens can produce movements where the carrying-them 'dispersing energy' inunens are dragged along, with this making light appear to be curving.

As 'dispersed energy' inunens meet other 'dispersed energy' inunens in vacuum space of Eternity, they begin to form enormously long and wide lightless 'streams' of 'dispersed energy'.

When two or more streams of 'dispersed energy' inunens meet, the stream flow begins to wrap themselves into one mass, due to the overall attraction between them, and to their 'larger-space-beyond'.

Gradually, these invisible streams of 'dispersed energy' become so vast that they begin to influence surrounding stars by attracting them to follow towards the central mass. Because the central mass of 'dispersed energy' inunens has no external boundary of influence (gravity) all the attracted suns and stars are able to come in direct contact with the surface of the 'dispersed energy' mass. This causes the 'uniting' and 'united energy' inunens to rapidly convert to 'dispersing energy' and from that to more and more 'dispersed energy' adding to its 'dark mass' resembling a black sphere with no surface shell, surrounded by all attracted stars and suns. So begins a formation known as a galaxy.

It is now opportune to provide an overview of the two systems, which are responsible for this whole Universe looking as it does.

THE TWO SYSTEMS OF THE PHYSICAL PROCESS OF CHANGE

Every day thousands of astronomers look at all that surrounds this planet in the endless vacuum space, and have no real idea of what they are actually looking at. For what is there before them is the physical process of change. This physical process of change comprises of two physical systems by which all that physically exists, as active inunens, is constantly cycled through their four phases. One of these systems is the Energy and Matter Attracting and Releasing System (EAMAARS) while the other system is the Energy and Matter Absorbing and Compressing System (EAMAACS).

EAMAARS (Energy and Matter Attracting and Releasing Systems) is the name for a period when 'uniting' and 'united energy' inunens form themselves into atoms, then into all objects and bodies, even as they are being converted into 'dispersing energy' inunens and then into 'dispersed energy' inunens.

These EAMAARS systems include the quite common occurrences of supernovas, which, as vast collections of 'uniting' and 'united energy' inunens, release these as light of 'dispersing energy' inunens and clouds of atoms that go on to unite into new suns. These suns go on to disperse their contents as 'dispersing' and 'dispersed energy' inunens, as well as a great deal of 'uniting' and 'united energy' inunens in large and small quantities.

Our Sun is also an EAMAARS and so are its planets. The Sun is a large structure with powerful internal and external directions of heat attraction, which account for its gravity ('boundary of influence'). But as with all EAMAARS gravities ('boundaries of influence'), it is incapable of preventing its 'united energy' chemicals and 'dispersing energy' light and heat from leaving it and dispersing into vacuum space due to their dispersals and attraction to the larger-space-beyond.

Our Milky Way galaxy is a compilation of well over 200 billion huge suns and stars, many with their own solar systems. All of these are EAMAARS, with a single EAMAACS attracting them all to its centre of the galaxy, along the streams of 'dispersed energy' inunens.

Then there are the Energy and Matter Absorbing and Compressing Systems (EAMAACS). EAMAACS represent a physical process by which streams of 'dispersed energy' inunens deliver the attracted stars and suns to the black spheres of solid space, where these are converted to 'dispersed energy' onto its total mass.

(Currently EAMAACS are incorrectly referred to as 'black holes': incorrectly because it is impossible to have any kind of a 'hole' or an orifice in the nothingness of vacuum space of Eternity.)

The nucleus of EAMAACS (Energy and Matter Absorbing and Compressing System) comprises of cold sphere, which can do nothing to prevent being attractive to all the attracted EAMAARS. Indeed, it is the EAMAARS suns and stars that are attracted to the EAMAACS by their very own gravities ('boundaries of influence'). For it is their own external directions of heat attraction that embrace an EAMAACS, no different to that of an individual 'uniting', 'united' or a 'dispersing energy' inunen embracing and carrying a 'dispersed energy' inunen.

Because EAMAACS comprise only of expanded bodies of 'dispersed energy' inunens, their contents behaves the same as that of the 'dispersed energy' inunens: they have no external boundary of influence (gravity) but only the inward direction of heat attraction and 'space-beyond-attraction. This mean they are not able to repel the boundaries of influence (gravity) of the attracted to them EAMAARS with 'grappled attraction. That is why the drawn to EAMAACS EAMAARS physical embrace them, in the process becoming an addition to an EAMAACS.

The overall pattern of all the EAMAARS being attracted to a central EAMAACS in vacuum space of Eternity is currently known as galaxy.

Despite that the EAMAARS that approach an EAMAACS stationed at the centre of a galaxy, their decreasing proximity to each other produces growing repulsions, which their individual inunens overcome due to their overall attraction to the an EAMAACS 'dispersed energy'. The closer the EAMAARS come to each other around an EAMAACS the more their growing repulsions accelerate the conversions of their combined mass to 'dispersing energy, and then to 'dispersed energy'. This only stimulates more and more 'dispersed energy' inunens to surround and press down onto the surface of the EAMAACS sphere.

Once there is no remaining matter left to convert and absorb, an EAMAACS becomes dormant as an extremely cold, lightless, black sphere, remaining indefinitely in the black vacuum space of Eternity. It may remain in a dormant state for a few seconds, or incalculable billions of Earth years, before dispersing its contents back into the vacuum space of Eternity.

While EAMAACS remain without rotation, the vacuum space of Eternity causes no friction, allowing a small EAMAACS sphere – in the process of absorbing surrounding stars and suns – to be dragged along by its galaxy, as it disperses as all Universe. Such occurrences allow EAMAACS spheres to become mobile: to move in space even after having absorbed all the contents from an EAMAARS.

Having formed a central sphere of an EAMAACS – in this, the fourth phase of their cycle – the 'dispersed energy' inunens experience a tremendous compression. The 'dispersed energy' inunens, especially those at the centre of the core, are unable to move forward beyond the 'dispersed energy' inunens directly facing them, while being pressed from behind by incalculable numbers of other 'dispersed energy' inunens.

At some point, the 'dispersed energy' inunens at the very centre of an EAMAACS' core become so compacted that their spherical bodies are pushed inwards, compressing them.

At that instant of compression, the 'dispersed energy' inunens become 'uniting energy' inunens once again. They become "switched on", with their internal directions of heat attraction restarted and their 'boundary of influence' re-activated, producing repulsion to all surrounding inunens. This produces an extra compressing strain on the 'dispersed energy' inunens, as from the core they are being repelled by the 'uniting energy' inunens (because they cannot be embraced by the 'uniting energy' inunens while being pressed from all sides by other 'dispersing energy' inunens), and from the other side they are also being compressed by those above them. This transformation at the core produces a 'ripple effect' on the 'dispersed energy' inunens, as layer after layer of 'dispersed energy' inunens revert to being 'uniting energy' inunens.

An internal conversion of 'dispersed energy' inunens into 'uniting energy' inunens continues until the top layers of 'dispersed energy' inunens are incapable of pressing down the repelling-from-the-inside 'uniting energy' inunens. The remaining layers of 'dispersed energy' inunens are eventually pushed out into vacuum space by the mass of 'uniting energy' inunens, which, due to their repulsion of each other, begin to disperse as layers of 'uniting energy' inunens.

Gradually, the whole structure of an EAMAACS sphere begins to silently disintegrate (disperse) three-dimensionally from all sides, as a Great Dispersal, with this outward movement comprising of the liberated 'uniting energy' inunens allowed to go on accelerating from the initial impetus of motion.

Almost immediately, and for billions of Earth years after being released outwards into vacuum space of Eternity, the sightless 'uniting energy' inunens begin to unite into ununens (united units of energy) and then into grununs (grouped units of ununens), on their way to forming 'united energy' atoms. The first to be formed are Hydrogen atoms: a single nucleus of 'united energy' inunens surrounded by a single 'energy body' (electron) 'uniting body' shell. That is why Hydrogen atoms are the building blocks of all existence.

Simultaneously incalculable numbers of 'uniting energy' inunens fail to unite from collisions and become converted to the 'dispersing energy' phase. These incalculable conversions produce vast volumes and areas of light and head dispersals, lighting up those regions of vacuum space of nothingness. The very same dispersals become responsible for development of complexity to incalculable numbers of Hydrogen atoms, which unite into more complex structures of Helium, and so on.

It is possible to witness such EAMAACS dispersals in many locations of our Universe. Unfortunately these events are not recognised for what they are: supernovas. While not all supernovas are that of dispersing EAMAACS, when an EAMAACS initially dispersed there is nothing to witness, as the newly released 'uniting energy' inunens invisibly spread out in dispersal. It is only when the 'dispersing energy' becomes active in vast numbers that the event becomes visible as huge clouds of "star dust" spreading as enormous quantities of heat and light.

In such way EAMAACS again produce the two systems of physical change, circulating all the individual units of energy (inunens) in their new cycle.

Our Universe had been a vast EAMAACS that had experienced a Great Dispersal, its contents now forming the EAMAARS suns and the stars with their gravities ('boundaries of influence'), coming together in galaxies that shall be absorbed and compressed by EAMAACS spheres at their centres.

Just as the Great Dispersal of an EAMAACS produced our Universe, so it is responsible for the fact that currently humans reside on Earth.

There is nothing to prevent an EAMAACS dispersal from taking place closer to Earth. After all, small EAMAACS are common enough. Considering that if there are well over 200 billion galaxies in our Universe, then there are at least that number of EAMAACS. And although they are nowhere near in size of the EAMAACS whose dispersal made this Universe possible, any of their dispersals would have a profound effect upon any region of this Universe. Not that this would matter to the released 'uniting energy' inunens, in the Phase 1 of their new cycle.

SUMMARY OF THE FOUR INUNEN PHASES

In a summary of the four phases of inunen one-directional cycle, it can be unequivocally stated that it is they, inunens (throughout their cyclical four physical body changes) who are responsible for all physical existence, due to their unions and dispersals of those unions. And the pivotal essence for all these unions is their internal direction of heat attraction, capable of producing their external direction of heat attraction: their external 'boundary of influence'.

It is no fluke that everything in physical existence resembles either the compressed, saucer-like shapes of 'uniting' and 'united energy' inunens or the expanded, ball-like spherical shapes of 'dispersing' and 'dispersed energy' inunens.

Take a look around you, and it becomes clearly apparent how everything in nature on Earth, whether body-brain (flora) or brain-body (fauna) replicate the compressed and expanded inunen shapes.

Examine any leaf, for instance, and note how the overall shape is that of compressed inunen, rounded with the sides slightly rolling over. The same applies any blade of grass, even if overall shape is elongated. Notice how all the stems of plants and trunks of trees are round, how the overall tree shapes with their leaf canopies are spherical and round. Similarly, this also applies to any kind of berry or fruit. They all resemble the spherical 'dispersing' and 'dispersed energy' inunens.

Many seeds of fruit, such as plums, apricots and mangoes, resemble two 'uniting' or 'united energy' inunens 'halves' clamped together into a single roundish shape. The same applies to the shellfish, their shells replicating the compressed shapes of 'uniting' and 'united energy' inunens. Even the elongated shape of a banana maintains the round shape with a curvature, while fruit like a pineapple are not only spherical-shaped but their surface is covered in hexagonal compressed inunens shapes.

Then there are all the rounded-shaped animal bodies (including fish), with rounded-shaped bones, rounded-shape organs, rounded-shaped cells, rounded-shapes everything: from eyes to hairs, to skin, to feathers, to internal rounded-shaped cells, bacteria, viruses and tumors.

And while humans have learnt to use flat surfaces to construct structures for their benefit, all they build and manufacture – from glassware to vehicles to architecture – replicates the curved or spherical inunen shapes, which, throughout their four cycles, are structured for one thing and one thing only: enclosure: enclosure of themselves and enclosure of all that they form in pursuit of larger-space-beyond (which, in the case of humans, had been their constructions of roads, landing strips and piers).

This also applies to all the EAMAARS celestial bodies, be they planets, suns and stars, just as it does to the highly condensed, lightless EAMAACS. They are all spherically shaped for enclosure, just like that of all inunens of which they are all formed. This is due to all such objects and bodies having an internal direction of heat attraction. And while EAMAARS structure develop internal heat movement with polarities that produce external boundaries of influence' (gravities), it the inward direction of heat attraction that causes them all to be spherical from that attraction moving inwards from all sides of the surface.

By having an understanding of how all the inunens provide an ongoing change to everything in existence by their existence, it becomes obvious how they, together with the nothingness of vacuum space of Eternity, formulate the constants of all physical existence.

The first constant of all physical existence is the physical existence of the vacuum space of Eternity – as previously explained.

The second constant of all physical existence is the physical process of change. This process represents the sum total of all inunen intelligence and activity that produces an eternal change from their constancy to influence, or to be influenced, in their eternal cycles comprising of four phases that are responsible for the EAMAARS and EAMAACS systems, which result in an endless process of unifications and dispersals.

The third constant of all existence is the simultaneity of all physical activity, everywhere. This constant means that irrespective of distance that separates all the inunens and all that they form, everywhere – in whichever phase they may be (not just in this Universe but throughout the vacuum space of Eternity) – all their activity of change occurs at the same instant of 'now'.

These three constants of all physical existence are not something that humans can devise and alter according to their opinions and perceptions. These three constants of all physical existence do not ever waver, or alter, or change; they are not just universal laws but eternal laws.

GRAVITATION: THE ATTRACTIONS AND REPULSIONS OF COMBINED MASS

When it comes to all EAMAARS (Energy and Matter Attracting and Releasing Systems), the common elements that apply to all that occurs within those systems, and makes it physically possible for them to be (be it temporarily), is their internal and external directions of heat attraction that provide an external 'boundary of influence'. While current science has no perception of 'boundary of influence', or of its different sizes and applications, they know one of its attributes, which is currently known as gravitation, or gravity.

Gravitation has puzzled humans for ages. It has been pondered and calculated and theorized. All because of one unknown attribute of gravitation. While its power of attraction had supposedly been easy to explain, it has been another effort altogether in trying to explain the reason why celestial bodies employing their attractions of gravitation manage to keep apart in vacuum space.

After all, if gravitation meant only attraction, then that would eventually result in all celestial bodies merging into one single mass. But such is not the case.

Of course, there has never been a lack of theories trying to explain this, including those proposing centrifugal forces or the famous current theory proposing grooves in vacuum space, along which the planets of our Solar System supposedly move around the Sun.

All such current theories maintained by current physics are not just wrong but are ludicrous. They fail to appreciate physical reality that the orbits of planets are never constant from being elliptical, with constantly changing ellipsis, and with the whole Solar System moving at a vast speed in vacuum space. All this would mean that even if grooves could be somehow be made in nothingness of vacuum space, these grooves could never be made accurately enough and quickly enough to provide for movement of all the planets about the Sun.

The actual reason for gravitational behaviour of the Solar System planets orbiting the Sun but never being able to merge with it – something that humans should have recognised ages ago, as it is so obvious – is that gravitation comprises of BOTH attraction AND repulsion. Just like with inunens, the principle of internal and external directions of heat attraction, which produce the external 'boundary of influence' that allows inunens to attract and to repel other inunens, applies similarly to all other bodies.

The fact of the physical process of change is that whatever occurs on the smallest scale is replicated on the largest possible scale. A difference in scale has no distinction in the fundamental processes comprising of internal and external directions of heat attraction. Be it an inunen or our Sun, the fundamentals of internal and external direction of heat attraction causing the external coils of direction of heat attraction, which is the 'boundary of influence', (or in current understanding: field of gravitational attraction), are exactly the same, with exception of their scale.

Where scale does matter, is in the incalculable numbers of inunens simultaneously involved in producing the internal and the external directions of heat attraction (boundary of influence) so powerful that huge objects, like our Sun, have the ability to possess influence over the most distant of planets, which have their own, reciprocal, 'boundaries of influence' (gravitations).

When it comes to celestial objects and bodies, their internal direction of heat attraction is the component that directs all the heat inwards. The reason for such structural characteristic of an external convex with an internal concave, is because this is the only way that combined solid mass can be achieved on a very large scale by the 'united energy' inunens. The following simplified graphic illustrates how a combined solid mass is derived.

A BRIEF LOOK AT ALL PHYSICAL EXISTENCE

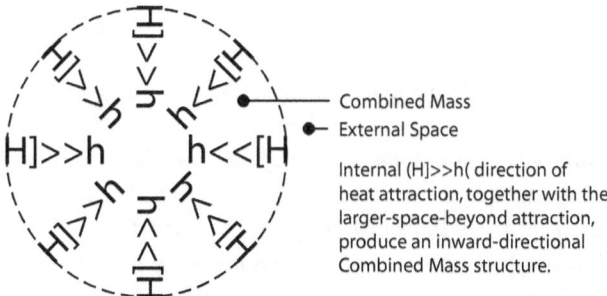

TWO-DIMENSIONAL REPRESENTATION
OF INTERNAL INFLUENCES DERIVING COMBINED MASS

Combined Mass
External Space

Internal (H]>>h(direction of heat attraction, together with the larger-space-beyond attraction, produce an inward-directional Combined Mass structure.

H]>>h = Higher heat repelling any oncoming heat, while being attracted to lower heat that attracts higher heat

From this graphic it can be seen how the loose components of matter (being formed from 'uniting' and 'united energy' inunens) are influenced to come together, their overall heat direction being influenced to move inwards towards internal core region of lower heat. But while this inward direction of heat attractions is directly responsible for the external 'boundary of influence', it is also, primarily, responsible for enclosure of its combined mass in its own right: producing the 'gravitational pull' with no repulsion on the surface of that combined mass. It is for this reason when anything that penetrates the 'boundary of influence' of any celestial body is bound to collide into the surface of that celestial body, as the united mass of that body attracts all on its surface into itself (or rather, to its internal direction of heat attraction's polarity stream.)

Actually the internal inward-direction of heat attraction does not form one solid mass but that of layers, because the build-up of combined mass occurs when layers of new material are positioned (or deposited) on top of existing layers.

Once the heat moves inwards to the core, the core heat has nowhere else to go, so it remains there, increasing in size, as it cannot move against the incoming heat direction from each layer above. It is only when the surface layers become much cooler than those beneath them that the core heat can reverse its direction of heat attraction and move towards the surface.

Still, this reverse direction is much slower than inward direction, as each above layer resists the approach of internal heat from the layer beneath it. It is for this reason that it takes longer to cool an object than to heat it. It is also for the same reason that it is taking our Sun much longer to release its contents that it took to compile them.

When there is a large spherical combined mass with a hot core, then there are large amounts of internal 'dispersing energy' inunens, and these are responsible for conducting a 'transfers of intent' throughout the whole combined mass. Wherever the internal 'dispersing energy' inunens can produce a substantial exit for themselves on the surface of that combined mass, this is where a concentrated movement of heat becomes formed, from the core towards the surface region. Once such a region is established – despite any internal heat being emitted at the surface – it becomes inevitable that most of the internal heat (with nowhere else to go) is influenced to move back internally, by looping around the internal polar heat movement as fields of internal heat direction, ending up at the rear of the heat movement. This self-propelling internal circuit of heat movement replicates the internal heat movement in compressed and expanded inunens, as shown in the diagram below.

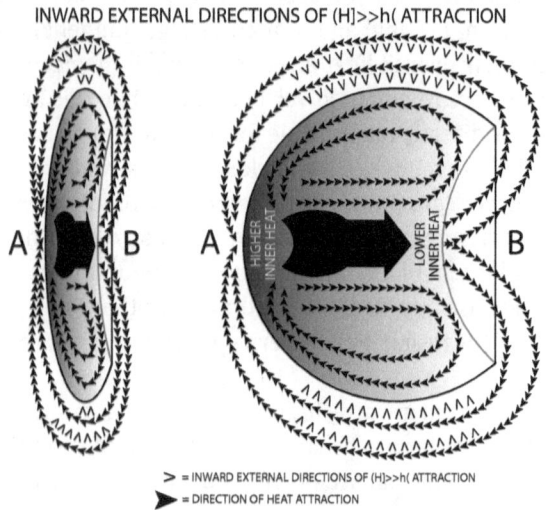

When the internal heat moves from its central core to region 'B' and then returns internally to the internal region 'A' (so as to re-join the core polarity and move forwards once again towards 'B' region) the external region 'B' becomes an exit polarity for any departing heat. As the external head direction moves back to external region 'A' – over the surface of the combined mass – it establishes a directional internal polarity between external and internal region 'A' and the external and internal region 'B'.

This circuit of internal/external heat movement, derived from the internal polar heat movement going through the central core, inevitably results in the three-dimensional fields of 'boundaries of influence'. In other words, the structure of these external 'boundaries of influence' coils of celestial bodies is the very entity that currently is known to be 'gravitation'. It is the external 'boundary of influence' of internal and external heat directions of combined mass that attracts foreign objects and bodies inwards, to become enclosed into the structure of its combined mass.

There is however, an important difference between the 'boundary of influence' (gravitation) of inunens and those of EAMAARS planets, suns and stars. When the 'uniting energy' inunens unite, their 'boundaries of influence' bring them together at a uniform distance from each other, while disallowing them to touch each other due to the repulsion. And while each 'united energy' inunen ('united' after having united with other 'uniting energy' inunens) vibrates from constant attraction and repulsion to inunens around then, they remain in place until the moment when they become 'dispersing energy' inunens.

This is not the case with celestial bodies. The planets of our Solar System, for instance, are in constant but differing directional movement around the Sun, which also rotates. And their individual varied distances from the Sun constantly fluctuate, due to their elliptical orbits of the Sun. And the speed of their orbits also fluctuates and differs between planets.

Such complexity may seem difficult to reconcile by mere 'boundary of influence' explanation. And yet, it does just that. That is because while attractions are all the same, be they those of inunens or huge celestial bodies, repulsions can be somewhat different. While on the miniscule scale of inunens, their repulsions are those of external directions of heat attraction, which hold them apart, while their mutually accommodating internal directions of heat attraction allow their unions. But when it comes to enormously sized 'boundaries of influence' (gravitation) coils of stars and planets, these 'boundary of influence' coils have another method of repulsion, which is classified as: 'grappled repulsion'. It is this 'grappled repulsion' that is fully responsible for the phenomena of gravitation: the ability to attract while preventing collisions and mergers between the attracting and the attracted celestial bodies.

In order to explain 'grappled repulsion' it is necessary to re-visit 'boundary of influence'. Each and every 'boundary of influence', be that of inunens or celestial bodies is a field, caused by internal and external directions of heat attraction, which externally extends some distance away from the physical body itself. These fields of 'boundaries of influence' can also be referred to as three-dimensional coils that emanate from the low-heat polar region of the physical body and externally moving back over that body, to the polar region of high-heat at the rear of that body.

The simplest way to illustrate the 'boundary of influence' coils is to have a sheet of paper covered with iron filings and placed directly onto a bar magnet. The metal filings will form a symmetrical pattern of curved lines joining one end of the magnet to the other, with these joined loops extending outwards on both sides of the bar magnet.

Of course, such coil patterns are currently classified as magnetic fields. But whether these coils are magnetic fields caused by magnets, electromagnetic fields caused by electromagnets, or gravitational fields caused by celestial bodies, they are all in fact the very same 'boundaries of influence', as they all comprise of tree-dimensional embracing coils that surround a physical body or an object, and are structured in layers one on top of another, spreading outwards from that physical body or object.

As the (H]>>h(equation of all heat direction indicates that heat moves from internal indent towards an internal bulge, so does the movement of general attraction. This attraction from the inner indent of every layered 'boundary of influence' coil invites anything outside it towards the next layer beneath it, closer to its structure of internal direction of heat attraction.

When it comes to celestial bodies, they all have some levels of 'boundary of influence'. Or, in current-speak, they all have some levels gravitational strength. Therefore, when two celestial bodies become located in vicinity of each other, their respective 'boundaries of influence' coils lock onto each other.

While this would appear as a definite reason for those two celestial bodies to physically collide and merge into one, it is the very same 'boundaries of influence' that prevent this from occurring.

The following graphic shall provide a visual explanation of 'grappled repulsion', whereby two separate 'boundaries of influence' (gravities) producing a simultaneous attraction also provides repulsion from attraction.

A BRIEF LOOK AT ALL PHYSICAL EXISTENCE

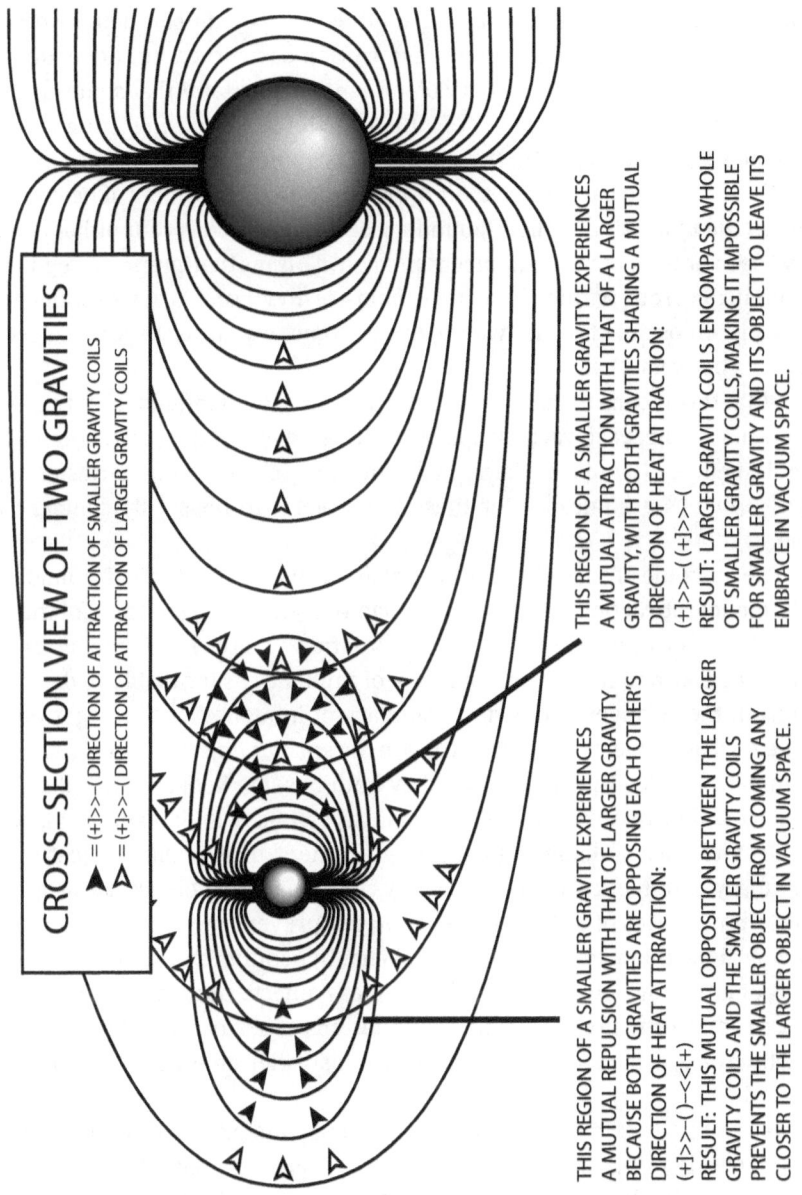

As this graphic shows, the three-dimensional fields of external directions of heat attraction ('boundary of influence') can be simplifies to appear as coils that extend out into external space from the polarity of the combined-mass. When two such 'boundaries of influence' (gravitations) come across each

PART 3. DISCLOSURE OF EVERYTHING

other they engage in attracting one another. As they do so, their respective 'boundaries of influence' (gravitations) overlap each other. But depending on individual levels of each 'boundary of influence' (gravitation) at some junction of their overlapping 'boundaries of influence' the two opposing 'boundaries of influence' (gravitations) find that they are in conflict.

While drawing each other, their opposing (+]>>− directions of heat attraction on the underside of their 'boundary of influence' coils cannot help but to repel each other by their 'grappled repulsion'. This mutual 'grappled repulsion' (a repulsion from within the curved fields of the 'boundaries of influence') makes it impossible for the two objects to come any closer beyond a specific distance between them.

What this means is that the external 'boundaries of influence' of two combined-mass objects, in vacuum space, produce an embrace of attraction, while simultaneously the mutual 'grappled repulsion' of their 'boundaries of influence' provides a hindrance for these two objects to physically contact each other.

The outcome of such circumstance also means that the mutual 'boundaries of influence' attraction (being greater than the 'grappled repulsion' of the two 'boundaries of influence') produces a situation whereby the smaller object, with a weaker and narrower 'boundary of influence, is incapable of disengaging from the stronger and wider 'boundary of influence' of the larger object, while the lager object is unable to release the smaller object from its 'boundary of influence' field. This means that they both become hooked into a distance-in-space union, unable to merge or separate, just like our Sun and its planets.

Despite such a distance-in-space attachment of a smaller object to the 'boundary of influence' fields, or 'gravity coils' of a larger object, these 'gravity coils' ('boundaries of influence' fields) do not prevent the smaller object from moving along them around the larger object. Depending on the inherent movement of either or both of the combined-mass objects can result in a constant, or occasional or non-existent speed of rotations of one or both objects, and the constant or varying speed of an orbit by the smaller object around the larger one.

While the 'boundaries of influence' of magnets and electro-magnets are symmetrical, and more-or-less equidistant from the body of a magnet, the boundaries of influence' fields of celestial bodies and objects are seldom such, due to their constant internal and external heat direction fluctuations. For that reason their three-dimensional 'boundaries of influence' (gravitations) can be elliptical. This causes many smaller celestial bodies to move about their larger hosts in ellipses rather than circles.

There is nothing in physical existence that remains forever the same. Change is inevitable. This applies as much to the four phases of inunen cycles as to the objects and bodies that inunens form. As objects and bodies experience physical change, their 'boundaries of influence' fields (gravitation) also undergo altered variance in levels of influence.

Whatever the structure of a combined mass of an object may be, its external 'boundary of influence' reflects that combined mass. The difference in 'boundaries of influence' depends on how porous or compacted they and their central cores are; how hot they are; and what they and their cores comprise of. This means, for instance, that an object whose core is porous or frozen is not going to produce very strong or distant level of 'boundary of influence' fields (gravitation).

Furthermore, the internal quantities of heat and their internal movements are never constant: some fluctuation always occurs. As stated earlier, these internal heat movement fluctuations directly affect the external 'boundary of influence'. This means that the three-dimensional fields of 'boundary of influence' attraction and repulsion are never absolutely steady. They can fluctuate from becoming weaker or stronger. This unsteadiness produces a movement for the whole combined mass structure. It also effects the movement in the objects attached to them by their 'boundary of influence' fields. Depending on these 'boundary of influence' (gravitation) fluctuations in all objects, these celestial bodies rotate (or not) on the axis of their internal polar heat directions, while orbiting along the 'boundaries of influence' of those combined-mass objects that they are attracted to the most.

An overall heat dispersal from a combined mass can also have an influence on 'boundaries of influence'. An object with a hot solid core, whose contents are allowed to leave its surface as 'dispersing energy' inunens of heat and light, as well as streams of atoms, grununs and ununens, will produce more than just a far reaching and highly concentrated 'boundary of influence' fields. Its emissions will dictate the distances to be maintained between itself and other objects and bodies that surround it. This means that levels of released 'uniting', 'united', and 'dispersing energy' inunens also have an influence on the distance at which the surrounding objects will remain. The higher the level of dispersing heat between two objects the further will that heat repel them apart.

Alternately, a frozen body comprising of a large core of highly compressed matter, such as that of metal like Iron (Fe) for instance, will have a pronounced 'boundary of influence' (gravitation), despite not having a hot central core.

PART 3. DISCLOSURE OF EVERYTHING

It is not just combined-body objects with a substantial mass that can activate and maintain a defined 'boundary of influence' field in vacuum space. Vast quantities of loose 'uniting' and 'united energy' inunens – forming various atoms ('united bodies') and particles located in vacuum space – can become captured within the coils or fields of the 'boundary of influence' of a solid mass object: unable to come closer and unable to leave. Such gathered atoms and particles can take on a form of a combined structure, in their own right. As a loose unity of external combined mass, these separate atoms and particles begin to exhibit their own 'boundary of influence' field, within the 'boundary of influence' fields of original highly compacted combined mass of a star or a planet.

Such a smaller 'boundary of influence' with token combined mass, within a larger one with an actual combined mass, produces two outcomes for the combined mass. Firstly, the smaller 'boundary of influence' produces some direct influence to the surface of the combined mass of the star, or planet; and secondly, it adds to the attributes of the larger 'boundary of influence' behaviour, by making it more defining in its attractions and repulsions.

This means that the smaller 'boundary of influence' within a larger one, becomes like an independent arbiter of what can penetrate to the surface of the combined mass, especially the incoming 'dispersing energy' inunens of heat and light – due to its own ability of providing close-range attractions and repulsions – and the physical presence of the loose union of atoms and particles acting as an obstacle to any approaching objects and bodies.

What is currently referred to as the Earth's electromagnetic field is, in fact, a 'boundary of influence' field (gravitation) within the Earth's larger 'boundary of influence' field (gravitation) – a form of gravitation within another gravitation.

Current science envisions gravitation as a mysterious and vacuum space-dependent force, from supposedly depending on the curvature of 'spacetime', resulting in 'gravitational time dilation', (while according to quantum gravity theory, gravitation has no role in determining the internal properties of everyday matter). Despite such elaborate presumptions – which grasp at fictitious make-belief while choosing to ignore physical reality – the actual physical function of gravitation (external 'boundary of influence'; derived from internal direction of heat attraction of combined mass of any astronomical body) is very precise and straightforward.

Any 'boundary of influence' (gravitation) needs no confirmation based on human imagination to function as an influence of attraction together with that of simultaneous 'grappled repulsion'. And if making a careful consideration of the two attributes of attraction and repulsion, it is directly thanks to the 'grappled repulsion' of every 'boundaries of influence' that EAMAARS (Energy and Matter Attracting and Releasing Systems) celestial bodies – including that of our Sun and all its planets – continue to exist at a distance-in-space from each other, rather than becoming one gigantic mass in space, were only gravitational attraction to prevail.

PHYSICAL BEHAVIOR OF 'ENERGY BODIES' (ELECTRONS)

Before attempting to explain the physical structures and behaviors of atoms, it would be prudent to give a more detailed explanation of particles called 'uniting energy' inunens, which are responsible for formations of 'energy bodies' (electrons).

As previously explained, once the newly compressed 'uniting energy' inunens are propelled forward, away from a dispersing EAMAACS (Energy and Matter Absorbing and Compressing System) in a process of a Great Dispersal, they immediately begin to unite with each other, thanks to their individual internal and external directions of heat attraction, which is responsible for external 'boundary of influence'.

At first the individual 'uniting energy' inunens form coiled strings of ununens (united units of inunens). Then these unite into sausage-like structures of grununs (grouped units of ununens). The grununs continue to combine into spherical structures of 'solid bodies' (protons) by the inward-direction of heat attraction that produces combined mass.

Meanwhile, many other 'uniting energy' inunens do not manage to unite into structures and so remain single. This does not prevent them from being attracted to the stronger 'boundary of influence' of 'solid bodies' (protons). As more and more unattached 'uniting energy' inunens surround from all sides the periphery of a 'solid body's' (proton's) external 'boundary of influence', their own individual external directions of heat attraction ('boundaries of influence') causes them to form unions with other 'uniting energy' inunens all around them, despite having personal repulsion to them. The following diagram illustrates how this is achieved.

PART 3. DISCLOSURE OF EVERYTHING

DISSECTED VIEW OF COMPRESSED 'UNITED' ENERGY INUNENS FORMING 'ENERGY BODY' ELECTRON MESHES BY SHARING THEIR EXTERNAL DIRECTION OF HEAT ATTRACTION WITH THOSE OF THEIR NEIGHBORS

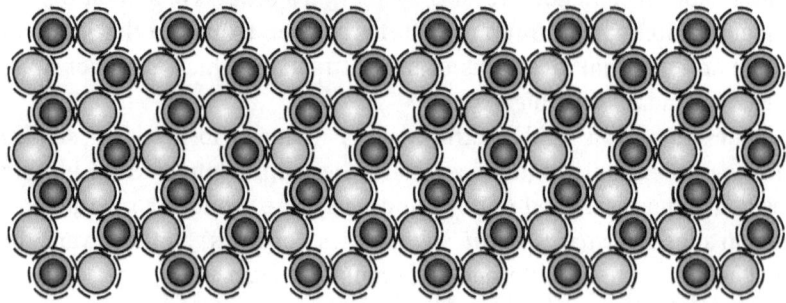

By alternating their bulge and indent facings towards the 'solid body' (proton) these 'uniting energy' inunens form a bounding mesh that encapsulates the 'solid body' (proton), with a mesh comprising of hexagons of three bulge facings being separated by three indents, as show by the following graphic.

BULGE AND INDENT 'UNITING ENERGY' INUNEN FACINGS FORMING A MESH TO ENCAPSULATE A 'SOLID BODY' (PROTON) AS AN 'ENERGY BODY' (ELECTRON)

The dashed line around each inunen represents their 'boundary of influence' at their sides, which prevents them from physically touching each other.

The gap within each hexagon is a space through which 'dispersing energy' inunens can travel, without touching the mesh itself.

Once such an 'energy body' (electron) encapsulates a 'solid body' (proton) as a shell, that 'solid body' (proton) and its 'energy body' (electron) form a lasting partnership, where the 'energy body' (electron) is both attracted to, and by, the 'solid body' (proton) while being repelled away by it to a very long distance.

Very often, however, a 'solid body' (proton) – due to a problem with its structure – is incapable of repelling its 'energy body' (electron).

In such a situation an 'energy body' (electron) shell collapses all over its 'solid body' (proton), covering its surface with layers of 'uniting energy' inunens, like a snow blanketing a surface. A combined structure of this kind is classified as a 'combined body' (neutron), and has a much smoother surface than the rough-terrain 'solid body' (proton). By having its 'energy body' (electron) wrapped all over itself, a 'combined body' (neutron) is larger and heavier than the 'solid body' (proton). The distinctive difference to a 'combined body' (neutron) is that by having its surface covered by its 'energy body' (electron), the 'solid body' (proton) underneath cannot activate its faulty 'boundary of influence'. This inability to engage actively in attractions and repulsions give it an almost non-responsive characteristic.

Such situation does not mean that a 'combined body' (neutron) remains as such forever. Depending on circumstances, the nucleus of the 'solid body' (proton) – beneath a layer of its 'energy body' (electron) of a 'combined body' – may become activated, so that it repels its 'energy body' (electron) away from its surface, and by that becoming a 'solid body' (proton) once again.

What remains to be said of the 'energy bodies' (electrons), is that while they are formed of individual 'uniting energy' inunens, everything physical is also formed from similar 'uniting energy' inunens. The only difference being is that the 'uniting energy' inunens that ended up forming 'solid bodies' (protons) have converted to the 'united energy' phase: they cannot freely leave the structures they form, unless it is as 'dispersing energy' inunens.

Such is not the case for 'uniting energy' inunens that form energy bodies' (electrons). 'Energy body' (electron) shells are not rigid, and therefore are more easily disrupted or disengaged, leaving the formations of 'uniting energy' inunens momentarily in disarray. It is in these moments of dislodgement that 'uniting energy' inunens display a behaviour that separated them from 'united energy ' inunens: their ability to have independent motion in the direction of heat attraction, but just as long as that direction is contained in a closed loop, or closed circuit.

Therefore, it is helpful to understand that any and every unattached 'uniting energy' inunen (responsible for everything in physical existence) is in fact the provider of 'electricity' when 'uniting energy' inunens, on mass, allow themselves to be converted into 'dispersing energy' inunens, by whatever means.

According to the current understanding of physics, the electrons ('energy bodies') are supposedly all-negative and the protons ('solid bodies') are all-positive. These all-negative and all-positive structures are supposed to produce a relationship of attraction between the electrons (energy bodies) and protons (solid bodies), with no repulsions and no other interaction occurring between them in vacuum space.

In actual fact, of course, the internal structure of proton ('solid body') has a central polarity of direction of heat attraction, and that means a simultaneous possession of opposing 'positive' and 'negative' sides. It is these individual 'boundaries of influence' that allows protons ('solid bodies') and their electrons ('energy bodies') to retain an attraction to each other, while simultaneously retaining a repulsive distance from each other. This mutual repulsion allows the electrons ('energy bodies') to encapsulate their protons ('solid bodies') without touching them.

The relationship between nuclei of atoms and their 'energy bodies' shells are often variable, depending on quantities of surrounding 'dispersing energy' inunens. This internal change (in the space between the surface of nucleus and the insides of the 'energy body' shell [electron]) and external change from any influences impacting onto the surface of 'energy body' shell (electron), causes 'energy bodies' shells to either contract or to expand to the limits of their combined 'boundary of influence'.

For instance, when a nucleus is releasing 'dispersing energy' inunens, this may cause fluctuation in its combined 'boundary of influence, which can cause the 'energy body' shell (electron) to expand or stretch out in certain direction, so as to release as many 'dispersing energy' inunens as possible without effecting or damaging the 'energy body' shell structure. Once this is accomplished and the nucleus settles down it 'boundary of influence', the 'energy body' shell will also diminish its size according to its combined 'boundary of influence'.

But while the 'boundary of influence' of 'energy bodies' (electrons) is slight – but sufficient enough to hold all the individual 'uniting energy' inunens together in a combined shell – it also performs an altogether different function that is unique to 'energy bodies' (electrons) of some metals. This extraordinary function is classified as being 'transfer of influence'.

'Transfer of influence' is a method by which the 'energy body' (electrons – that is: net-type shells made up of 'uniting energy' inunens that surround an atomic nucleus) – of some metal elements use their combined 'boundary of influence' to communicate with other's 'energy bodies' (electrons). This occurs when all the 'energy bodies' (electrons) of a particular group of atoms, such as that of Copper (Cu), combine their 'boundaries of influence', allowing such overall 'boundary of influence' to instantly transfer information from one side of the grouped atoms to the other. This 'transfer of influence' ability basically allows an 'energy body' (electron) at one end of a metal structure' to know exactly what another 'energy body' (electron) is doing at the opposite end, even if these two opposite ends are part of a structure that is of a vast distance in length.

In order to explain how this comes about requires another look at the 'uniting energy' inunens.

As mentioned previously, a 'uniting energy' inunen will remain inactive until there is an approach to it by some other 'uniting' united' or 'dispersing energy' inunen. Till then its internal direction of heat attraction remains almost at rest: all present but almost inactive. This internal stillness equates to no external direction of heat attraction taking place, so that there is virtually no 'boundary no influence' to speak of.

But while such inactive 'uniting energy' inunen is dormant, it retains an ability of awareness. For as soon as it senses the 'boundary of influence' of approaching inunen (that is: its external direction of heat attraction) it immediately activates its internal direction of heat attraction, which, in turn, activates its external direction of heat attraction – its external 'boundary of influence.' Once the internal direction of heat attraction is turned on, it remains active for the duration of being in touch with the 'boundary of influence' of another inunen.

If this 'uniting energy' inunen forms a bond with another 'uniting energy' inunen – and by that converting to 'united energy' inunens – their internal and external directions of heat attraction movements prevent them from having the freedom to depart the union, until they become converted to 'dispersing energy' inunens.

But should a 'uniting energy' inunen again be left alone (before any unions with other inunens), it will again reduce its internal heat activity, doing the same with its external direction of heat attraction ('boundary of influence'), and by that remaining a 'uniting energy' inunen: free to experience whatever comes its way.

When singular 'uniting energy' inunens form into 'energy body' meshes (electrons) surrounding a nucleus, instead of becoming 'united energy' inunens, they remain 'uniting energy' inunens. This means that should their 'energy body' mesh rupture or disperse, for whatever reason, they have a freedom of movement.

And yet there are restrictions on this free movement by such disbanded 'uniting energy' inunens. This is because when an 'energy body' mesh is caused to leave its nucleus, it is not done so voluntarily but caused by some very powerful influence of direction of heat attraction, such as that of a lightning, or 'energy body' (electron) harvesting for electricity production, or even electricity usage. But whatever the cause, when the 'uniting energy' inunens are no longer forming an 'energy body' mesh shell (electron), they, as individual 'uniting energy' inunens are not given a chance to unite into 'united energy' inunens, This is because the action that caused the 'energy body' (electron) to come apart further influence them all to move forward as a uniform file in the direction of heat attraction: that is, towards the region of low heat; but only if that forward direction is a closed circuit. And only if the low heat region in that closed circuit draws to itself the advancing individual 'uniting energy' inunens, because it is at that region they become converted to 'dispersing energy' inunens after having performed 'effort' of 'work'. This function is currently termed as 'electricity'.

All these loose 'united energy' inunens that are forced to perform 'electricity' can be described as being: 'ebitransits'. Ebitransit (singular) or ebitransits (plural) is an abbreviation of: '**e**nergy **b**ody' inunens in **transit**. Of course, ebitransits are not only extracted from 'energy body' shells (electrons) of copper by human technology. They are constantly obtained under pressure from Nitrogen, Oxygen, and Hydrogen from water in the Earth's atmosphere, presenting themselves in circuited movement as lightning.

Irrespective of how ebitransits are produced, when they are caused to move forward, they can only do so within confines of chemical elements that are accommodating of ebitransits. The reason for that is because these chemical elements and atoms need to be constructed of 'solid bodies' (atoms) whose 'energy bodies' (electrons) can also be easily influenced to leave their nuclei to become ebitransits. And it is within such environment that 'energy bodies' (electrons) can perform their unique action of 'transfer of influence' over long distances.

In forming a long circuit of copper wire, for example, the 29 'energy bodies' (electrons) of every copper atom (Cu) maintain an individual united 'boundary of influence' for their atom. Once a foreign entity of ebitransits is felt by any of the 'energy bodies' (electrons) within the copper wire structure, all the individual 'energy bodies' (electrons) of all the copper atoms, in the whole copper wire, become combined into a mutual 'boundary of influence' that involves each and every atom. At this juncture, the region of high heat would be at the region of the stationary ebitransits presence, with the remaining circuit of copper wire having no direction towards lower heat.

This situation immediately changes when a region of low heat becomes active anywhere along the circuit of the copper wire, be it at the furthest distance from where the stationary ebitransits are located. This region of low heat is where ebitransits are invited to enter, so as to be converted to 'dispersing energy' inunens and do 'work' by expanding their compressed bodies.

Once the region of low heat becomes active, two unique functions occur.

Firstly, the combined 'boundary of influence' of all copper atoms produces a 'transfer of influence' function, which comprises of showing the moving forward ebitransits the shortest path to the region of the lowest heat.

Secondly, and even more astonishingly, even as ebitransits begin to move forward at astonishing speed, but long before they can reach that region of low heat, the 'energy bodies' (electrons) at the region of low heat, begin to disperse, sending their disengaged 'uniting energy' inunens as ebitransits into the region of low heat, to be converted to 'dispersing energy' inunens.

This function has been called: 'movement by proxy'. This 'movement by proxy' occurs because the 'transfer of influence' function instigates that as the ebitransit forward movement is taking place, it becomes necessary that the region of low heat is filled, irrespective of where the moving ebitransits (as electrical current) may be.

While the entering ebitransits move towards the region of low heat – as directed by the 'transfer of influence' direction – this action is responsible for an overall stronger 'boundary of influence' that can attract other appropriate metals to it. This overall 'boundary of influence' is currently known as a 'magnetic field'.

Any ebitransits that are not converted to 'dispersing energy' inunens at the region of low heat, continue to move forward in a circuit, back towards the region where the initial ebitransits entered the copper wire.

Due to the functions of 'transfer of influence' and 'movement by proxy', it is possible for entering ebitransits to be where they are not. As a practical application of this, envisage that an electric light bulb is attached by a copper wire circuit to an electrical cell, or battery, with an on-off switch, and the distance between the light bulb and the electrical cell being that of a hundred miles. When the electric switch is turned 'on', the electric light bulb instantly lights up. And it does so simply because the 'energy bodies' (electrons) next to the light bulb (as the region of low heat) are attracted to disperse, and send the displaced 'uniting energy' inunens, as ebitransits, racing towards the light bulb as 'movement by proxy' (from the ebitransits in the battery cell), long before the ebitransits (electric current) from the battery cell, moving along the 'transfer of influence', reach the light bulb.

These principles of combined 'boundary of influence' of all relevant 'energy bodies' (electrons) and their ability to provide a 'transfer of influence' direction to ebitransits between the 'energy bodies' (electrons), on their way from region of high heat to that of low heat within a closed circuit, apply to both alternating current (AC) and direct current (DC) arrangements.

It may appear that just because all atom surfaces are covered by 'energy body' (electron) shells, and these 'energy body' (electron) shells are made up of 'uniting energy' inunens, that these 'uniting energy' inunens can be harvested from all atoms, so as to be used as ebitransits (electric currents). This is not the case.

The removal (or not) of 'uniting energy' inunens from the 'energy bodies' (electrons) depends on the strength of the 'boundary of influence' a nucleus structure develops and maintains, so as to manage the combined 'boundary of influence' of the 'energy body' (electron). This depends on the numbers of 'solid bodies' (protons), and 'combined bodies' (neutrons) forming the composition of a nucleus, and by that have either a stronger or a weaker 'boundary of influence' with which to prevent not just its closest 'energy body' (electron) shells from departing, but also the distant ones, as well.

This 'boundaries of influence' relationship between the nuclei and their 'energy body' (electron) shells is also responsible for the strength of the bonds maintained between atoms when they form molecules and compounds. The stronger the combined 'boundaries of influence' of atoms are, the closer these atoms come to each other, while the weaker combined 'boundaries of influence' result in weaker bonds between atoms.

The entities that can disrupt and weaken any 'boundary of influence' are the 'dispersing energy' inunens. The more and larger the volume of 'dispersing energy' of heat and light, the greater and faster is the reduction in the 'boundary of influence' of that which they contact. To illustrate this, one only needs to heat up any metal to see them melt. A heating of a magnet also shows how the internal 'frozen' 'transfer of intent' ceases to function.

Due to the Sun's constant stream of 'dispersing energy' of heat and light (which is the reason for Earth's overall temperature), most elements and molecules on Earth inevitably experience this 'dispersing energy' and are affected by it. With a decrease of this level of 'dispersing energy' inunens (which constantly produce 'dispersing energy' chain reactions), elements increase their gravitational attraction to each other as their 'boundaries of influence' enlarge and strengthen. With an increase of 'dispersing energy' the elements decrease their 'boundaries of influence'. This reduces and weakens their attractions, making it possible for solids to become liquids and even gasses. Water is a good example of this.

As a final point in regards to how 'energy body' (electron) shells maintain their 'transfer of influence' communication with all other 'energy body' (electron) shells, this concerns the stability and permanence of the 'transfer of influence'.

Many elements act as resistors to ebitransits (electrical current) because their atomic configuration does not facilitate their 'energy bodies' (electrons) shells to produce strong and distinct 'transfer of influence' direction to any ebitransits.

Other elements, such as iron (Fe), have an atomic structure whose 'electron body' (electron) shells provide very accommodating 'transfer of influence'. This is evident by how a passage of ebitransits (electrical current) through a metal can permanently 'freeze' the direction of 'transfer of influence'.

This kind of unmoving 'transfer of influence' produces a rigid polarity of heat direction: from high heat region towards the lower heat region. Such a combined rigid 'transfer of influence' has the ability to provide a more expanded and defined 'boundary of influence' attraction that is part of every permanent magnet, until the permanence of the static 'transfer of influence' is nullified by large volumes of 'dispersing energy', in a form of heat and light.

By representing the higher heat region with a positive (+) and the lowest heat region with a negative (−), it is possible to map and establish in any permanent magnet the regions where ebitransits (electrical current) had originally entered and where they had departed.

VISUALIZING ATOMS ('UNITED BODIES')

For all their so-called achievements derived by way of incalculable trial-and-error experiments, a physical reality remains that all those involved with current sciences are still in dark about how atoms function and how they look.

For instance, it is still considered that an 'energy body' (electron) is an all-negative particle and a wave that is structured as a cloud. It is therefore thought that an electron is a solid structure, rather than being a composite of countless 'uniting energy' inunens forming a mesh shell, with an internal and external directions of heat attraction producing an internal polarity. And the 'energy body' shell itself that can expand and contract according to the limitations of the combined 'boundary of influence' and the 'boundary of influence' of the nucleus itself.

According to current science, these electron clouds satellites maintain an order and distances away from their nucleus. These distances are referred to as: 'levels' and 'sub-levels', where each subsequent level (distance from the nucleus) comprises of two – and only two – electrons, while each separate sub-level within each extending level can comprise of more numerous groupings of electrons. Each electron of the more-distant-from-nucleus orbital level and sub-level positions had been calculated to have a higher energy level, which means that it would require less energy to attract more distant electrons away from their nucleus.

The separate levels had been classified as: 's' (for: sharp – referring to spectral lines), starting with s1 (which atoms of all elements have), then proceeding onto s2, s3, s4, and so on. Between each 's' level, there are numerous sub-levels, starting with 'p' (principal), then sublevels of 'd' (diffuse), 'f' (fundamental), and onto to: 'g', 'h', etc.

In such orbital configurations the electrons are supposedly able to jump to higher energy levels when the atoms supposedly receive extra energy (because currently 'energy' is thought to be a commodity which can be passed around with no ill or diminishing effects) and then return back to their original level – while releasing an equal amount of energy that was initially received, in a form of a photon (a photon being light, which is supposed to be either a particle or a wave).

An electron, now referred to as a 'cloud', supposedly take on shapes of a sphere, or a double doughnut, or a doughnut inside a larger doughnut, all depending on the electron's state of energy levels. Considering that all these fancy configurations are supposed to apply just to the humble Hydrogen (H), with its single electron, it is no wonder that with such a complexity no physicist is attempting to describe the appearance of an atom with multiple electrons, such as Uranium (U) with its 92 electrons.

Furthermore, it is also now thought that the electron clouds spin, and simultaneously, do not actually move, from being a stationery wave. An electron is even thought not be in any particular place when not being searcher for, so that it can only be located where it is not expected to be.

So what then, do the 'energy bodies' (electrons) shells actually look like?
'Energy body' (electron) structures actually are:

- Comprised of a net-like shell formed by 'uniting energy' inunens, which have an internal and external directions of heat attraction that produce an internal polarity, with the external direction of heat attraction being the external three-dimensional fields of 'boundary of influence' (gravitational or electro-magnetic attraction with a repulsion)...
- ... which can provide in certain elements a mutual external 'transfer of influence' (electric field) 'shiver' or 'wave' movements on the surface of shells, with each one moving in an opposite direction to that of its neighbor shell...
- ... where loose 'uniting energy' inunens are capable of moving forward in closed circuits as ebitransits (electricity current).

For all that, the actual shapes of atoms look different to all current perceptions. This is because, with exception of Hydrogen (H), the structures that 'energy bodies' (electrons) form comprise of two halves. With two halves making a whole, these structures strive to be balanced for their nuclei, despite them being symmetrically opposite units. There are physical reasons for that:

Firstly: these 'energy body' (electron) shells replicate the internal and external shapes of inunen bodies; that is: they produce an external bulge and an internal indent, so that internally and externally the direction of heat attraction apply in order to assist the nucleus in maintaining its central polarity.

Secondly: the structures that the 'energy body' (electron) shells form are defined by their intent to have as much symmetrical balance as possible.

Thirdly: these shells provide protection to the nuclei they encase from external contacts with 'uniting', 'united', and 'dispersing energies', and from escalations of dispersals within the internal space.

Fourthly: as with all functions that physically take place everywhere, they comprise of cycles, be those also called systems or circuits. All cycles are concerned with providing a balance. And that is exactly the concern of 'energy body' (electron) shells: a balance of internal space for their nucleus, by attempting to achieve a balanced opposing symmetry of internal space. The following diagram illustrates the three-dimensional external 'energy body' (electron) shells architecture of a small sample of elements.

PART 3. DISCLOSURE OF EVERYTHING

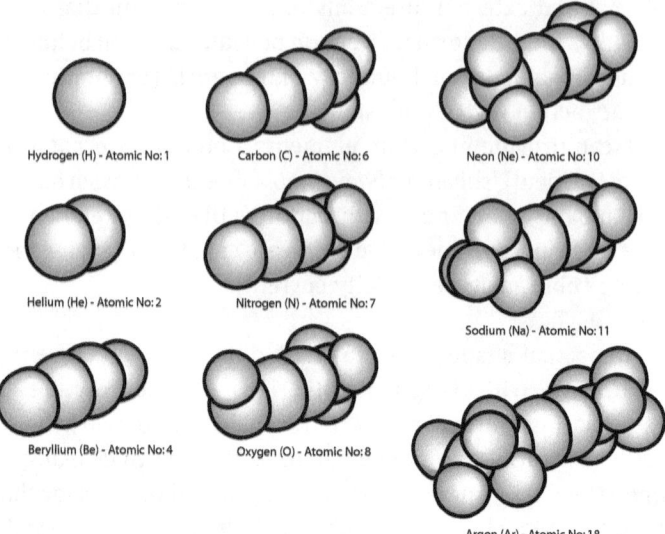

EXAMPLE OF 'ENERGY BODY' (ELECTRON) SHELLS STRUCTURES

What can be seen from this graphic is that the levels (the distances away from their nucleus) are divided into two symmetrical sides. Where the current understanding of electrons ('energy bodies') presumes that the levels and sub-levels take place simultaneously all around a nucleus, they are actually separated into two halves, with these halves uniting into symmetrically-opposite structures.

It is perfectly correct that all the levels 's' consist of two electrons ('energy bodies'), with the 'p' sub-levels capacity being that of six electrons ('energy bodies'), the 'd' sub-levels capacity of 10 electrons ('energy bodies'), and 'f' sub-levels capacity of 14 electrons ('energy bodies'). However, the way that the 'energy bodies' (electrons) actually come together is in entirely different formations. The architecture that the 'energy bodies' (electrons) form for their nuclei comprises of shells increasing in symmetry from one, then to three in each of the two 'p' sub-levels, then five in each of the two 'd' sub-levels, and finally seven in each of the two 'f' sublevels.

It is these increments of shell numbers, on both sides, that produce the difference in the physical behaviour of all elements. The more symmetrical and simpler the structure, the more it is stable. The more complex and unbalanced – the more it is unstable and even volatile, with a shorter half-life.

A BRIEF LOOK AT ALL PHYSICAL EXISTENCE

To provide more insight into 'energy body' (electron) structures, the following pages depict diagrams of elements in a flat-plan format, beginning with Hydrogen (H), with atomic number of one, all the way to Copernicium (Cn), with an atomic number of 112.

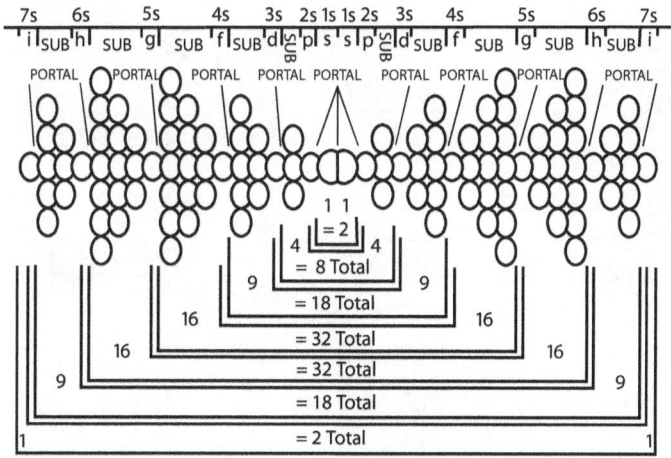

HOW TO READ THE TWO-DIMENSIONAL DIAGRAM OF COPERNICIUM (Cn) ELECTRON ('energy bodies') STRUCTURE: ONE OF THE LARGEST ELEMENTS IN THE PERIODIC TABLE

1s, 2s, 3s, etc. = SHELLS THAT ARE PORTALS TO SUB-LEVELS
s, p, d, etc. = LEVELS
SUB = SUB-LEVELS

Grand Total = 112 'energy body' (electron) shells, which is the Atomic Number of 112 for Copernicium (Cn)

PART 3. DISCLOSURE OF EVERYTHING

1: H: Hydrogen
1 : 1s¹
○

2: He: Helium
2 : 1s²
⊙

3: Li: Lithium
2, 1 : 2s¹
⊙○

4: Be: Beryllium
2, 2 : 2d²
⊙○○

5: B: Boron
2, 3 : 2s² 2p¹
⊙○○○

6: C: Carbon
2, 4 : 2s² 2p²
⊙○○○○

7: N: Nitrogen
2, 5 : 2s² 2p³

8: O: Oxygen
2, 6 : 2s² 2p⁴

9: F: Fluorine
2, 7 : 2s² 2p⁵

10: Ne: Neon
2, 8 : 2s² 2p⁶

11: Na: Sodium
2, 8, 1 : 3s¹

12: Mg: Magnesium
2, 8, 2 : 3s²

13: Al: Aluminum
2, 8, 3 : 3s² 3p¹

14: Si: Silicon
2, 8, 4 : 3s² 3p²

15: P: Phosphorus
2, 8, 5 : 3s² 3p³

16: S: Sulfur
2, 8, 6 : 3s² 3p⁴

17: Cl: Chlorine
2, 8, 7 : 3s² 3p⁵

18: Ar: Argon
2, 8, 8 : 3s² 3p⁶

19: K: Potassium
2, 8, 8, 1 : 4s¹

20: Ca: Calsium
2, 8, 8, 2 : 4s²

21: Sc: Scandium
2, 8, 9, 2 : 4d¹ 4s²

22: Ti: Titanium
2, 8, 10, 2 : 3d² 4s²

23: V: Vanadium
2, 8, 11, 2 : 3d³ 4s²

24: Cr: Chromium
2, 8, (13, 1) : 3d⁵ 4s¹

25: Mn: Manganese
2, 8, 13, 2 : 3d⁵ 4s²

26: Fe: Iron
2, 8, 14, 2 : 3d⁶ 4s²

27: Co: Cobalt
2, 8, 15, 2 : 3d⁷ 4s²

28: Ni: Nickel
2, 8, 16, 2 : 3d⁸ 4s²

29: Cu: Copper
2, 8, (18, 1) : 3d¹⁰ 4s¹

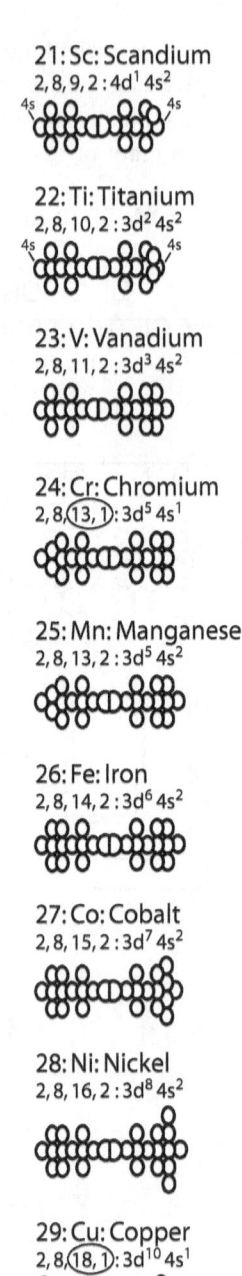

A BRIEF LOOK AT ALL PHYSICAL EXISTENCE

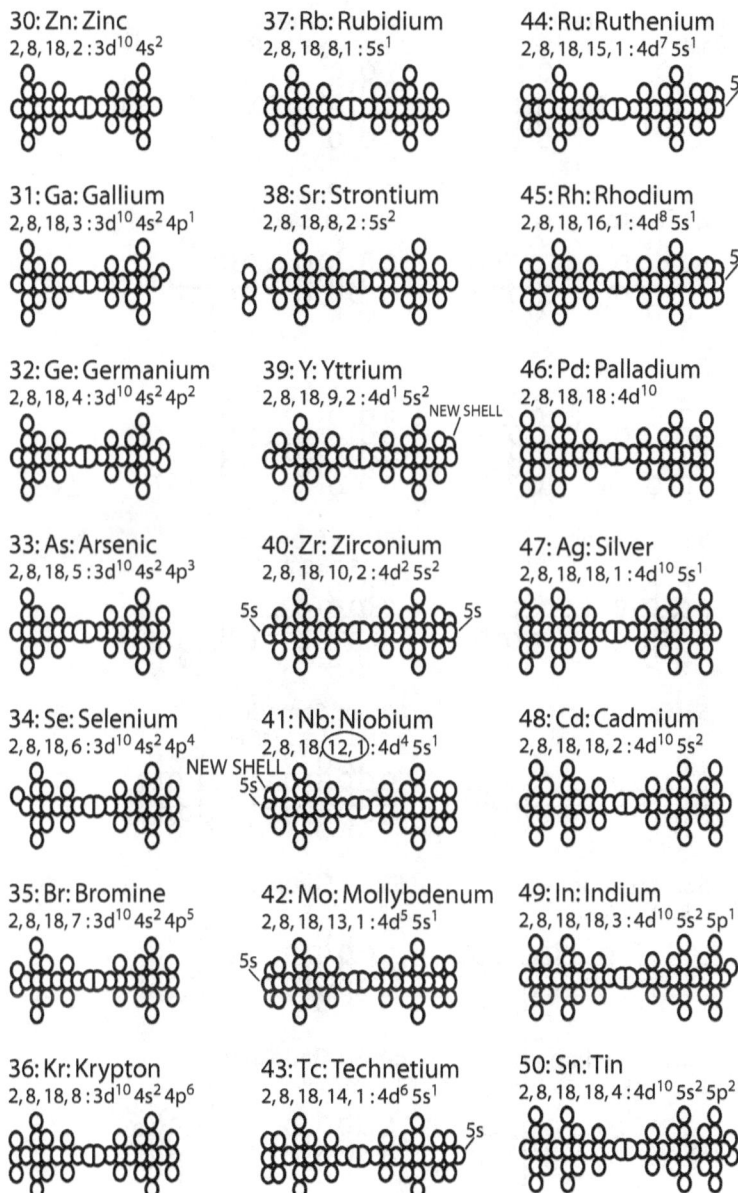

30: Zn: Zinc
2, 8, 18, 2 : $3d^{10} 4s^2$

31: Ga: Gallium
2, 8, 18, 3 : $3d^{10} 4s^2 4p^1$

32: Ge: Germanium
2, 8, 18, 4 : $3d^{10} 4s^2 4p^2$

33: As: Arsenic
2, 8, 18, 5 : $3d^{10} 4s^2 4p^3$

34: Se: Selenium
2, 8, 18, 6 : $3d^{10} 4s^2 4p^4$

35: Br: Bromine
2, 8, 18, 7 : $3d^{10} 4s^2 4p^5$

36: Kr: Krypton
2, 8, 18, 8 : $3d^{10} 4s^2 4p^6$

37: Rb: Rubidium
2, 8, 18, 8, 1 : $5s^1$

38: Sr: Strontium
2, 8, 18, 8, 2 : $5s^2$

39: Y: Yttrium
2, 8, 18, 9, 2 : $4d^1 5s^2$ NEW SHELL

40: Zr: Zirconium
2, 8, 18, 10, 2 : $4d^2 5s^2$

41: Nb: Niobium
2, 8, 18, (12, 1) : $4d^4 5s^1$ NEW SHELL

42: Mo: Mollybdenum
2, 8, 18, 13, 1 : $4d^5 5s^1$

43: Tc: Technetium
2, 8, 18, 14, 1 : $4d^6 5s^1$

44: Ru: Ruthenium
2, 8, 18, 15, 1 : $4d^7 5s^1$

45: Rh: Rhodium
2, 8, 18, 16, 1 : $4d^8 5s^1$

46: Pd: Palladium
2, 8, 18, 18 : $4d^{10}$

47: Ag: Silver
2, 8, 18, 18, 1 : $4d^{10} 5s^1$

48: Cd: Cadmium
2, 8, 18, 18, 2 : $4d^{10} 5s^2$

49: In: Indium
2, 8, 18, 18, 3 : $4d^{10} 5s^2 5p^1$

50: Sn: Tin
2, 8, 18, 18, 4 : $4d^{10} 5s^2 5p^2$

PART 3. DISCLOSURE OF EVERYTHING

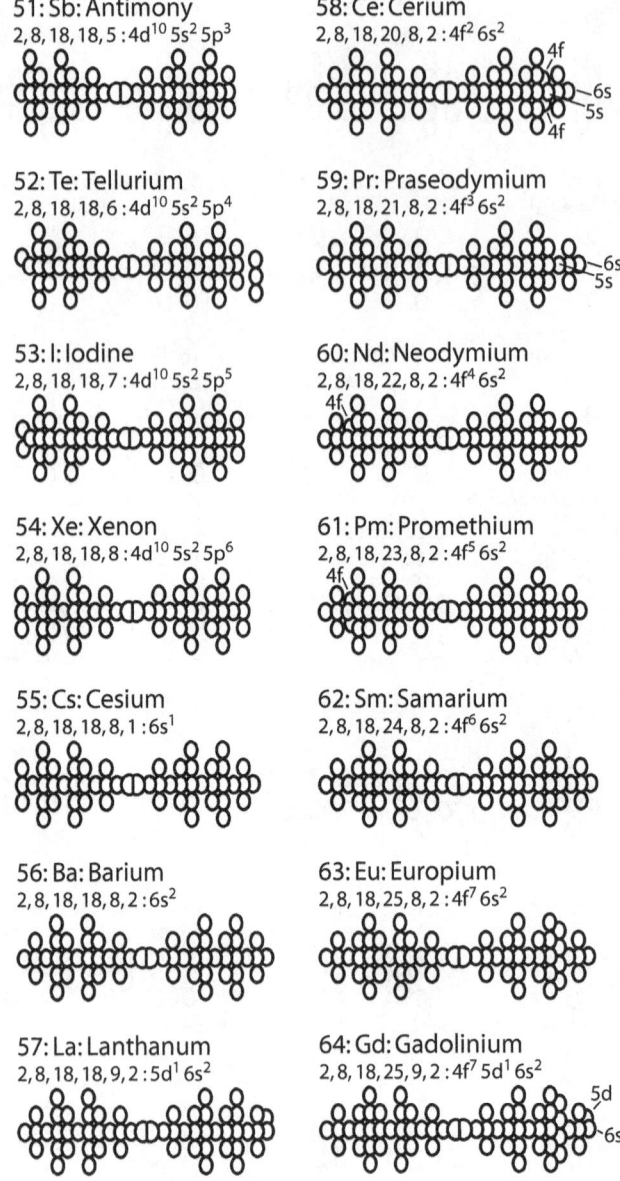

51: Sb: Antimony
2, 8, 18, 18, 5 : $4d^{10}\ 5s^2\ 5p^3$

52: Te: Tellurium
2, 8, 18, 18, 6 : $4d^{10}\ 5s^2\ 5p^4$

53: I: Iodine
2, 8, 18, 18, 7 : $4d^{10}\ 5s^2\ 5p^5$

54: Xe: Xenon
2, 8, 18, 18, 8 : $4d^{10}\ 5s^2\ 5p^6$

55: Cs: Cesium
2, 8, 18, 18, 8, 1 : $6s^1$

56: Ba: Barium
2, 8, 18, 18, 8, 2 : $6s^2$

57: La: Lanthanum
2, 8, 18, 18, 9, 2 : $5d^1\ 6s^2$

58: Ce: Cerium
2, 8, 18, 20, 8, 2 : $4f^2\ 6s^2$

59: Pr: Praseodymium
2, 8, 18, 21, 8, 2 : $4f^3\ 6s^2$

60: Nd: Neodymium
2, 8, 18, 22, 8, 2 : $4f^4\ 6s^2$

61: Pm: Promethium
2, 8, 18, 23, 8, 2 : $4f^5\ 6s^2$

62: Sm: Samarium
2, 8, 18, 24, 8, 2 : $4f^6\ 6s^2$

63: Eu: Europium
2, 8, 18, 25, 8, 2 : $4f^7\ 6s^2$

64: Gd: Gadolinium
2, 8, 18, 25, 9, 2 : $4f^7\ 5d^1\ 6s^2$

A BRIEF LOOK AT ALL PHYSICAL EXISTENCE

65: Tb: Terbium
2, 8, 18, 27, 8, 2 : $4f^9\, 6s^2$

66: Dy: Dysprosium
2, 8, 18, 28, 8, 2 : $4f^{10}\, 6s^2$

67: Ho: Holmium
2, 8, 18, 29, 8, 2 : $4f^{11}\, 6s^2$

68: Er: Erbium
2, 8, 18, 30, 8, 2 : $4f^{12}\, 6s^2$

69: Tm: Thulium
2, 8, 18, 31, 8, 2 : $4f^{13}\, 6s^2$

70: Yb: Ytterbium
2, 8, 18, 32, 8, 2 : $4f^{14}\, 6s^2$

71: Lu: Lutetium
2, 8, 18, 32, 9, 2 : $4f^{14}\, 5d^1\, 6s^2$

72: Hf: Hafnium
2, 8, 18, 32, 10, 2 : $4f^{14}\, 5d^2\, 6s^2$

73: Ta: Tantalum
2, 8, 18, 32, 11, 2 : $4f^{14}\, 5d^3\, 6s^2$

74: W: Tungsten
2, 8, 18, 32, 12, 2 : $4f^{14}\, 5d^4\, 6s^2$

75: Re: Rhenium
2, 8, 18, 32, 13, 2 : $4f^{14}\, 5d^5\, 6s^2$

76: Os: Osmium
2, 8, 18, 32, 14, 2 : $4f^{14}\, 5d^6\, 6s^2$

77: Ir: Iridium
2, 8, 18, 32, 15, 2 : $4f^{14}\, 5d^7\, 6s^2$

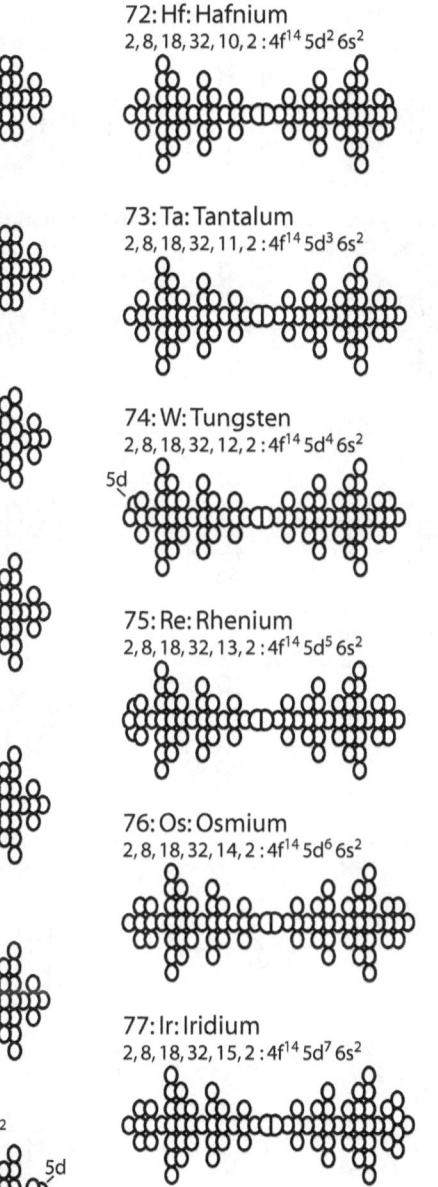

PART 3. DISCLOSURE OF EVERYTHING

78: Pt: Platinum
2, 8, 18, 32,(17, 1) :4f^{14} 5d^9 6s^1

79: Au: Gold
2, 8, 18, 32, 18, 1 :4f^{14} 5d^{10} 6s^1

80: Hg: Mercury
2, 8, 18, 32, 18, 2 :4f^{14} 5d^{10} 6s^2

81: Tl: Thallium
2, 8, 18, 32, 18, 3 :4f^{14} 5d^{10} 6s^2 6p^1

82: Pb: Lead
2, 8, 18, 32, 18, 4 :4f^{14} 5d^{10} 6s^2 6p^2

83: Bi: Bismuth
2, 8, 18, 32, 18, 5 :4f^{14} 5d^{10} 6s^2 6p^3

84: Po: Polonium
2, 8, 18, 32, 18, 6 :4f^{14} 5d^{10} 6s^2 6p^4

85: At: Astatine
2, 8, 18, 32, 18, 7 :4f^{14} 5d^{10} 6s^2 6p^5

86: Rn: Radon
2, 8, 18, 32, 18, 8 :4f^{14} 5d^{10} 6s^2 6p^6

87: Fr: Francium
2, 8, 18, 32, 18, 8, 1 :7s^1

88: Ra: Radium
2, 8, 18, 32, 18, 8, 2 :7s^2

89: Ac: Actinium
2, 8, 18, 32, 18, 9, 2 :6d^1 7s^2

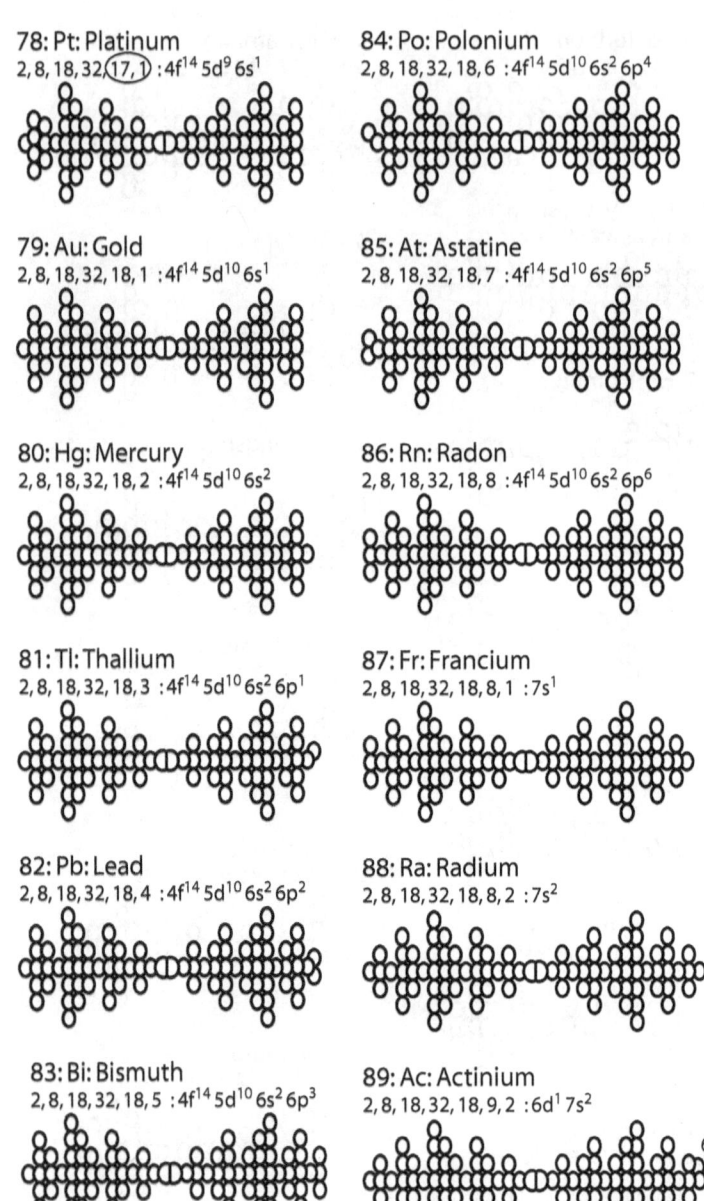

310

A BRIEF LOOK AT ALL PHYSICAL EXISTENCE

90: Th: Thorium
2, 8, 18, 32, 18, 10, 2 : $6d^2\ 7s^2$

91: Pa: Protactinium
2, 8, 18, 32, 20, 9, 2 : $5f^2\ 6d^1\ 7s^2$

92: U: Uranium
2, 8, 18, 32, 21, 9, 2 : $5f^3\ 6d^1\ 7s^2$

93: Np: Neptunium
2, 8, 18, 32, 22, 9, 2 : $5f^4\ 6d^1\ 7s^2$

94: Pu: Plutonium
2, 8, 18, 32, 24, 8, 2 : $5f^6\ 7s^2$

95: Am: Americium
2, 8, 18, 32, 25, 8, 2 : $5f^7\ 7s^2$

96: Cm: Curium
2, 8, 18, 32, 25, 9, 2 : $5f^7\ 6d^1\ 7s^2$

97: Bk: Berkelium
2, 8, 18, 32, 27, 8, 2 : $5f^9\ 7s^2$

98: Cf: Californium
2, 8, 18, 32, 28, 8, 2 : $5f^{10}\ 7s^2$

99: Es: Einsteinium
2, 8, 18, 32, 29, 8, 2 : $5f^{11}\ 7s^2$

100: Fm: Fermium
2, 8, 18, 32, 30, 8, 2 : $5f^{12}\ 7s^2$

101: Md: Mendelevium
2, 8, 18, 32, 31, 8, 2 : $5f^{13}\ 7s^2$

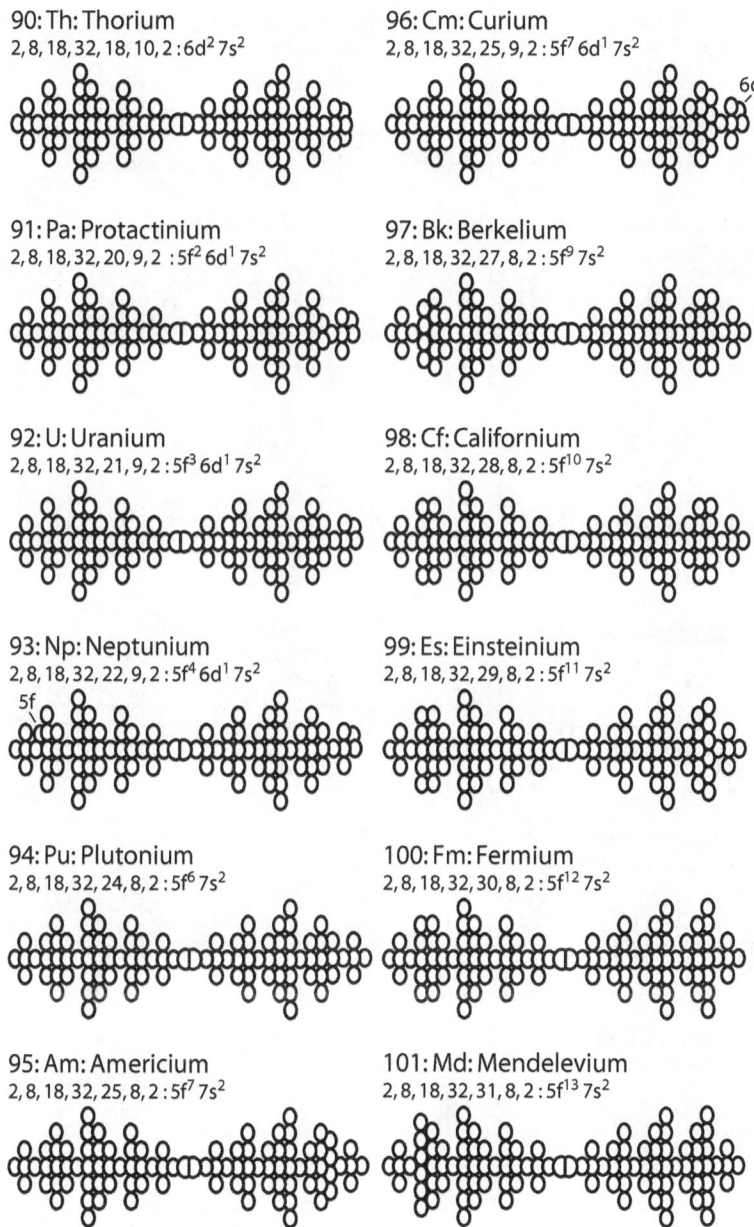

PART 3. DISCLOSURE OF EVERYTHING

102: No: Nobelium
2, 8, 18, 32, 32, 8, 2 : $5f^{14} 7s^2$
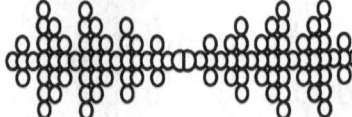

103: Lr: Lawrencium
2, 8, 18, 32, 32, 9, 2 : $5f^{14} 6d^1 7s^2$
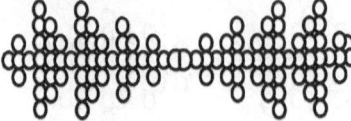

104: Rf: Rutherfordium
2, 8, 18, 32, 32, 10, 2 : $5f^{14} 6d^2 7s^2$
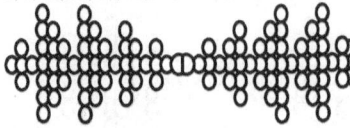

105: Db: Dubnium
2, 8, 18, 32, 32, 11, 2 : $5f^{14} 6d^3 7s^2$
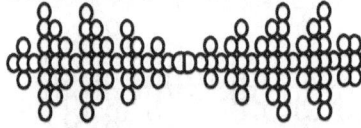

106: Sg: Seaborgium
2, 8, 18, 32, 32, 12, 2 : $5f^{14} 6d^4 7s^2$
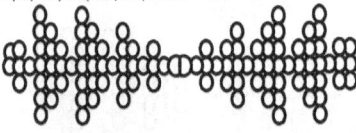

107: Bh: Bohrium
2, 8, 18, 32, 32, 13, 2 : $5f^{14} 6d^5 7s^2$
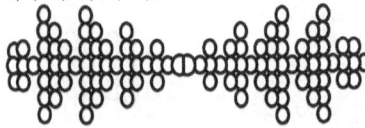

108: Ha: Hassium
2, 8, 18, 32, 32, 14, 2 : $5f^{14} 6d^6 7s^2$
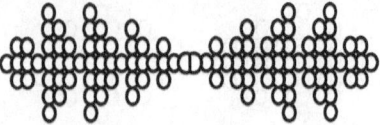

109: Mt: Meitnerium
2, 8, 18, 32, 32, 15, 2 : $5f^{14} 6d^7 7s^2$
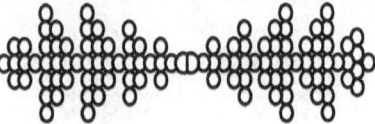

110: Ds: Darmstadtium
2, 8, 18, 32, 32, 16, 2 : $5f^{14} 6d^8 7s^2$
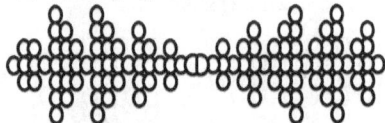

111: Rg: Roentgenium
2, 8, 18, 32, 32, 17, 2 : $5f^{14} 6d^9 7s^2$
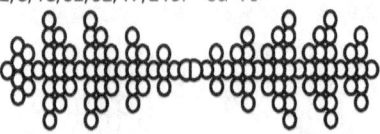

112: Cm: Copernicium
2, 8, 18, 32, 32, 18, 2 : $5f^{14} 6d^{10} 7s^2$
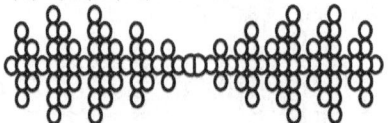

A BRIEF LOOK AT ALL PHYSICAL EXISTENCE

What has to be understood about all the listed elements is that a large number of them do not freely exist in nature on this planet. Even gasses like Hydrogen (H) and metals like Aluminum (Al) have to be obtained as by-products of manufacturing and chemical processes, while other – even more complex and larger elements – are actually produced in laboratories, and due to these factors have a very short half-life. But despite all the artificially produced element structures, there are still many combinations of 'energy body' (electron) structures that have not been realized.

In the following diagram just two of such structural compositions are shown, together with a representation of just how complex and large the structures of some elements can be – if only for a short while.

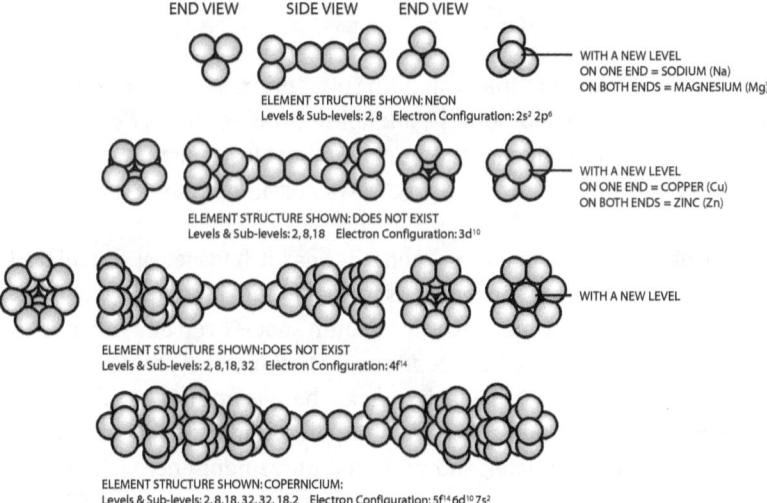

The purpose of all the 'energy body' (electron) structures is to produce an internal space, within which a nucleus resides. This internal space is necessary for the nucleus to keep itself away from needless exposure to any outside dispersing activities, to other nuclei, and to facilitate its movements within the internal space. By such means a nucleus attempts to extend its existence for as long as possible, no different to any cell or bacteria, which actually replicate the structures of atoms from which they are made.

Using an Oxygen (O) 'energy body' (electron) structure as an example, the following diagram depicts its internal space and the influences that take place in it.

PART 3. DISCLOSURE OF EVERYTHING

INTERNAL STRUCTURE OF AN OXYGEN (O)
(atomic no. 8)

THIS SIDE: 1 'p'
= INCOMPLETE SUB-LEVEL

'S' LEVELS = 'ENERGY BODY' (ELECTRON) SHELLS
FUNCTION AS PORTALS

THIS SIDE: 3 'p'
= COMPLETE SUB-LEVEL

'p' no. 4

s2   s1   s1   s2

'p' no. 1
'p' no. 2

NUCLEUS

'p' no. 3

+]>>– = INTERNAL DIRECTION OF HEAT ATTRACTION
⎯⎯ = CORRIDOR PASSAGE THROUGH 'ENERGY BODY' (ELECTRON) SHELL PORTALS
+ – = INTERNAL AND EXTERNAL POLARITIES

According to current understanding of atoms, a nucleus comprises of all-positive protons ('solid bodies') and all-neutral neutrons ('combined bodies'), grouped together like a little bunch of marbles, remaining stationery in one spot, while being surrounded, in one way or another, by a number of electrons ('energy bodies').

The problem with such atom model is that it had never considered that there is nothing to prevent any nucleus from moving in vacuum space, as nothing stands still. If everything moves in vacuum space, even our galaxy, and with it our Sun, why would there be exceptions for the nuclei of atoms? There would be none, as there are none. In which case, as there is nothing to prevent the nuclei from moving in vacuum space, then the surrounding electrons would need to follow their moving nuclei, and with that movement encroach onto the rotating electrons of other atoms; causing constant mayhem from collisions. The result of such behaviour would be constant sporadic dispersals taking place everywhere, with nothing, anywhere, being stable. This is not what takes place in physical existence.

The factor that allows any combined mass to remain as combined mass, is the fact that the 'energy bodies' (electrons) of atoms control the distance from their surrounding neighbors by means of their 'boundaries of influence', which holds them together with attraction, while holding them apart by repulsion. The same 'energy body' (electron) shells encase their nucleus within their internal space, allowing their nuclei to move in that internal space along the corridors provided by the portals between the shells. And with each 'energy body' (electron) shell having an internal concave indent, which directs all heat directions inwards: (+]>>–', these internal, inward directions of heat attraction prevent the nuclei from coming too close to the 'energy body' (electron) shells. By such means a nucleus is free to move about within the internal space of its combined 'energy body' (electron) structure, while remaining encased.

The more an 'energy body' (electron) shell structure is symmetrically composed and uncomplicated, the calmer and more static the nucleus is, while a higher degree of dissymmetry and complexity in a shell structure causes the larger nucleus to be more agitated in constant movements within the internal space of its shell domes.

In managing the unavoidable need to move about, nuclei do not constantly remain in a bunch, but can also unfurl themselves into double helix spirals, by means of the rough-surfaced spherical 'solid bodies' (protons) using their slightly larger smooth-surfaced spherical 'combined bodies' (neutrons) as hinges, which they hold together. Such change of form assists nuclei to produce a more defining polarity with which to move within the internal space, and to provide maintenance to the 'energy body' (electron) shells.

However, for many element structures the composition of their nucleus can be as unbalanced as their 'energy body' (electron) structure, as shown by the following graphic illustrating an Oxygen (O) nucleus.

OXYGEN (O) NUCLEUS STRUCTURE

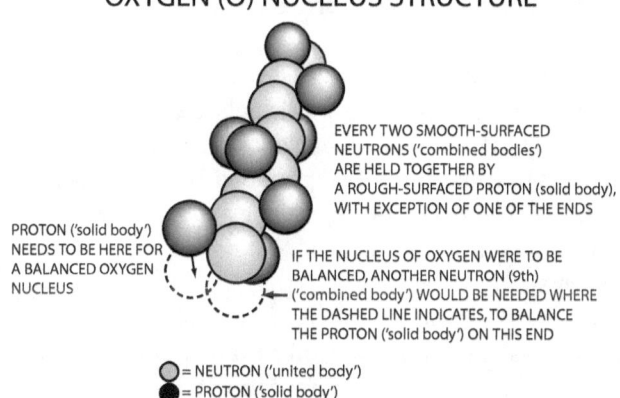

What can be seen in the Oxygen nucleus is that the 'solid body' (proton), at one of the ends of the group, has no other 'combined body' (neutron) to hold, and so it remains unfulfilled from being just outside of direction of heat attraction flow, as shown by the following illustration.

PART 3. DISCLOSURE OF EVERYTHING

POLARITY OF OXYGEN (O) NUCLEUS STRUCTURE

PROTON ('solid body')
ON THIS END
IS NOT ALIGNED
WITH THE OVERALL
DIRECTION OF HEAT
ATTRACTION,
CAUSING A BREAK IN
THE SYMMETRY
OF THIS NUCLEUS

OVERALL CIRCUIT DIRECTION OF
OXYGEN (O) NUCLEUS' POLARITY
BETWEEN THE '+' and '−' ENDS,
WHICH PRODUCES THE OVERALL
'BOUNDARY OF INFLUENCE'
(ELECTROMAGNETIC FIELD)

○ = NEUTRON ('united body')
● = PROTON ('solid body')

By having little idea of how atoms and elements actually look like, the current method of depicting chemical structures continues to be that of little colored balls. But atoms are not little colored balls. They are not merely spheres of different size, but are varied physical structures, and it is these structures – with their relevant internal and external directions of heat movement that provide the reason why they, as chemicals, behave as they do.

Take water, for instance: H2O. Currently always depicted as two Hydrogen balls attached to a single Oxygen ball. This kind of depiction shows that there is nothing complex about such structure. But then take a proper look at the structure of a water molecule.

Two Hydrogen (H) atoms, each comprising of a single 'solid body' (proton) encased in a single 'energy body' (electron) shell, is simple enough.

But then there is the element of single Oxygen (O), which is no colored ball, as it comprises of a nucleus with eight 'solid bodies' (protons), and eight 'combined bodies' (neutrons), encased in eight 'energy body' (electron) shells. Everything eight: and seemingly balanced. But not according to how elements compose themselves.

The structure of the nucleus, with its eight 'solid bodies' (protons) and eight combined bodies' (neutron), is balanced and placid. But not the structure of the 'energy body' (electron) shells. First comes the structure of two central 'energy body' (electron) shells of two symmetrical halves of level s1, which provide an internal portal for the nucleus. But from there the symmetry is not retained. Despite that there are further two 'energy body' (electron) shells of level s2, on either side of those on level s1, there is a complete sub-level s2 set of three 'energy bodies' (electrons), and on the opposite side there is only one sub-level 2 'energy body' (electron) shell. Such 'energy body' (electron) shell composition can now be clearly seen as being out of symmetry. One end of the

A BRIEF LOOK AT ALL PHYSICAL EXISTENCE

Oxygen atom is obviously in need of two 'energy bodies' (electrons), so as to balance out its structure.

But what for? What is so important for Oxygen to have a symmetrically-opposite balanced structure?

The reason for that is both external and internal. Having a lopsided external shape with an unbalanced internal nucleus structure results in Oxygen having a poor 'transfer of intent' and a lopsided 'boundary of influence'. This overall imbalance provides Oxygen with volatility of response to contacting 'dispersing energy' of heat. This is why Oxygen 'assists' any action of burning, while that combustion is surrounded by Oxygen.

An addition of two 'energy bodies' (electrons) and one 'combined body' (neutron) would resolve that problem for Oxygen, and provide a balanced polarity – but then this would produce an atomic structure that does not exist.

Oxygen structural imbalance could be solved externally by an addition of two separate Hydrogen atoms, or a single Helium atom, with its two 'energy body' (electron) shells. Helium, however, with its two 'solid bodies' (protons) and two 'combined bodies' (neutrons), within two 'energy bodies' (electron) shells is perfectly symmetrical and balanced, and so has no need to bond with any other elements.

So this leaves only Hydrogen atoms to fulfill an Oxygen's external requirement of a more balanced 'boundary of influence', by combining its 'boundary of influence' with that the two Hydrogen atoms, even if the internal space will not achieve a balance. The following diagram illustrates how the two Hydrogen atoms become united with an Oxygen atom, which results in a water molecule.

ELECTRON ('energy body') STRUCTURE OF WATER
Oxygen (O) + 2 Hydrogens (H)

Presumed water structure

A water molecule is presumed to be a tetrahedron, with two Hydrogen atoms being present at two sides of the Oxygen atom.

Actual water structure

2 separate Hydrogens

Oxygen Structure: comprising of 8 combined 'energy body' (Electron) shells

The covalent bonds between Hydrogen and Oxygen are not very strong at the relatively high levels of 'dispersing energy' experienced at 'room temperature' or even cooler. This results in water not always sustaining a constant bond with both Hydrogen atoms.

PART 3. DISCLOSURE OF EVERYTHING

Of course, it is not just elements like Oxygen that utilise the support of Hydrogen atoms in their desire to rectify the lack of symmetry and balance in their natural structure. The following diagram shows a structure for Methane, a very common type of gas, which results from Carbon attracting four Hydrogen atoms in a covalent bond.

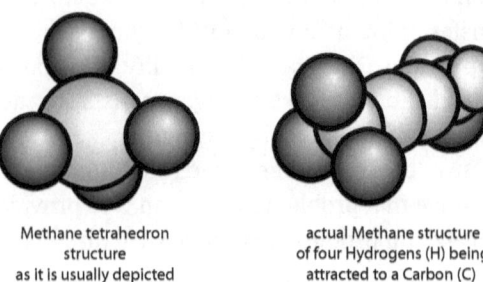

Methane tetrahedron
structure
as it is usually depicted

actual Methane structure
of four Hydrogens (H) being
attracted to a Carbon (C)

When the levels of 'dispersing energy' drop within the spaces between the 'energy body' (electron) shells, elements and molecules increase their levels of 'boundary of influence', so that they, as compounds, come closer together, or 'shrink'. This applies equally to water. And yet, in freezing, the water expands its combined mass as ice. The reason for that is due to the Hydrogen atoms in the H2O combination, experiencing a strengthening of a mutual intersecting polarity between them and Oxygen, brought about by the increase of their 'boundary of influence'. This results in a more pronounced bi-polarity between the two Hydrogens towards other Oxygen pairs, as the following diagram illustrates.

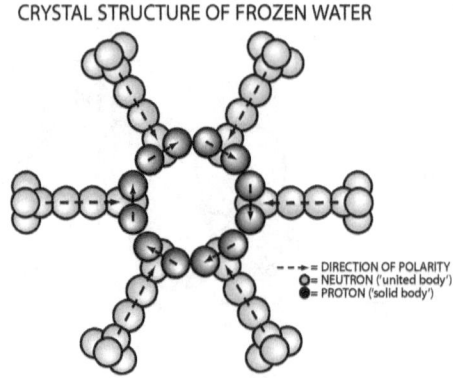

CRYSTAL STRUCTURE OF FROZEN WATER

- - - + = DIRECTION OF POLARITY
○ = NEUTRON ('united body')
● = PROTON ('solid body')

Such intersecting bi-polarity provides an incentive for Hydrogen atoms of individual molecules to merge. As the levels of 'dispersing energy' decreases (that is: as it freezes) the Hydrogens lock into circuits, usually in hexagon formations, although other formations are also possible. As the water molecules adopt such opened-out rigid structures, their formations may be thin layered but they nevertheless manage to expand the volume of water as ice. Being made of Oxygen, being surrounded by Oxygen as air, and having a great deal of Oxygen atoms within it, allows frozen water to be supported by water surface gravity (to float on it).

But what of the so-called chemical bonds, which supposed to be a strong force attracting and holding atoms together in molecules, achieved by means of transferring or sharing of electrons?

While current science may refer to bonds between atoms and molecules as electromagnetic fields occurring between the electrically charged all-negative electrons and the all-positive protons in the nuclei, the only physical means by which separate atoms form compounds and combine to produce molecules are by the internal and external directions of heat attraction, which produce 'boundaries of influence'.

In some unions between different type of elements, as with water for example, the two Hydrogens uniting with a single Oxygen combine their separate 'transfers of influence' into a single movement across all their combined structure, no different to that of their combined 'boundaries of influence', which results in water having a double polarity. This kind of bonding is currently classified as 'covalent'.

Other kind of bonding between elements is known as 'ionic' bonding, an example of which is shown below.

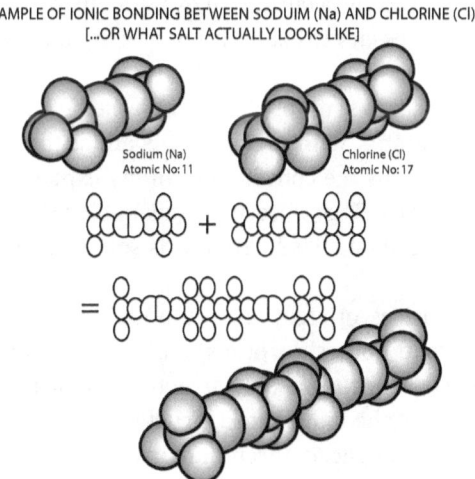

EXAMPLE OF IONIC BONDING BETWEEN SODUIM (Na) AND CHLORINE (Cl)
[...OR WHAT SALT ACTUALLY LOOKS LIKE]

Sodium (Na) Atomic No: 11

Chlorine (Cl) Atomic No: 17

This is when two separate elements – due to their mutual 'boundary of influence' end-attractions – manage to exchange one or more of their 'energy body' (electron) shells between them for their mutually improved symmetry and balance. While continuing to maintain independent external 'transfers of influence, their individual 'boundaries of influence' become mutually combined, and this holds the two newly structured elements together, but without touching. The result of such 'energy body' (electron) shells transformation produces an altered interior architecture for both of the unchanged nuclei and their overall exterior 'boundary of influence'. This kind of chemical change can transform, for instance, the physical attributes of poisonous Sodium metal and Chlorine gas into harmless common salt.

According to current understanding of chemistry, chemical bonding is supposed to take place between two, or more, individual atomic element structures directly involved with formation of a molecule. Taking water as an example, it is supposed to be a formation produced from a covalent bonding between two Hydrogen atoms and one Oxygen atom. Pure and simple! No other considerations are given as to why the two Hydrogens remain with the Oxygen, considering that the H2O can only remain as water while being surrounded by Oxygen, the very same Oxygen that makes it possible for combustion to continue while being surrounded by Oxygen.

Furthermore, while water is surrounded by Oxygen, it maintains its own combined 'boundary of influence', regardless of Earth's 'boundary of influence' (gravitation). And from that, it can be said that while and where water is surrounded by Oxygen, its surface forms its own gravity. While currently unheard of, this is perfectly physically accurate. To explain this so that it all makes perfect sense, it is first necessary to examine Oxygen in more detail.

While some 20% of Earth's atmosphere comprises of Dioxygen, which is a union of two Oxygen atoms, with a formula O2, and is the colorless and odorless gas we all breathe, for the convenience of my explanations I have used a singular Oxygen structure. Therefore, it is reasonable to assume that in reality the influence of Dioxygen on all it surrounds is somewhat more substantial to that of just Oxygen.

Oxygen (O) comprises of eight 'energy body' (electron) shells, encasing a nucleus of eight 'solid bodies' (protons) and eight 'combined bodies' (neutrons). As explained earlier, this nucleus structure has an imbalanced direction of heat attraction, while the imbalanced structure of 'energy body' (electron) shells produces an imbalanced three-dimensional 'boundary of influence' (electromagnetic field), as shown on the following page.

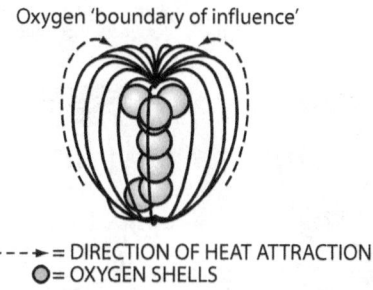

‑ ‑ ‑ ▸ = DIRECTION OF HEAT ATTRACTION
◎ = OXYGEN SHELLS

The results of all these imbalances means that the Oxygen external direction of heat attraction is constricted at the region of exit and rapidly spreads out on returning to the region of re-entry. This represents a relatively weak 'boundary of influence', because it is stymied throughout its external movement. There is also a small 'boundary of influence' protrusion at the constricted end of the element.

As it is uneven from one end to the other, the Oxygen 'boundary of influence' causes the whole Oxygen structure to wobble as the uneven sides attempt to attract other similarly unbalanced 'boundaries of influence' of other Oxygen atom. This results in Oxygen atoms being unable to form cohesive combined mass with each other unlike that of water, where for instance, inside a space station, outside Earth's 'boundary's of influence' (influence of gravity), water will form a floating combined mass (due to its own gravity) while the Oxygen filling the interior of the space station remains spread throughout, without concentration in any particular region.

This evenly-spaced detachment between Oxygen atoms means that they are very good conductors of 'dispersing energy'; allowing 'dispersing energy' of heat and light to be easily spread by them, using their 'transfer of intent'. This is why the transfer of Sun's heat is felt more prominently near the ground, at Oxygen level, rather than in the higher, freezing region of Nitrogen surrounding the Earth, despite it being closer to the Sun, and therefore, by all rights should be hotter. This is also the reason why the presence of Oxygen around a fire will encourage it, while an elimination of surrounding Oxygen will stifle the fire.

But then these Oxygen properties become absolutely reversed once two Hydrogen atoms become connected to it. Why? From what reason would Oxygen – as part of water – extinguish combustion when added to a fire, instead of maintaining it as Oxygen?

Part of the reason for that is in the combined 'boundary of influence' that H2O produces, as shown on the next page.

PART 3. DISCLOSURE OF EVERYTHING

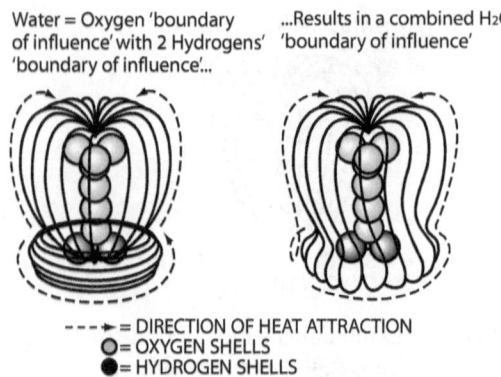

Water = Oxygen 'boundary of influence' with 2 Hydrogens' 'boundary of influence'...

...Results in a combined H_2O 'boundary of influence'

- - - ▶ = DIRECTION OF HEAT ATTRACTION
○ = OXYGEN SHELLS
● = HYDROGEN SHELLS

But why then is this combined 'boundary of influence' is only a part of the reason for water behaving as water on Earth?

Because, while the uniform polarity of two Hydrogens – in unison with the single sublevel 'd' 'energy body' (electron) shell – provide a correction to the unbalanced Oxygen 'boundary of influence'. This allows the molecules of water to have a weak, but sufficiently strong enough 'boundary of influence' to attract and temporarily hold each other, but not sufficient for water molecules to hold onto their two Hydrogens, especially at the surface. The only factor that allows the surface of water molecules to retain their state is the presence of the surrounding Oxygen. Without surrounding Oxygen water molecules become unstable, incapable of retaining the two Hydrogens, and in that process appearing to boil, as the Oxygen and Hydrogens separate, breaking their bonds while converting to 'dispersing energy' of heat. This is why water cannot remain as water in vacuum space. It disperses while some remnants of it freeze.

How then does water remain as water?

Water molecules under the thin very top layer of water are very capable of remaining as water molecules, as each higher layer of water presses down on those beneath it, in response to the Earth's 'boundary of influence' (influence of gravitation fields). But the very top layer of water has nothing to prevent them from losing their Hydrogens, if not for the surrounding Oxygen layer, which blankets the Earth's surface. It is these Oxygen atoms that make it possible for water to remain water.

This Oxygen, surrounding the water molecules, does not actually succeed in containing Hydrogens in water by pressing water down. Instead, Oxygen combines its 'boundary of influence' with that of the water molecules, so that its added attraction to a water molecule provides a combined 'boundary of influence' sufficiently strong enough to hold the Hydrogens in place, in a manner shown by the following diagram.

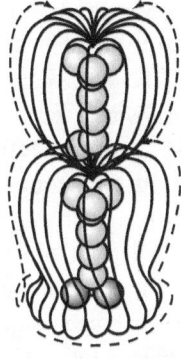

At the boundary between the Oxygen layer and the water layer Oxygen 'boundary of influence' comes in contact with a water molecule's H_2O combined 'boundary of influence'...

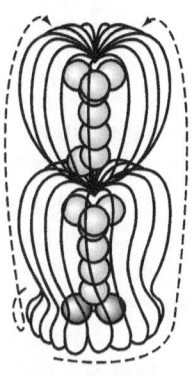

...producing a temporary, larger, combined 'boundary of influence', which assists the Oxygen in an H_2O water molecule to hold onto its 2 Hydrogens

---→ = DIRECTION OF HEAT ATTRACTION
○ = OXYGEN SHELLS
● = HYDROGEN SHELLS

There is, however, much more to the process of Oxygen atoms and those of water molecules becoming momentarily attracted, and in doing so produce a combined 'boundary of influence' that allows the molecules of water to retain their bond with their Hydrogens.

The first of these is the ability of the combined Oxygen and water molecule 'boundary of influence' to make a stronger, ongoing – if always temporary – unions with other combined Oxygen and water molecule 'boundaries of influence' that form all surfaces of water; and secondly, to have all of these combined Oxygen and water molecule 'boundaries of influence' directed inwards, at right angle to any surface of water, as an inward direction of heat attraction.

To simplify this, it can be said that water has a strong surface unity (currently referred to as 'surface tension'), due to the surrounding Oxygen and its own gravity.

Yes, indeed, these surface properties of water – including the refraction of light, cohesion, surface unity (surface tension), and adhesion – are brought about by nothing else but the relationship between Oxygen in the atmosphere and the Oxygen of the water molecules producing a water gravity. The physical proof of this is evident everywhere there is water on and off this planet.

For example, the water gravity can be witnessed in a large spherical mass of water floating inside a space station (outside Earth's influence of gravity ['boundary of influence']), and in every falling spherical droplet of water within Earth's 'boundary of influence' (gravity); in every curving-on-the-underside ocean wave; rounded from all sides running water; and in every water surface that is always rounded; while always being surrounded by Oxygen.

Even the bubbles on the water-surface are formations of combined Oxygen and water molecules 'boundaries of influence' coming from both the surface and from within the Oxygen bubble, producing an overall, thin film-layered round shell that remains temporarily on top of water molecules.

The combustion-eliminating ability of water is totally due to its water-surface 'membrane'. There are two factors to this. The first factor, is that the water-surface 'membrane' – temporary as it is constant – with its strong-enough, combined, side 'boundaries of influence' (providing strong surface unity), covers the flames with itself as a sheet, preventing the Oxygen atoms above it to be contacted by the 'dispersing energy' of combustion beneath it. The second factor is that the combined 'boundary of influence' of Oxygen and water molecules is stronger than that of individual Oxygen atoms. Being stronger means its ability repel advancing 'dispersing energy' inunens is also stronger. This slowing down of approaching 'dispersing energy' inunens result in its responsive 'transfer of intent', with its 'dispersal code' being slower.

When these two factors are applied to combustion, the 'dispersing energy' inunens, producing combustion, fail to reach surrounding Oxygen so as to be transferred in continuum, and the rate and speed of their 'transfers of intent' are substantially reduced by the water-surface 'membrane'. The outcome is that all these physical restrictions make it very difficult for the process of combustion to continue.

This water-surface 'membrane' is the reason why shower water feels firm, while bath water – beneath the water-surface 'crust' – feels soft. For there is a distinct difference between them, due to the below-the-water-surface water molecules – having a weak combined 'boundary of influence' – behaving as Oxygen bodies moving freely on their two Hydrogens, as if on hinges; and on many occasions temporarily losing them. This makes the below-the-water-surface water more pliable: more influenced by the Earth's 'boundary of influence' (gravitation).

The water's inward-directed 'transfer of influence' that influences its 'boundary of influence' (gravity) can also be seen in water's behaviour as water, despite the overall influence of the Earth's internal and external 'boundary of influence' (gravitation).

Unlike the Earth's above-the-ground cover of Oxygen, or rather Dioxygen,

A BRIEF LOOK AT ALL PHYSICAL EXISTENCE

which allows acceleration to a falling object, the two-Hydrogens-one-Oxygen combination of water behaves differently, despite experiencing the same influence from the Earth's gravity as that of Oxygen.

It may seem as incredible that the difference of behaviour between water and Oxygen comprises of two, minute Hydrogen atoms. But that is exactly the reason of water being so much 'thicker' than air. The addition of Hydrogens to Oxygen produces a more defining 'boundary of influence' not just to every water molecule, but to the grouped 'boundary of influence' (water gravity) of all combined water molecules as a whole, wherever water is amassed on the planet's surface or as water vapor held up by the Oxygen layers, forming clouds in the sky.

For these reasons water prevents the acceleration to objects moving from its surface towards the 'bottom' of earth's surface underwater.

In actual fact, because of its united 'boundary of influence', which provides a so-called 'weight' to its mass, it actually hinders any progression to its depths, despite the increasing pressure, as each layer of water presses down on those below them.

This can easily witnessed with constancy of 'air bubbles' rising from any depth to the surface. While this expression seems logical, lighter-than-water gasses rising by themselves through water, this is just another inaccuracy.

'Air bubbles' or gasses do not rise to the surface of water: they are actually 'pushed up' or 'lifted' by water molecules. When there are grouped gasses in water, the surrounding water molecules attracted inwards by their combined 'boundary of influence' (gravitation) and the Earth's gravitation, move down beneath these gasses, and by doing so provide an elevation to these gasses with their own physical presence. And because of constancy to their 'boundary of influence' (gravitation), the gasses are raised by the water molecules at a constant rate, irrespective of water's depth or pressure.

In visualizing atoms and other structures they compose, either with or without human (conscio) involvement, it seems inconceivable that current science continues to view atomic and chemical structures as some simple colored balls. They either ignore or dismiss the shear complexity and intricacy of atoms' external and internal architecture that causes their particular behaviour.

For instance, right now it is still accepted in science that the Moon's 'boundary of influence' (gravitation) can influence a minute tide in a glass of water, considering that it is been also accepted by all humans (conscios) that the Moon controls the ocean tides.

And yet, how can it be explained by current science that low tides are possible when both the Sun and the Moon are directly overhead? Or why it is possible for an ocean tide to be extremely low at the height of a Solar Eclipse, where both the Moon and the Sun are directly overhead?

FROM INTELLIGENCE TO LIFE

Water on Earth, with its surrounding layer of Oxygen, has another particular attribute that no other element or molecule has: it is, and was capable of not just supporting life but actually becoming life.

But how did life come about, anyway?

The most fundamental intelligence is that of inunens – the individual units of energy. This intelligence is not that of reasoning, but it does comprise of knowledge of function, of response to function of others, and memory of self. This intelligence allows inunens to know what to do when it is expected of them, at the instant of a physical contact with another inunen, and as 'dispersing energy' inunens to calculate their response to that physical contact. The use of memory allows them to know what structure they are part of, so that when the moment arrives for them to depart as 'dispersing energy' they know precisely what kind of distinct forward movement they should make.

The next step to intelligence is the combined intelligence of all inunens forming a complete element, whatever that atom may be. This combined element intelligence comprises not so much as of an understanding of survival but of knowing ways of extending their united existence. A 'united body' (atom) intelligence of the combined whole structure understands how to replicate the internal and external directions of heat attraction, so as to have some control over any approaching 'dispersing energy'. The same applies to any internal, departing 'dispersing energy'. To do this, the elements utilise 'energy body' (electron) shells to protect their nuclei from exposure to external 'dispersing energy', and from rapid deterioration caused by internal 'dispersing energy' inunens.

As 'dispersing energy' is unavoidable (considering that everything must eventually end up becoming 'dispersing energy' and 'dispersed energy') some molecules, more than others – because of their structural combinations – ended up using their combined intelligence to become more adapt at utilizing 'dispersing energy' to their advantage.

On Earth, water molecules turned out to be the most successful at this, not because of water's own structure, but from the relationship it could have with the surrounding Oxygen atoms and the relatively manageable levels and quantities of 'dispersing energy' coming from the Sun and from within Earth.

The Oxygen and water had filtered and evened-out the Sun's 'dispersing energy'. The structure of water caused any penetrating 'dispersing energy' to do so at a refracted angle, slowing it down and allowing the moving 'boundary of influence' of water molecules to pass it around, without actually making immediate physical contacts. This kind of 'dispersing energy' capture is not unique to water molecules, but the ability of water molecules to use it for a gradual, long term 'transfer of intent' is.

As a result of such ability to control 'dispersing energy', heat could be drawn down with water molecules, who are attracted inwards by their 'boundary of influence (gravitation and the Earth's 'boundary of influence' [gravitation]), without this causing a great deal of dispersing disturbance within water.

A relatively warm temperature of water had meant that the 'boundaries of influence' of all other chemicals in water were weak, with this circumstance making it possible that various flexible individual chemicals, such as Hydrogen (H), Nitrogen (N), Carbon (C), and Oxygen (O), could end up being inside an Oxygen bubble within water. Many such bubbles – all of them having the surface-water 'membrane' (where the internal side of the bubble is a surface no different to that of the region between water and the Oxygen layer above it) – would 'cook' the chemicals within.

A chemical reaction always requires an impetus of 'dispersing energy' of heat (or light) to produce 'effort' of 'work' of stripping the 'energy body' (electron) shells and exposing the nucleus within. Such 'dispersing energy' impetus produces a reaction of further dispersals – providing more effort of work – when the nuclei begin an ongoing cycle of converting their 'solid bodies' (protons), and 'combined bodies' (neutrons) into 'dispersing energy', while the remaining 'energy body' (electron) shells collapse into being unattached 'uniting energy' inunens, which can move in circuits as ebitransits (electricity). The result of these activities could alter the original chemicals into new structures, such as amino acids.

These chemicals undergoing physical change within the air bubbles of Oxygen in the water eventually devised an 'understanding' that to extend their duration of activity they would require an encasement more durable than that of water-surface 'membrane'. So that is what come to be, when the early chemicals learned how to replace the temporary water-surface 'membrane' with a more permanent boundary, within which they could store an ongoing supply of chemicals, with some of these being converted into ebitransits (electricity) and 'dispersing energy' heat cycles; and by such physical processes becoming short-span early life forms.

PART 3. DISCLOSURE OF EVERYTHING

By gradual efforts of attempting to exist for longer periods, the initial temporary life forms had made an evolutionary leap: they resolved to replicate themselves by division, instead of continuing to be a brief-living primitive one cell organisms. But even this form of sustained existence by division was often temporary.

Eventually a more permanent existence was achieved by cell-dividing organisms. This was done by a development of a chemical representation, with which all the membrane and internal contents were precisely recorded in sugar of their protein, as a 'transfer of life' code, and passed on from one dividing cell to the next. Such early 'transfer of life' code eventually evolved (that is: developed by life forms) into what is currently classified as DNA (Deoxyribonucleic Acid): a double helix shaped nucleic acid structure, along and between which chemical codes are attached to instigate specific chemical functions that lead to a reproduced internal watery space bound by a membrane, to replicate its maker.

The shortcoming of reproduction by exact duplication – known as cloning – consists of absence of diversity, which is necessary for life's adaptation to a new environment. This, the early living cells resolved by two solutions.

The first of these was to unite the intelligence of all single dividing cells into one unit, with that united structure being able to 'think' as a unit. These multiplying-through-division living cells developed into life forms currently classified as bacteria. These bacteria had developed a sophisticated intelligence of adapting to different environments by evolving their ability to consume different 'united energy' chemicals surrounding them, so as to convert them into 'dispersing energy', the process of which is the stimulus of life and reproduction. And should bacteria find themselves without further nutrients of foreign cells to consume (that is: without the necessary water and 'dispersing energy' of heat), their intelligence instructs them to stop multiplying, so as to remain dormant for thousands – if not millions – of years.

Every bacterium (all of which comprise of atoms formed of inunens) have achieved this ability of survival – and continue to do so – by their own intelligence; without the assistance or involvement of anything but the most basic of elements of salts and minerals, water, and the surrounding 'dispersing energy', mainly coming from the Sun.

The bacterial intelligence had brought about the second, more advanced method of reproductive duplication. This method comprised of two separate bacteria merging their separate 'transfer of life' codes into one depository, with that container holding and merging the two separate 'transfer of life' codes into one new composition. This new 'transfer of life' composition would then result in new physical structure, which could go on to deposit its 'transfer of life' code so as to be mixed again with other 'transfer of life' codes.

With such technique, a new process of gender reproduction came into existence on Earth, replicating the behaviour of atoms and 'dispersing energy' inunens, where 'dispersing energy' inunens penetrate the 'energy body (electron) shells in order to reach the nucleus. This led to a reproductive system where one cell could physically penetrate another cell, at which instant the two separate chromosomes unite and produce a new diverse trait, suited better for its existence.

From this point reproduction became divided into two factions: the body-brains and the brain-bodies.

The body-brains were those organisms whose total body structure maintained an intelligence derived from all the cells forming that body. These combined-into-one-intelligence cells (basically comprising of water) were content to rely on external 'dispersing energies' of heat and light for their main form of sustenance, together with supplemented nourishment obtained from chemicals and water. But with such living process being incapable of obtaining large quantities of 'uniting' and 'united energy' inunens, so as to be converted to 'dispersing energy' in order to provide 'effort' of 'work', this has meant that such body-brains could never increase their capacity for higher intelligence. And so, as body-brains, these life forms chose the path of either bacteria, or as phytoplankton and algae that evolved into marine grasses and plants, and then the land-based flora.

Alternatively, the brain-bodies had resolved to produce unions of water-based cells, which would form groupings dedicated to producing specific calculations. All these separate groupings would also be united into one single unit, within which the separate groupings would communicate with each other by means of ebitransits (electric current) code pulsing, moving along connecting passages of nerve cells with synapses. These combined groupings – forming a single structure of a brain – would then decide the method by which they would move in space not just once, as seeds and pollen do (always dependent for that on currents and winds, etc.) but at any moment when there was a need to do so.

Such earliest brains were minute and very basic. Nonetheless, they had managed to produce for themselves bodies of their own design, with which to move their bodies, so as to acquire sustenance to feed the bodies and them, the brains.

These brain-bodies never discarded 'transfer of life' body designs. Instead they reapplied them when they felt it necessary. This frugal attitude had gradually allowed brain-bodies to develop eyesight for their body structures. The process of eyesight came about from the efforts of earlier life forms (such as shellfish) attempting to feed on the 'dispersing energy' of sunlight penetrating water.

In everything that the brain-bodies devised for themselves and their bodies, they had always replicated the inunen structures and the 'united body' (atom) structures. Their own shape, as brains, is reminiscent of 'expanding energy' inunen shape, with the divided balanced symmetry present in 'energy body' (electron) shell structures. In fact, every body that brains had ever produced have comprised of this divided symmetry, be that the skeletal structures, the organ structures, and the symmetrical balance of opposite-sided limbs, horn or teeth protrusions.

Despite all the evolutionary developments undertaken by the brain-bodies and the body-brains, one physical fact had always been present for all of their existence: all of if depended and depends on water, Oxygen, and constant, moderate amounts of 'dispersing energy'. Without these main ingredients there would have been no chance for any body-brains and brain-bodies to come into physical existence, irrespective of the fantasies humans (conscios) devised for themselves, or care to believe.

THE ABSOLUTE OF ALL ACTIVE PHYSICAL EXISTENCE

Humans think themselves very clever simply due to having some intelligence. Indeed, the intelligence of their brains is most impressive. Not that they ever use a great deal of it, as their conscios obstruct that intelligence with interference of their 'wants' and 'desires'. Which is a pity, because were they to reduce their dependence of wanting immediate fulfillment of their wants, their brains could have led them to empowered and wondrous lives. As it is, conscios rule: this resulting in all the deceptions by which humanity lives and suffers.

The problem with humans (conscios), including those in sciences, is that they tend to believe that everything that surrounds them has to be complicated. If it is straightforward and relatively uncomplicated they refuse to take it seriously.

For them energy has to involve all that is fanciful and non-existent, rather than be that individual units of energy (inunens), which are eternal, cannot be penetrated, and change with phases within singe cycles. This energy description is evident to anyone who cares to look about, and recognise inunen presence in, and as, everything that physically exists and undergoes change in the physical process of change.

For that reason, should energy, as inunens, require an expression of an equation, then that equation would need to have a consideration of everything: the three eternal constants comprising of the vacuum space of Eternity, physical process of change, and the simultaneity of all physical actions, which are responsible for all the Energy and Matter Absorbing and Compressing Systems (EAMAACS), and all the Energy and Matter Attracting and Releasing Systems (EAMAARS); and even the lives of each and every human on this planet.

Perhaps that equation could begin with the expression that applies to all active physical existence: [H]>>h(or: (+]>>–(

This can be further simplified to its most fundamental function, which represents all physical existence as a physical process of change, where all the physical structures are combined (united) with all the physical actions (dispersals):

$$E = \left(\!\left(\pm\right)\!\right)$$

This equation, therefore, reads: energy 'E' equals "to union, to dispersal, in a repeating cycle, with all of this being a three-dimensional activity."

Where: the 'up' curved arrow represents a direction to 'union', while the 'down' curved arrow represents the direction to 'dispersal'; where the plus '+' sign represents 'union', while the minus '–' sign represents 'dispersal; and the surrounding curved arrows represent an eternal cycle.

If this expression of energy makes a poor equation, it, nonetheless, is the very absolute expression of all active three-dimensional physical existence.

DIRECTIONS OF HUMAN SPACE TRAVEL

In order to complete this chapter of explaining of the basics of all physical existence, here is some information to show that we, humans (conscios), reside on this planet as human space navigators.
Living blissfully on this planet, we, humans, have little physical sense of where we are. Oh, sure, we know that we live on a planet that rotates and orbits our Sun; which is part of a Milky Way galaxy; which is part of this expanding Universe. But we do not feel this. Thanks to the Earth's strong polarities that produce a substantial 'boundary of influence' (gravitation) – which is responsible

PART 3. DISCLOSURE OF EVERYTHING

for Earth's 'boundary of influence' within a 'boundary of influence' (electromagnetic field), by which a great deal of that 'dispersing energy' is reflected and deflected – there is no physical sensation of what is actually occurring to us.

But just because humans feel no change of movement that they constantly experience, this does not mean that their change of location-in-space is not taking place – for it does.

Therefore, here is an outline to convey some directions of space travel, in which all humans (conscios) participate. Some space travels are beyond human control, and by that are involuntary for humans. Others are caused by intentional human actions, as part of human ability to navigate space – and from that are voluntary forms of navigation.

- **First direction of space travel**

All the 'uniting energy' expelled at the dispersal of the EAMAACS (Energy and Matter Absorbing and Compressing Systems) – currently referred to incorrectly as Big Bang – in the Great Dispersal, shall continue to move out three-dimensionally into vacuum space of Eternity, away from the location-in-space of the Great Dispersal. Everything physical – that is this Universe – which had been formed from inunens since the , will continue to constantly radiate outwards into the vacuum space of Eternity, at an accelerating rate of velocity – that acceleration originating with the impetus of the Great Dispersal. This is the first involuntary direction of all human space travel.

- **Second direction of space travel**

After quintillions of Earth years, galaxies – comprising of EAMAARS (Energy and Matter Attracting and Releasing Systems) surrounding central EAMAACS (Energy and Matter Absorbing and Compressing Systems) – came into existence. The total sum of these galaxies is called the Universe. All galaxies in the region of the Universe had developed their own individual directions of space travel. The EAMAACS – the makers of galaxies – may be caused by the structure of their galaxy to spin and tumble. So apart from the overall outward direction caused by the Great Dispersal, they also move in various, haphazard, secondary directions, causing many collisions and ricochets between galaxies, which further alter their individual directions.

Our Milky Way galaxy is an enormous collection of some billion of billion of EAMAARS (stars, and suns, and sun systems), resembling a bright vortex of glowing, spiraling arms. While the Milky Way galaxy is dispersing as a unit, while being absorbed by an EAMAACS at its centre, it is moving at over a million miles an hour in its own specific direction as well as away from the centre of the Great Dispersal. This is the second involuntary direction of space travel that humans undergo, due to their Milky Way galaxy.

- **Third direction of space travel**

Apart from the individual direction of the Milky Way galaxy, the entire collection of individual stars, suns, and other particles and objects also rotate as a unit around the central EAMAACS.

As the galaxies rotate about the EAMAACS, they also cause attractions among each other, forming into groups that appear as individual tentacle arms of the whole vortex. The outer ends of the tentacle arms are moving slower than those at the centre where an EAMAACS acts as a pivot, absorbing the energy and matter of the galaxy. Eventually, the tentacles of a galaxy will wrap themselves, as veils, around each other, into a mass of hundreds of billions of stars and suns surrounding the EAMAACS. After quadrillions of Earth years, the tentacles may again reappear, as the stars on the periphery are pushed out by repulsions from the centre heat, repeating the vortex appearance.

In the present moment, while the Milky Way galaxy still remains in its present shape, the unified spin of its whole structure provides the third involuntary form of space travel for our whole Solar System; a movement of millions of miles an hour.

- **Fourth and Fifth directions of space travel**

Without going into how our Sun is influenced in its direction within the whole Milky Way galaxy by other surrounding stars – which is the fourth, involuntary, direction of space travel – the Earth revolves about the Sun at close to a million miles a day, presently taking 365 and a quarter Earth days to complete an orbit. This movement about the Sun is the fifth, involuntary, space travel that humans experience every moment of their lives.

- **Sixth direction of space travel**

As the Earth orbits the Sun it also spins daily on its geomagnetic axis. What this means is that while orbiting the Sun at an odd angle that produce seasons on the planet, the planet also rotates on its north-pole-south-pole axis of polarity, resulting in the day and night cycles. This, too, is an involuntary direction of space travel.

- **Seventh direction of space travel**

Humans are mobile. They have a physical ability to use their body as means of moving from one location to another. This mobility is a voluntary seventh direction of human space travel, where space is not that of vacuum but the expanse of Earth's surface.

- **Eighth direction of space travel**

While humans can move about freely on their legs they can also choose to utilize animals and vehicles of transport to speed-up the transfer of their bodies, from one location to another (as navigators in space). With some modes of transport they are restricted in movement, where they virtually merge their body with that of their transport vehicle, and by that remain in the seventh direction of space travel.

However, there are also the large vehicles of transport, such as: buses, ships, trains, and large aircraft, all of which allow human passengers to retain their personal mobility of the seventh direction of space travel while being able to physically move in a separate direction within that space. Such transport vehicles allow humans to experience an eighth direction of space travel.

These eight directions of human space travel represent various processes of dispersion, as dispersion is actually what occurs to all the objects in vacuum space. They also indicate that any physical object or a body in existence – be this a human or the Sun – are all subject to space travel due to the influence of enormous 'uniting', 'united', 'dispersing', and 'dispersed energies' that are them and surround them and their ongoing physical movements in space of Eternity.

All these voluntary and involuntary movements in space signify some very important facts:

1. In the first place, all these eight involuntary and voluntary movements in space signify that in all physical existence, all movements in space are all caused by physical actions and reactions that are unions and dispersals of physical objects and bodies in space. Therefore, it is these physical influences that compel everything physical to constantly move, as they do, and not any supernatural entity of a god, or spirit.

2. The constant movement of everything physical in vacuum space means that no physical object or body can ever remain at the same stationary point. This indicates that there is no fixed permanency to anything in physical existence, where even objects on Earth are in constant involuntary movement.

3. While all movement of objects and bodies in vacuum space, and even on Earth, may have variables of individual velocity, all these actions of movement

take place simultaneously, no matter where they may be. Therefore, everything, everywhere, occurs in the instant of present-moment: not before and not after.

4. No doubt that the credit to the eight movements in space would be given – especially by those in science of physics – to an entity of 'time'. The fact of the matter is that 'time', just like any god, has no involvement from having no physical existence.

This claim may surprise if not shock many of those who believe in existence of 'time'. So in order to provide a compelling explanation of a present-moment, it is necessary to debunk one of the greatest deceptions that all humans (conscios) hold so dearly in folly.

25

When science deceives as much as religion

DECEPTION OF 'TIME' AND 'TIME TRAVEL'

There is one important factor that humans (conscios) should be fully aware of: they all live with notions earlier humans invented. Despite that these notions have nothing to do with physical realities that constantly surround them, humans chose – and continue – to reject the physical realities, preferring to embrace the notions obtained from their ancestor's imaginations. Not only are all cultures and religions are based on these human brain invented notions for their conscios, but much of science as well.

It must also be understood that these invented notions are not some miniscule superstitions, but are wide and far-reaching directives on whose premise one human generation after another continues to uphold what they presume to be factual absolutes that constitute physical reality.

Such human self-devised deceptions, be they religious or scientific, do not harm physical existence. After all, their beliefs in physically impossible entities and events are not, in any way, influencing all that is the physical existence. Physical existence has no need of naïve human fabrications to continue behaving as it had, does, and always shall. And any erroneous scientific make-beliefs that discard physical reality, with which humans attempt to explain physical behaviour of physical existence, shall never make physical reality be what it is not, and can never be. Nor, it can be said, do human fantasy beliefs harm many individuals directly. What harm they do bring is to humanity as a whole. But not just to humans – as a species – but to nearly all species on the planet.

The basic means with which harm is done is by having their deceptions of physical reality hide from view all the physical limitations that apply to all that represents physical existence. This means that while humans choose to ignore and avoid understanding all the physical limitations that apply to them and this planet – just as it applies to all physical existence – they shall go on deceiving themselves into believing that they can continue to overpopulate and utterly exploit this planet, without this directly impacting Earth and all future generations of all life forms.

It is for this reason that beginning with their scientific nonsense, all their main self-deceptions should be exposed and shown for what they are: conscio-desired absurdity.

Before confronting current scientific and religious beliefs, it is worthwhile to present an example of just how irrationally humans (conscios) have treated the history of their own existence.

As members of human species, they have existed on Earth for around two million years. Of that, they have existed as members of Homo Sapient Sapient branch of the species for some tens of thousands of years. And yet, the current calendar indicates their latest date of existence as that of a twenty first century. Just over two thousand years. The reason for this is because a 6th century Catholic Church 'Spiritual Male' monk, Dionysius Exiguus, had instructed other 'Spiritual Males' that the beginning of Christian calendar be based on birth of Christ Jesus, (another human brain invention of a fictitious spiritual idol who remains revered as a god). In doing so, these 'Spiritual Males' had assigned all of human existence before that date into a reversed order, simply because all humanity born prior were classified as 'pagans', living, presumably in the BCE (Before Common Era), whereas all those living after the year 'zero' would be doing so in CE (Common Era or Current Era).

It would have been an enormous effort for all the historical texts and records to be recalculated, revised and rewritten. After all, no prior historical records calculated dates backwards.

The Egyptians, for instance, did not record the early days of their country's existence as that of six thousand BCE running down to zero. Instead, the progression of their lives and culture was recorded accumulatively, as is done these days, with one year being added to those of the past.

Nor did the Greek and Roman historians record their leaders' and emperors' births to be larger in number to that of their deaths. All the recorded dates were accumulative according to their calendars. Gaius Julius Caesar, for instance, did not live 100 to 44 BCE, but in the progressive calendar dates of that period. No one was living towards year 'zero', which was but a human invention.

Even the events and lives occurring before the Exiguus pronouncements, in the so-called 6th century, were all recorded progressively, according to practice of that period, as they were from the beginning of Roman Empire, without being divided into BC (BCE) and AD (CE). That is: dividing the so-called 'time' of human existence into two segments of 'ancient' and 'modern' or 'current'.

And yet, this disfigurement of historical human existence was obeyed. Now this 'Spiritual Male' human intervention in human progression is accepted without any question or consideration, as if this is how it had always been. Even those in science and historical research accept this historical dating divide as a natural order, without referring to how the ancient scholars actually recorded their historical dates.

The above example is not an isolated case of human intentional insertion of their opinion into the actual physical actions of this planet or the whole cosmos. This results in an acceptance of illogical and unreasonable view of all that should be seen as logical and reasonable.

For example, when a Roman Emperor desired to be considered a god, and be allocated an additional day to the month of his birth, this was done by removing a day from the month of February. To the very present, the calendar in use has failed to be devised on a realistic distribution of days in a year, but on how earlier 'Dominant Males' and 'Spiritual Males' had dictated their will to their subordinates, with these fallacies never again being questioned by logic and reason.

This is exactly how and why humans (conscios) have come to accept all their other delusions, including that of 'time'.

In physical reality 'time' does not exist just like, for instance, a mile and a kilometer do not physically exist. The physical land surface on Earth exists, which may be measured in miles and kilometers, but the measurements of miles and kilometers are only human-devised forms of accepted units of measure: a convention which represents a specific physical distance between two separate physical points in space. In the same way, kilograms or pounds do not physically exist, for they are also human inventions of measure: in this case of weight, based on Earth's internal and external 'boundary of influence' (gravity).

'Time' is therefore but a conventional devise – derived from Earth's ongoing daily exposure to the Sun, and its yearly orbits of the Sun – by which it is possible for humans to maintain two functions: a management device to manage human daily activities on Earth, and a navigational devise to manage human locations on Earth.

As a management devise it assists by indicating to humans what they should be doing in their life at any particular period of a day (or night), and gives indication of how long they spend performing a task or a function. These are possibly expressed as: "Time for this, and time for that," and the likes of: "I spent a lot of time in the garden."

As navigational devise, every period of a day (or night) provide indication to human life of locations (on Earth – be that the same city, or the even the same house but different rooms) where they are required to perform some tasks or functions at any particular periods, throughout the day. These circumstances are possibly expressed as: "I have to be at such and such address, for my appointment at 10 o'clock," or, "On that date we shall be on the other side of the world, in New York."

Such management and navigational assistance is provided by mechanical and digital devices, in the form of clocks and watches.

Similar directive and navigational devises are used for human activities requiring not just daily but also seasonal assistance, by means of yearly calendars.

This stipulates that all the units of years, days, hours, minutes and seconds are not meant to define some mysterious and intangible 'time' but that of human daily lives dedicated to spending segments of it performing for some periods some tasks and functions, at some locations.

There is a vital factor that applies to all this human based activity that enjoys the assistance of Earth's relationship with the Sun for its daily life management: it is all, with no exceptions, derived from human choices, with no influence of any 'time'.

Whether the decisions are based on group or individual choices, everything that humans do, whenever and where they choose to do it, and for how long, depends on human decisions and choices, and no other entity. There are no gods or 'time' that tell humans when, where, and what to do, and for how long they want to do it. Only humans themselves are responsible for this. They may be influenced by other humans in forming their choices, but this still remains a human interaction and involvement amongst humans and not that of 'time'.

This very principle applies not just to the process of human lives but to Earth itself, in its relationship with the Sun.

While improbable, it is not impossible that the Earth's daily rotations around the Sun may possibly alter. Should the physical process of change cause the Earth to dramatically speed up or slow down its daily rotations – to make, let us say, a complete daily rotation in 23 hours, or 25 hours, instead of the usual 24 hours – then undoubtedly, by mutual consent for convenience of calculations, humans would alter the scale of their 'time' measurements. The 23 or 25 hour Earth day would be reconfigured back to 24-hour units.

This would not mean that 'time' on Earth had sped up or slowed down, but that some physical change in their gravitational 'grappled attraction' has caused a slowing down, or speeding up of Earth's daily rotations. Were the

PART 3. DISCLOSURE OF EVERYTHING

planet's daily rotations to slow down or speed up, then that would probably influence the calendar measurements, where the Earth's calendar would be represented by a year comprised of either more days or less days. These changes, derived from altered periods that the Earth faces the Sun, would simply mean that the cause was the physical process of change and nothing else.

As a matter of fact, every few years the atomic clocks (which actually are tabulations reached from observing a physical dispersal occurring to a chemical element, such as Caesium [Cs]), need to be adjusted by a few fractions of a second. This is necessary because the planet's rotation is affected by its physical internal seismic activity, and the slight physical fluctuations in all 'boundaries of influence' (gravitation) that is produce by a change in the Earth's physical relationship with the Sun, and not from any influence of 'time'.

So the situation is: all chosen human activity on Earth, and its daily management and navigation obtained from calculated movement of Earth before the Sun, are completely and totally derived from physical actions and reactions that have nothing to do with any influence from 'time'. For were 'time' to extend any influence, then there would be unexplained fluctuations on Earth between expected durations in common activities.

For instance, if a dropped spoon fell to the floor in slow motion, then that would undoubtedly be an influence of 'time'. Or if a slowly strolling couple suddenly increased their movements without any intent of doing so, that would also be an obvious influence of 'time'. And if the Moon suddenly made brief, accelerated movements, then that too would be attributed to 'time'.

Furthermore, if 'time' is supposedly responsible for aging, then how is it that aging is one directional: from the cradle to the grave? Such natural progression has no stamp of 'time's' influence on it. If 'time' were to be involved with aging, then there would be no impediment for 'time' to function in reverse on a human body: from a has-been to a teen.

Considering that no such deviant events had ever taken place on Earth, this would indicate that physical existence has no need of 'time'. And by not being needed this would more than justify 'time's' nonexistence.

It would seem, however, that such logic and reason are insufficient to convince those who continue to believe in 'time'.

The notion of 'time' is much older to those of religions. Once human brains had launched their secondary consciousness, their conscios (commanding override negotiator, selectively controlling input-output) became consciously aware of change taking place in surrounding space (environment) and their own bodies. The unusual aspect of this is that all animal brains are aware of prolonged environmental change, which may cause them to consider activating evolutionary change to their bodies. But in this instance it was the external consciousness, outside of the brain, that was noticing and pondering what the surrounding change was all about. And because the early human brains could not directly explain what change actually meant – a physical process of change that comes about from actions and reactions constantly occurring between atoms and chemicals that form everything, including human bodies – conscios began to conclude (using the ability of their brains) that it had to be some form of invisible vapor which they breathed and which covered everything. But because it supposedly produced change to their bodies and to surrounding environment in a consistent progression, where they and animals aged and seasons followed one another, this invisible presence of change could not be spiritual or magical, something that spirits could be.

So the conscio consciousness of change became developed by them parallel to religions. 'Time' became a representation of change taking place on and around Earth, and applicable to physical things – including the movements of planets – while religions came to represent spiritual, metaphysical, and superhuman beings, supposedly capable of producing extraordinary change, whose change-ability and power were beyond that of 'time'.

But religions were always aware of the 'time' entity, gradually incorporating the notion of 'time' into their doctrines. 'Time' became a spiritual pathway that merged the past, present and future, so that it could – according to Judaic and later Christian religions – deliver great religious warriors, such as messiahs, to 'times' when they are needed, and according to Christian religion, would cause gods to judge all men at its end.

As human (conscio) understanding of themselves and of their surrounding expanded with the development of the so-called science, instead of gaining a rational and logical understanding of what 'time' represents – a calculation mechanism based on Earth's relationship with the Sun – humans (conscios, especially those involved with science) chose dispense with the attributes of rational and logic. Instead, not only had they retained the irrational and illogical abilities of 'time' to connect all past, present and future events, they incorporated this nonsense into their scientific doctrines: in effect, basing their scientific principles on fictitious 'time'.

'Time' was given an indispensible position in formation and ongoing function of all existence, where everything in existence owes its existence to 'time'. 'Time' was allocated physical attributes that would allow it to accelerate or slow down, to expand or contract, and to even bend. 'Time' became an element that was combined with the nothingness of vacuum space to become the "space-time continuum", which supposedly controls the expansion of the Universe with its own expansion.

All these attributes to 'time' were made, and continue to be made, by what could be described as reasonable humans (conscios) who considered, and consider themselves as being members of highly educated scientific community. All these humans (conscios) continue to view 'time' as being a physical reality, despite that not a single human (conscio) in all of human existence had ever presented even a shred of physical evidence that 'time' actually, physically exists, from having either heard it, or physically felt it, or actually seen it.

As with invisibility and indefinability of gods, no human (conscio) had ever declared: "Here I present to you God for your examination!" Just as no one had ever declared: "Here I present to you 'time' for your examination!"

Of course scientists make experiment by which they attempt to prove the existence of 'time'. But they do this without realising that all their experiments involve physical attributes of physical actions and reactions of inunens in their various phases of change, rather than any physical 'time', for were 'time' to exist it would need no physical actions and reactions to produce change. It alone would be responsible for all change.

Instead of acknowledging this truth, those in science continue to uphold the fiction of 'time' with such enthusiasm that most, if not all, scientific community can rightly scoff at the notions of gods, while the very same people find no problems with accepting a physically impossible notion of 'time'.

Sadly for all of humanity there is no future in 'time', as it is nothing but yet another great big hoax that humans (conscios) had perpetrated on themselves.

The main reason why humans, on the whole, deceive themselves with various notions – such as that of 'time' – which have no connection to physical reality, is because they have no knowledge of inunens. Those in science have notions of minute particles being vital for all existence, but these particles are not given the responsibility of inunens. Without knowing the inunen behaviour of their four phases, by which everything is physically formed before it is eventually dispersed, means that the current understanding of physical actions and reactions is dependent not on concrete knowledge of physical limitations but on ancient superstitions and modern fantasy of unlimited possibilities

Unlike the current understanding of energy, inunens (individual units of energy) are simultaneously energy and mass – the only true mass because it cannot be divided or penetrated by anything, and energy because it performs 'effort' of 'work'. Inunens are eternal because energy (energy/mass) can neither be created nor destroyed – it can only perform two functions: energy/mass can converts its state from active to dormant (as is explained in the last chapter), and it can change its phases in its ongoing cycle of existence.

As explained previously, an inunen cycle comprises of a physical body change when compressed 'uniting' and 'united energy' inunens expand their bodies in the process of producing 'work' and dispersing as 'dispersing energy' inunens of heat and light. And when the 'dispersing energy' inunens contact any other 'uniting' and 'united energy' inunens, they pass on their 'dispersal code' instructions and convert to spherical 'dispersed energy' inunens, incapable of any further effort but that of being attracting to all other inunens.

Their limitation is in that once the compressed 'uniting' and 'united energy' inunens have performed their function and expanded their shape to that of 'dispersing energy' inunens, they have no physical ability to compress themselves back into a compressed shape in the same cycle. This inability is the limiting factor that makes it impossible for energy, once used, to convert back into active energy/mass of any structure. It is also this limiting factor that makes it impossible for that which was to go back to what it was. All inunen change can only be as a simultaneous, ongoing present-moment of 'now', when change continues to proceed to further change – infinitum – with all the ongoing change being different to all previous change in manner of how it occurred, to what extent it occurred, and at what location it occurred, with all these factors being a physical process of change.

What this signifies is that physical change – in the manner it is produced by inunens, and occurring in the present-moment of 'now' – which occurs simultaneously everywhere – has no ability to revert to what it was. So no matter how change is perceived and examined, the fact remains that it is physically impossible for change to be anything but altered new change, which can never revert or reverse its direction from being ongoing change.

As such is the factual situation concerning change brought about by inunens in their simultaneously ongoing one-directional change, the 'past' instances of all present-moments of 'now' cannot be physically repeated or returned to. For once the present-moment of 'now' had taken place and been replaced by the next ongoing present-moment of 'now', it no longer physically exists anywhere in physical existence. It can only remain in the memory of human brains or as recordings of what had been the present-moment of 'now' when it was just that: the present-moment of 'now'.

As an example, consider an Egyptian pyramid. This pyramid was built in ongoing present-moments of 'now'. It stood and deteriorated always in the instant present-moment of 'now'. And now it is viewed by archaeologists and tourists, also in the present-moment of 'now'. And while the archaeologists and tourists can examine the current state of the pyramid and the very location where it had stood for thousands of Earth years, it impossible for them to visit the 'past' present-moments of 'now' when the pyramid was being built.

It is possible to presume that the pyramid shall continue to remain in existence for further thousands of years, but all of this duration shall be in the present-moment of 'now', and no other instant. Not even that of 'future', as 'future', like that of 'past' has no existence because nothing can exist before the present-moment of 'now' had taken place. But because the present-moments of 'now' replace one another simultaneously everywhere, this continuum of present-moments of 'now' prevents any 'future' events to take place anywhere in existence.

It is not only 'time' that became incorporated into scientific doctrines because humans (conscios) have no knowledge of inunens and the physical limitations of their functions. The vacuum space of nothingness has also been awarded physical abilities it never could or can possess, by human (conscio) perceptions obtained from human brain imaginations.

Unlike fictitious 'time' the vacuum space of nothingness does exist. It is everywhere: infinity with no end or a beginning. It is comprised of absolutely nothing and as such cannot do anything but allow all that is physical to share its existence. If the vacuum space of nothingness were not to exist (as a nothingness of nothingness) then there would be nothing for anything physical to exist in, including the nothingness of vacuum space. Without the vacuum space of nothingness there could be no existence whatsoever.

And yet, those currently in science seriously infer that this vacuum space of nothingness has physical attributes of physical objects and bodies, because it is supposedly capable of expanding, bending, stretching and shrinking, and being totally combined with 'time' in the form of "space-time continuum".

Without even the simplest of logic and reason, those in science soberly assure others in their lectures, scientific papers and books that not only was matter released from a mass the size of pinhead in the so-called Big Bang, but that physical matter had emerged from the absolute nothingness of nothingness, after which the nothingness of vacuum space, 'time', and matter came into being. As if physical matter was hidden in the absolute nothingness of nothingness, only to infiltrate – together with nothingness of vacuum space – through some kind of pores in the nothingness of nothingness.

Without an understanding of physical limitations, those in science are free to ignore physical reality, where no physical object or body can increase mass of its own volition – that is: mass cannot simply enlarge its mass without a physical addition of other mass to its own. This means inunens were and are of the same size, and cannot enlarge or reduce in size (to accommodate a reduction of this Universe into the size of a pinhead).

But just as physical limitations of inunens functions are responsible for physical limitations that apply to all physical existence, the vacuum space of nothingness has its own limitation: the limitation of being the only nothingness in existence.

As explained previously: for there to be a nothingness within another nothingness would represent absolute nothingness: a nothingness that would comprise of no nothingness of vacuum space or of any physical existence derived from inunens. There would be nothing at all. And such nothing-at-all would certainly not support the existence of any pinhead of any future universe because there would be nothing: not the pinhead, and not the vacuum space of nothingness inside that pinhead.

Furthermore, such nothing-at-all would not support any expanding universe with its expanding space-'time', because, once again, there would be nothing for anything to expand in, just as there would be nothing that could expand.

So when the current scientists and academics presume that their version of existence comprises of an expanding Universe within an expanding space-time, within a nothing-at-all, they are either bereft of any logic and reason, or are living in their own version of reality, or both.

However, fortunately for all who live with physical reality, the proof that there is no nothingness of nothingness is in the existence of vacuum space of nothingness within which everything physical exists and can be witnessed to exist. So if the vacuum space of nothingness exists it has nothing to expand in – because it is already everywhere. Irrespective of any human (conscio) speculations, the vacuum space is not about to change from what it actually was, is, and forever shall be: a vacuum space of nothingness, which attracts with its nothingness, but in no way impedes, hinders, or restricts any physical existence, simply because it is incapable of doing so as a nothingness.

In allowing everything physical to be as it will, the vacuum space allows everything to move systematically or randomly, while leaving behind no physical markers as to what precise locations all the objects or bodies had previously occupied in vacuum space.

This physical fact of all physical existence, including that of vacuum space of nothingness, proves that the human (conscio) theory of expanding space-time is also a nothingness: worth nothing and deserving nothing.

Nonetheless, 'time' remains such a popular concept – on which theories, like the 'time dilation', are based – that many scientists seriously search for ways to achieve 'time travel'. After all, according to the current acceptance of "special relativity", there supposed to be "symmetries of space and 'time'" – where the plateau of "hypersurface of the present space" is penetrated by 'time' comprising of "future light cone" and "past light cone": all this inferring that light can have a past and future existence. And if light can have a past and a future, then that would mean energy and mass also share the same attributes of past and future.

After all, what is the difference between the present-moment of 'now' to that which occurred a second ago? It must all be within the same vicinity of 'time'. So if the 'time' slows down, as it can according to "special relativity", then the instant of the present-moment of 'now' must slow down, by stretching out and being wider. And that would have to be a way of entering the 'past' and 'future' 'time' zones.

Using such flowed rationale, those in science and science fiction have long perceived machinery that would allow them to navigate 'time' in the 'past and the 'future'. Merely twist and turn some knobs, press some buttons, and the 'time machine' would accelerate not just into the 'past' or 'future' but different locations on Earth, obviously aided in this by the various flows and currents of 'time'.

Unfortunately for such fiction the physical reality representing all physical existence, with the limitations of their abilities, cannot produce such impossible feats.

Basically there are three fundamental reasons for that:

Firstly, as explained earlier, it is only the instant of present-moment of 'now' that exists.

Secondly, nothing can be stored or retained permanently in the vacuum space of nothingness. And for past to exist, then that is exactly what would need to be.

Thirdly, when the instant present-moment occurs, this represents a simultaneous involvement of every action in existence. So in order to revisit the past all of the present-moments of 'now' of all the simultaneous involvements of all actions in existence would need to reverse in its totality.

Apart from these insurmountable physical handicaps to the notion of "time travel", here are few more.

If 'time' of the present-moment of 'now' were to exist, and with that be inclusive of the 'past' and the 'future', then the 'past' would presumably be full of imagery of everything that had occurred on Earth in the past, together with actions of all the people and animals, together with all the sounds and smells. The 'future' would presumably also comprise of imagery and actions of all that is yet to come.

The difficulty that applies to these presumptions is in ascertaining how exactly all this 'past' and 'future' imagery and actions had been relayed to the past and future and retained there, and where exactly are they stored for posterity? Considering that astronomers can view all the vacuum space surrounding this planet, where there are no 'past' or 'future imagery of any kind, where are they then hidden? But apart from all that takes place on Earth, where is the location of all that had taken place in the 'past' of the whole Universe, and all that is yet to occur in the 'future' of the Universe? Is it all somehow burnt into the fabric of the 'space-time continuum' or hidden by a cloak of 'time'?

Furthermore, by what or whose means had all of the 'past' been structured as the 'past', and by what or whose means would all the 'future' images and actions of the whole Universe be devised and implemented before they actually occur?

Beside all these unanswerable physical impossibilities there are more. If 'time' were to exist, then the present-moment of 'now' would be between the past-moment and the future-moment. So in order to penetrate into the 'past' or 'future' it would be necessary to slip into these zones from the present-moment of 'now'. But because the present-moment of 'now' is without a pause, the past-moments and the future-moments are never given a chance to be realised, even if there was some method of gaining entry.

Still there are more physical impediments to the concept of "time travel". In order to revisit the 'past', would require everything that had occurred to reverse, not unlike a rewinding of a projected movie film. This means that every single action of every single inunen, object, and body everywhere in the Universe would need to revert to what it was previously.

But even if this physical impossibility were to be achieved, the "time travel" into the 'past' would be nothing like that depicted in science fiction. If 'past' existed, it would not be an unrestricted and free-flowing environment of the present-moment of 'now' existence, where actions and reactions take place as they occur. In a 'past' zone everything would be rigid and strictly regimented, with no intervention of any kind being possible: the past cannot be altered or undone.

What this means is that no insertion of foreign bodies would be possible, especially not how science fiction depicts "time travellers" freely moving about and producing new change within the 'past' 'time' zone. The only function that "time travellers" could experience in the 'past' 'time' zone would be a conscious observation without any participation. "Time travellers" would simply be invisible in surrounding 'past' environment, because they were never part of original present-moments of 'now', and from that cannot be real in that which once was real.

So there you have it: a testament of physical reality, which confirms that a physical impossibility shall remain a physical impossibility irrespective of how it is approached.

Humans (conscios) may continue to imagine that imaginations may lead them to transcend physical reality into realms of physical impossibility with their beliefs in religions and scientific fiction, like their notions of 'time' and 'space-time', but they are wrong: for there is no entity in existence that can ever produce physical reality from unreality.

This, of course, is not what current scientists choose to accept or understand. Instead of seeking real physical causes to physical actions and reactions, they have simply used notions, such as: 'space-time continuum' as a way of explaining physical phenomena they fail to resolve. A slight disparity in the expected outcome to any length of duration from any physical activity involving velocity and vacuum space? No problems. Use the usual explanation: attribute it to 'time'! Why bother with using all the observations and reasoning to determine the real reason for those inconsistencies, when there is the 'time' and 'space-time' to be used as an answer?

Such attitudes had resulted in development of theories, such as 'time dilation', with which those in science attempt to prove the existence of 'time'. It should, therefore, be worthwhile to explain the possible reasons for physical discrepancies that amaze those in science, as well as to dispel some of their currently favorite theories.

In the theory of 'time dilation', either 'gravity', or 'velocity', supposedly cause 'time' to differ at different points in space, while appearing to be correct at those different points in space. There are two parts to this theory. There is the "relative velocity time dilation" and the "gravitational time dilation".

The "relative velocity 'time' dilation" is often illustrated with a hypothetical situation where two space ships, A and B, are passing each other at fast speed in vacuum space. As they do so, the occupants of both space ships notice that the clocks and movements of those opposite are slower to their own, while seeing their own clocks and movements being normal. This would be "time dilation", where either those on ship A or B would age less because their clocks and movements are slower to those on the passing space ship. But if there was a third space ship C flying alongside space ship B, those on B and C would see each other's clocks and movements synchronized with each other, and therefore would retain their respective ages without 'time' change.

With the "gravitational 'time' dilation" the premise is that 'time' is slower the closer it is to a gravitational mass, and becomes faster the further it is away from that gravitational mass. This suppose to mean, for instance, that astronauts on a space station circling Earth experience slower 'time' to that which is on Earth, and by that are supposed to remain younger. Such gravitational 'time' dilation is supposed to be responsible for clocks on International Space Station (ISS) to run slower, as shown by atomic clocks, while those on Global Positioning System (GPS) and various other satellites to run faster.

What these theories, and all those who devised them – as well as those who accept them – fail to recognise, is that what they describe as 'time' dilation is actually the rate of speed at which a physical influence affects a physical object or a body: whereby all entities involved are physical, and only physical, needing no involvement from any 'time'.

To explain this, it is necessary to once again apply the knowledge of inunens and their functions, of which current science should know but had been too backward to obtain.

It is the 'uniting' and 'united energy' inunens (individual units of energy: heat) that form all the atoms (united bodies), which then form all the physical objects and bodies in all existence; only to be eventually dispersed as 'dispersing energy' inunens as heat and light (otherwise known as electromagnetic wave particles of the electromagnetic spectrum).

Consider also the two factors that apply to inunens: As 'uniting' and 'united energy' inunens they are influenced by the overall internal and external 'boundary of influence' (gravitation) of vast objects and bodies in vacuum space. And as 'dispersing energy' inunens they leave behind their 'transfer of intent', and retain their responsive 'dispersal code' as they depart their previous location having produced 'effort' of 'work' at having expanded their spherical bodies.

What the first of the two factors represents, is that as 'uniting' and 'united energy' inunens, they are influenced by the strength of a so-called 'gravitational field' of an object or body they are in presence of, and less so when moved further away from the source of that gravitational field. This means that the closer a 'uniting' or 'united energy' inunen is to the external 'boundary of influence' field or shells (gravitational field or shells), be it as part of an atom, the more 'tense' or 'rigid' it will be ¬– just like all other inunens forming that atom: both the 'energy bodies' (electrons) and the nucleus.

Alternatively, the further away a 'uniting' or 'united energy' inunen is from the external 'boundary of influence' field (gravitational field) in vacuum space, the more 'relaxed' or 'flexible' it becomes, together with all other 'uniting' and 'united energy' inunens forming all the atoms of that object or body. (But just because atoms in vacuum space and with less influence from Earth's gravitation become more flexible, they do not develop any increase in their activity as they would with influence of high heat; for high heat can also influence atoms to become 'relaxed', but not without becoming highly 'emotional' and reactive.)

It is because of inunen reaction to 'boundary of influence' (gravity) influences in vacuum space that astronauts, residing in zero gravity on a space station orbiting Earth, rapidly begin to lose their body functionalities despite all their efforts to maintain physically exercise regimes. With none of Earth's usual 'boundary of influence' (gravitational) influence causing the inunens forming all the atoms to retain their conformity, the more relaxed inunens allow the atoms forming astronauts' bodies – including their bone structures, organs and all that constitutes a body – to become more 'relaxed' and pliable, altering their body definition and function, which leads to loss of body mass and overall body malfunction.

This fact is very apparent when those living in zero gravity (well, not quite zero gravity but reduced gravity, considering that they remain within the boundaries of Earth's and the Sun's gravitational fields) return to Earth. Their bodies are initially incapable of self-support, needing long periods of not just recuperation but of rejuvenation, which only comes about when their inunens forming all their atoms becoming less relaxed and more rigid, as imposed by the levels of Earth's 'boundary of influence' (gravity).

When the factors of 'uniting' and united energy' inunens being influenced by Earth's and the Sun's boundaries of influence (gravitations) – together with the Earth's electromagnetic field – is combined with the factor of 'dispersing energy' inunens using their intelligence to decide their 'dispersal code' and the rate of velocity of their dispersal, and applied to astronauts in vacuum space, living in zero gravity, it can be said that their overall body metabolism minutely slows down. **But this has to be viewed as being harmful to the body** and in no way benevolent, in the way that supposed "reduced aging" suggests, **and** this process has absolutely nothing to do with involvement of 'time'.

Because Earth rotates before the Sun once every complete day (as a physical event and not due to any 'time'), the Earth's surface was divided, by convention, into twenty-four equal segments – each segment partitioned from the

North Pole down to South Pole – with intention for each slice of Earth's surface to represent one hour. This one-hour-per-segment is meant to be the same one-hour at the widest part of the slice at the Equator region as it is at the thinnest regions of the two Poles. (This, of course, means that at the very centres of the two Poles – where the thin edges of the slices merge into one pivotal centre of the axis – the twenty-four hour segmentation cannot apply because the area is less then on human footprint. This means that if 'time' were to exist according to their partition of Earth's surface, it would still not exist at those North Pole and South Pole regions.)

This device of 'time' zones partitioning has given humans (conscios), just as with their watches and clocks, a practical ability to establish a position of the Sun in its relationship with Earth, and with that the appropriate indication of the hour in each segment. And by such allocation of hours it is possible for travellers to know their exact hour of the day according to their location on Earth (and vice versa).

In order to achieve a higher degree of accuracy for such assessments of allocated hours, a new method had been devised to those of mechanical and digital watches and clocks: the use of atomic clocks.

Due to their accuracy, atomic clocks had become the essential tools for assessing the duration of any physical activity, mistakenly attributed to passage of 'time'. So in order to test the 'time' dilation theories, atomic clocks had been consistently used. Atomic clock had been used aboard the ISS, they had been used in satellites, and in various experiments using aircraft. Not surprisingly, the verdict had been to confirm the theory of the "gravitational time dilation", while the "relative velocity time dilation" is accepted to be valid as well.

However, having just explained how 'uniting' and 'united energy' inunens – which form all atoms, and which in turn form everything else – are directly influenced by boundaries of influence (gravitations), it should not be surprising that the atoms used for atomic clocks are likewise influenced by their inunen's influence of 'boundaries of influence' (gravitations) in zero gravity.

There is however another element to the 'boundaries of influence' (gravitations) fields. On Earth, for instance, the 'boundary of influence' (gravitation) field is not one continuous extension, but are like that of Earth-encompassing shells, with each shell extending further away from Earth, and by that encompassing the smaller shell thin it. The shells closer to Earth are smaller but much stronger in influence to those that are further away from the planet. These shells extend out into vacuum space as far as the Earth's 'boundary of influence' is capable of maintaining their presence.

Such 'boundary of influence' invisible architecture applies to all the planets in Solar System, including the Earth's Moon, and to the Sun itself.

PART 3. DISCLOSURE OF EVERYTHING

As explained earlier, there is a vital element to the 'boundary of influence' shells (fields) where the 'boundary of influence' shells of one large object in vacuum space merge with those of another: at some region of some of the shells a repulsion occurs, in a form of 'grappled attraction'. This is where the undersides of the interlocked shells produce situation of preventing the two objects in space from coming any closer together, despite of their individual 'boundary of influence' shells attracting each other. Were it not for the 'grappled attraction' function, all objects in space would merge into one big mass, irrespective of any current 'relativity' theories presuming that objects in space roll around in grooves, supposedly caused by their own weight, in the nothingness of vacuum space. As Earth rotated before the Sun, the region of 'grappled attraction' constantly changes, but the crossover of their respective 'boundaries of influence' fields or shells (gravity fields or shells) remains active. Where this occurs, the two shells produce a momentary varied region of attraction.

Just as the Moon integrates its 'boundary of influence' (gravitation) shells – or gravitation fields – with those Earth, the Earth does the same with those of the Sun. In this way, Earth 'boundary of influence' (gravitation) shells are in constant crossovers with those of the Moon and the Sun. And whenever the 'boundary of influence' (gravitation) crossovers occur, there exist the constantly changing, stronger pulses of (gravitational) influence on all the inunens of all the atoms in Earth's vicinity of space.

What this, therefore, signifies, is that no matter what objects there may be in space around Earth, be it space stations, satellites, or atomic clocks, not only are their physical structures affected in some subtle way by Earth's 'boundary of influence' (gravitation) shells – be it of reduced strength – but also by the crossover 'boundaries of influence' shells from the Moon and the Sun.

No doubt atomic clocks are highly accurate, with the oscillation of cooled gases being probed by electronic transition of high frequency. But irrespective of their construction and the atoms employed, any departure from Earth's surface (where all atomic clocks function without deviation) into regions of vacuum space, will inevitably result is some slight variance of function, and by that: duration of function. And what needs to be understood is that variance and duration of function applies not just to the atomic clocks, but also to the container in which the atomic clocks are incased and to everything that is the spacecraft. It is just that the atomic clocks are more sensitive to physical change than all other structures making up the spacecraft.

All this signified that the so-called "gravitational 'time' dilation" is nothing but physical activity that has nothing to do with the entity of 'time'.

Similarly, any muon experiment asserting that there is some kind of 'time'-involved variance of muon (particle of a grunen: grouped units of ununens, in the shape of short sausages that combine to produce the structure of a proton) decay (a conversion) should review that perception to that of a purely physical activity involving actions and reactions, including those of gravitational and electromagnetic influences.

So why then the atomic clocks aboard International Space Station and flying aircraft show delay in their function while those on Galileo and other satellites indicate an increase, despite being in higher orbits and from that further away from Earth centre of 'boundary of influence (gravity).

The answer to this situation is quite simple: it a matter of interior environment. The interiors of aircraft and that of ISS are pressurized and heated, to replicate the conditions on Earth's surface, while the satellites have no need of this. So while the 'uniting' and 'united energy' inunens forming chemicals of human bodies and those used in apparatus, such as atomic clocks and electrical battery cells, are allowed to become 'relaxed' and minutely slower acting, the apparatus with active chemical activity aboard unmanned satellites remain in a cold interior environment. This causes the 'uniting' and 'united energy' inunens on unmanned satellites to be more firm and more precise in their function due to being less contacted by surrounding heat of 'dispersing energy' inunens, by that avoiding a conversion to 'dispersing energy', which would alter the chemical actions of any apparatus.

It is for these reasons that large, mainframe computers are housed in sunless, very cool locations, which prevents any generated heat from affecting the microchips and the magnets of the processors.

No doubt it would be difficult for all those who became reliant on assigning any unexplained deviation in research to that of 'time' to remove the notion of nonexistent 'time from their equations. After all, it is all so convenient to nominate 'time' as the cause of all that is hard to explain when discrepancies of duration arise between physical actions taking place on Earth and those outside Earth, with all the involved influences from the Sun's gravitation – as well as other influences derived from the Sun, in the form of Van Allen belt surrounding Earth. It maybe convenient when studying muons deterioration and particles accelerated along electromagnetic fields in the Hadron colliders to simply allocate any variances to 'time', but if physical puzzles are to be resolved then it is appropriate now that those in science turn their attention to the domain of physical reality of all physical existence, by discarding the notion of nonexistent 'time'.

As to the question of "relative velocity 'time' dilation", this theory surmises but proves absolutely nothing, because velocity – at least any velocity in vacuum space – produces no physical change to any object or body involved. Any physical variance that may occur to any object and body in vacuum space does so from duration of its presence at a particular location that may exhibit a particular physical influence, but not from any movement, no matter how rapid.

The concepts of "relative velocity 'time' dilation" and of "gravity 'time' dilation" had been derived from a theory of "special relativity". The problems with these concepts is that the theory of "special relativity" had always been based on two mistakes:

First mistake is that there is no such entity as 'time', and the second mistake is in an assumption that there is a relationship between separate entities of mass and energy, whereas mass and energy are one and the same entity. Because inunens (individual units of energy: individual units of heat) cannot be penetrated or divided by any other inunen – in whatever physical phase – their individual structures represent the only form or mass that can possibly exist. It can be said that as mass, inunens go on to form other mass of atoms, and these form chemicals that form objects and bodies. The fact remains that the only real mass are the bodies of inunens. And because all inunens follow a one-directional cycle of converting their bodies (mass) from 'united' and 'uniting energy' inunens to those of 'dispersing energy' inunens, and finally to that of 'dispersed energy' inunens, that means the same mass goes from being building mass to dispersing mass (currently thought of as energy) of heat and light.

This means that the equation of $E=mc^2$ is wrong. E does not equal mass, for energy is mass, just as mass (m) does not equal energy, as it is energy in its dispersing phase, while the constant (c), which is supposed to be the speed of light squared (multiplied by itself) is the mass in its dispersing phase of 'dispersing energy', in process of dispersing as heat and light (where heat is light and light is heat). In other words: mass is energy (mass in dispersal) where the same mass in dispersal is light and heat (moving at speed of light).

This reality of physical existence has no need of any "special relativity", "rest energy", "invariant mass" or "symmetry of space and time", and many other imaginary notions on which so many had wasted so much of their lives.

This includes the irrationality of the "velocity 'time' dilation", which argues that any traveller approaching the speed of light does not noticed a slowing of 'time' for oneself, but can notice this in others moving past in an opposite direction. These kinds of erroneous deliberations had been brought about by lack of understanding of physical reality. By accepting the notion of 'time', humans (conscios) – including those in science – had fallen for an assumption that there is a physical past and a physical future, whereas in physical existence there is only a simultaneous instant present-moment of 'now' and nothing else. And while that instant present-moment of 'now' – otherwise known as the 'present' – results in ongoing existence, that ongoing existence always remains in the instant present-moment of 'now'. This means:

Fact 1: the notion of past and future are human brain inventions for their conscios, because the human brains can maintain past recall due to memory-ability, and future projections thanks to its imagination-ability.

Fact 2: once the instant present-moment of 'now' had been replaced by the next, and the next, and the next instant present-moment of 'now', representing increments of motion in physical existence, that which had taken place exists no more, while the future is physically unattainable as it does not yet exists.

Therefore: were 'time' to exist, it could only do so in the instant present-moment of 'now', as the 'past' and 'future' have no physical existence. But because the instant of the present-moment of 'now' merely represents all the simultaneous movements, actions, and reactions of inunens – which form everything physical – their intelligence of knowing when and what to do when it is necessary has no need of 'time's' assistance.

To repeat this physical fact, the instant present-moment of 'now' represents a **simultaneous**, seamlessly incremental actions and reactions of all physical objects and bodies in all physical existence. If this understanding is applied to an example of two spaceships passing each other at speed, the occupants of each spaceship could see each other absolutely still in a single instant present-moment of 'now'. Were their respective clocks to be synchronized and if each crew on respective spaceships **photographed** each other's clocks, when later compared, both clocks would show exactly the same 'time': no fasted and no slower.

Therefore, there would be no 'time' dilation on either of the passing spaceships, due to simultaneity of the instant present-present-moment of 'now' (and nonexistence of 'time').

There is another way to show that velocity produces no 'time' dilation. If 'time' were to slow down within a speeding spaceship, it is currently accepted that the slower 'time' somehow influences the occupants to age slower to those on Earth. That then would mean the same slower 'time' would fill the whole of the spaceship's interior. As such, the slower 'time' would obviously have the same slower aging affect on every body and component of that interior space of the spaceship. This would obviously include all food supplies, allowing them to remain fresher, longer. And if there were any fruit and vegetables in storage, then these would also age slower.

PART 3. DISCLOSURE OF EVERYTHING

In which case, the same logic can be used in application to Earth, which is moving in vacuum space at millions of miles every hour. Therefore it would be reasonable to assume that 'time' has some similar overriding influence over all aging on Earth, if planet Earth were to be considered being a huge spaceship.

To examine this circumstance, an experiment can be set up, comprising of a working refrigerator being placed in a room constructed in a desert, with another working refrigerator placed in the direct sunlight outside the room. By all expectations – according to slowing 'time' producing slowing aging in the same interior space of a spaceship – the insides of these refrigerators, whether in a room or outside would have the same influence on their contents, considering that the two refrigerators are at the same location of Earth 'time' zone and the insides of the two refrigerators would be filled by the same Earth 'time' of that particular location.

Now two similar fresh, green leaves are produced, with one being placed inside a refrigerator inside the room and the second leaf on top of that refrigerator inside the room. Simultaneously the same is done with the outside refrigerator: one fresh green leaf placed inside it and the second similar leaf on top of the refrigerator. Additionally, four identical and synchronized clocks are distributed: one each inside the refrigerators with the leaves, and one each on top of refrigerators, next to the green fresh leaves. The premise being that if Earth 'time' were to exist, as the clocks would supposedly show, it would be affecting all the leaves identically, or at least similarly, considering that all leaves are at close proximity to each other; not unlike all the vegetables and fruit on the speeding spaceship being similarly affected by the same slowing 'time' on board.

A week later the leaves are examined.

The result of the experiment shows that all the clocks continue to indicate the same Earth 'time'. The leaves, however, are far from being in same condition. The leaf inside a refrigerator in the room is still crisp and green. The leaf inside the refrigerator outside the room has withered but still in good condition. The leaf on top of the refrigerator inside the room has withered, while the leaf on top of the refrigerator outside the room has dried out, bleached and disintegrating.

The inevitable conclusion of exposing similarly fresh green leaves to a presumed same Earth 'time' in the same location but different environments shows that it is the physical conditions that directly influence the state of the leaves and not any Earth 'time'. The supposed 'time' in no way had produced any similarity of aging influence on any of the leaves, not even those inside refrigerators, and therefore can be deleted from any consideration of being an effective agent of controlled change. Which, in fact, it cannot be, as it has no existence.

Were Earth 'time' to exist, it, as a constant presence of the same 'time' in a precisely same location, would have influenced a similar speed of aging on all leaves, or similar 'time' dilation, be those leaves encased in refrigerators or not. That is why everything that deteriorates, ages, and becomes decrepit anywhere, has nothing to do with 'time' but with the physical process of change.

There is yet another example of 'time's' non-existence, which is there for anyone to witness. The physical proof of this comprises of a physical process that is so well known and documented that it would be laughable to all those involved to dedicate any of it to 'time', as an entity that intentionally manages controlled change. This process is that of wine making.

Every wine maker (and many wine drinkers) know how wine is made and the precise durations of physical process of change required to convert grape juice to that of wine, and the further durations of physical process of change required to convert the contents of wine into more palatable liquid. And while those involved with wine may call the overall duration of the process as that of 'time' – as it is a current human custom – none of them would even imagine that some active entity of 'time' was physically having a direct influence on the winemaking process.

But how are the durations of these physical process of change are actually measured? By years and months: which represent a measuring device based on duration of the physical process of change comprising of Earth's single, complete elliptical orbit of the Sun, with this being divided, by human convention, into twelve months. So in physical reality, any measurements of any wine's 'life cycle' is based on physical process of change as it applies to Earth's relationship with the Sun, and not on any mystical or mythical invisible 'time'.

It is the physical process of change that applies to every single activity required to convert grape juice into wine, from harvesting at a particular period of Earth's relationship with the Sun (season) to crushing, pressing and fermentation, blending and refining, adding of preservatives, and finally to that of bottling – all being physical activities that are measured by an accepted convention of months and years.

Likewise, the conversion of grape juice and the following maturing of wine comprises exclusively of various chemical actions and reactions occurring between each other, be that at the initial stage of fermentation or the maturing of wine in bottles are measured by months and years: an accepted convention representing the Earth's relationship with the Sun, with no contribution from any mythical 'time'.

Wine making is simply a physical process of change, where yeasts are used to change sugars (in the grapes) into alcohol. Initially this takes during fermentation, (timed by days: Earth's daily rotations), during which the heat released is physically controlled so as to achieve a precise grape aroma, colour and flavor, as well as formation of yeast by-products. Then the physical process of change continues in barrels that are stores at a particular temperature: a physical environment of heat to physically control that physical process of conversion (change). After that the wine is bottled, with the process of change continuing in the bottles. That means the so-called 'aging of wine' has nothing to do with 'time' but the physical actions and reactions taking place between various chemicals that constitute wine.

If the contents of a wine bottles are not consumed within an allocated period of years (based on Earth's individual orbits of the Sun) the wine structure will eventually deteriorate due to continued chemical actions and reactions taking place in the wine bottle, especially if oxygen penetrates the cork. Even when such wine is discarded, by being poured into a drain or a sink, the physical process of change continues when the spoilt wine is blended with many other chemicals in the drains, eventually (be that months or years, which are based on Earth's orbits of the Sun) being converted into water by a waste recycling plant; all of this being achieved with physical activities of chemical additives and physical processes of filtration: none of which represent any involvement of mythical 'time'.

As wine is produced all over this planet at various locations (be they dry, wet, cool or warm climates) and at various elevations, all these physical process of change – without any exceptions – comprise of physical efforts and physical chemical actions and reactions beholden to no 'time' but to the necessary duration of the physical process of change as required by those physical and chemical activities.

There are multitudes of such physical activities that easily prove the absence of any 'time' involvement. Even the scientific experiments, including that of 'time dilation', that supposedly support the existence of 'time', if examined objectively and with the knowledge of the physical process of change, can be shown to be nothing but physical reactions occurring from physical actions. If properly examined, instead of proving 'time' they actually confirm that in physical process of change there are all kinds of variable durations that can, and do occur in functions.

So the absolute difference between the physical process of change and 'time' is that the physical process of change is physically evident everywhere, whereas 'time', which is supposed to be everywhere, is physically evident nowhere but in human and scientific beliefs.

It is surprising how all the past theories continue to be accepted despite all the available physical evidence that dispute their presumptions. Even with the concept of light currently not being understood that it is actually 'mass/energy' in a dispersing phase of heat and light, its physical abilities and limitations can still be worked out using logic and reason, so as to dispelled the fiction attributed to it by the "special relativity".

Because there is no 'time' dilation due to all actions and reactions being only conducted simultaneously in the instant of the present-moment of 'now', here is a further explanation of how physical functions can influence the speed of light, something that 'time' could never do, were it to exist.

Consider yourself, hypothetically, being on a spaceship that is travelling in vacuum space at full speed of light. Consider also that there are powerful spotlights mounted at the front, the sides and the rear of your spaceship. You have decided to switch on the forward spotlight. You turn on the spotlight, and... the light beam from the spotlight instantly projects forward at speed of light, even as your spaceship is moving at the speed of light. What does this mean?

Firstly, this means that the incremental forward movements of your spaceship's instant moments of 'now', similarly applies to the instant moments of 'now', which are the incremental forward movements of light emitted by the forward-facing spotlight, as those on Earth.

Secondly, this means that the light from the spotlight is moving forward at a combined speed of its speed of light plus the full speed of light of your spaceship. This would equal to twice the speed of light, because the vacuum space – as nothingness – has no ability to hinder any movement of any object, or body, or 'dispersing energy'.

This physical fact can be proven by an experiment of projecting forward an object from a forward moving vehicle, with the result being that the projected object would maintain (no matter how briefly, due to the restrictions of impetus of the projection and the influence of Earth's gravity) its own velocity combined with the velocity of the moving vehicle.

This also means that when a vehicle travelling at night has its headlights switched on, for the driver to see the road ahead, this represents a physical fact that the light emitted by the headlights is travelling faster than the speed of light, from having the additional assistance from the velocity of a moving vehicle, and all this occurring with the present-moment of 'now' applying simultaneously and similarly to the speed of light of the headlights, the vehicle

PART 3. DISCLOSURE OF EVERYTHING

and its velocity, and Earth with its velocity. So much for the concepts that light cannot surpass the speed of light.

Now you decide to turn the spotlights on the sides of the spaceship. Instantly the light beans issue forward in a straight trajectory, perpendicularly to the spaceship with the same principle of incremental instant moments of 'now' applying. But because of the speed of your spaceship, the side-projections of light beams will take an appearance of light wings spreading backwards, but with a straight edge at the front, on either side of the spaceship.

Finally, you switch on the rear spotlight, but instead of an instant projecting stream of light shooting forth behind the spaceship, you see... nothing. The spotlight is lit but no beam of light is being emitted. This would not mean that 'time' had stopped at the rear of your spaceship that is moving forward at the speed of light, because the present-moment of 'now' continues, but that the forward direction of your spaceship at speed of light had cancelled-out the opposite-directional movement of speed of light at the rear of your spaceship. An object moving at a particular velocity cancels the momentum of anything released from it, in the opposite direction at the same velocity. This fact can also be verified by experiments on Earth.

These are not semantics but physical facts derived from physical actions, requiring no involvement of any fictitious 'time'.

So what does this hypothetical experiment disclose? That if 'time' were to exist it would not be slower with increased velocity, as presented by the "special relativity". The speed of light can increase and still remain light, while the present-moment of 'now' remains unchanged no matter how fast an object travels in vacuum space. If light – which is dispersing 'mass/energy' – remains unchanged in volume, then neither will any other mass at any velocity.

This fact of physical existence totally contradicts the opinions presented in "special relativity". "Special relativity' was devised, and is now maintained, by all those lacking knowledge that mass and energy are one and the same entity of inunens, which can be converted from being 'uniting' and 'united energy' inunens into 'dispersing energy' inunens.

Instead, it was thought – and still is – that no object or body could attain, and even exceed the speed of light because mass and energy are interchangeable. This was determined by a presumption that mass has a resistance to energy attempting to accelerate it: the larger the mass the greater amount of energy required.

(No doubt that this perception had been based on movement behaviour on Earth's surface, where this notion is valid due to influences of friction caused by resistance from ground, air and water, and influence of gravity. But it does not apply to a mass in vacuum space, as vacuum space provides no resistance to acceleration of any mass.)

So what would prevent an accelerating mass from exceeding the speed of light? The explanation was that mass of an object or a body must increase in size as it travels faster. And the way this is supposed to occur is by energy being transferred back into the very mass it was propelling forward, so that with increasing mass the energy producing the effort of work diminishes, by that preventing that enlarging mass from reaching the speed of light. The function of energy converting back into mass would also be interchangeable, where mass could produce energy and become reduced.

The problem with this theory is that energy can never be converted back into mass, as only mass can be converted into energy. This is because, as had been previously explained on many occasions, 'uniting' and 'uniter energy' inunens – that form everything in existence – can only be converted into 'dispersing energy' inunens, which, as 'dispersing energy' produces 'effort' of 'work' before dispersing away from an object or a body they were part of as radiating heat and light. If and when 'dispersing energy' inunens do physically contact any object and body comprising of 'uniting' and 'united energy' inunens, these 'uniting' and 'united energy' inunens instantly become converted to 'dispersing energy' inunens, while the contacting 'dispersing energy' inunens become converted to 'dispersed energy' inunens and become dormant.

Therefore, once mass converts any of its atoms – such as propulsion fuel of a spaceship – to 'dispersing energy', that 'dispersing energy' departs as heat and light, reducing the size and the weight of the mass, and if any of that 'dispersing energy' were to be directed back into the mass (of the propulsion fuel of a spaceship), then that mass would not expand but would further deteriorate in its mass size and weight, by converting to 'dispersing energy'.

This means that the mass of a spaceship accelerating in vacuum space towards the speed of light cannot possibly convert the exhaust heat, light and gasses (force) that are released by the engines of a spaceship back into the mass of the spaceship, making that mass "infinitely great", so as to prevent its acceleration.

Besides, how could the exhaust gasses, heat, and light (force) possibly convert back into the mass of the spaceship and expand it? This can only be done in imaginations of those who are bereft of logic.

Furthermore, if mass were to expand with approaches to the speed of light – as it does according to "special relativity" – then light, all light, would need to expand as it travels as light. In which case all light would expand the further it travelled, instead of diminishing (as there is inevitable change in numbers of 'dispersing energy' inunens to that of 'dispersed energy inunens, from physical contacts) as it does in physical reality. So this physical impossibility is just another observation how science continues to be based on nonsense and irrationality.

And as far as the size of mass in vacuum space is concerned, it is absolutely irrelevant how large a mass may be in travelling in vacuum space, and how fast it may be travelling, because the vacuum space of nothingness has no ability to impede, hinder or restrict anything physical, of any size or weight from moving at any velocity whatsoever.

In conducting this brief examination of the "special relativity" the only conclusions that can be reached is that it is not just wrong but naïve of physical reality. For were "special relativity" to hold true, then it would have been possible mass to convert some of itself into energy, recoup that released energy back into its mass, again convert some of its mass into energy and again recoup that released energy back into its mass... infinitum, and by that being a self-perpetuating mass/energy that could virtually travel forever. As it is, the only element that had been capable of performing a self-perpetuating function is this theory of physical impossibility.

So why had those in science devise a belief that an increase in velocity instigates a slowing of 'time'? The answer to this question is quite simple: it was recognised that there could not be two "universal constants".

The speed of light had been calculated as being a universal constant at having 186,000 miles per second velocity. Therefore, if that was the fastest velocity applicable to all trajectories, then 'time' had to be somehow restricted in speed. And the way to do that was to make it slow down in proportion to any acceleration towards the speed of light.

But just for argument's sake – for those who shall continue to believe in existence of 'time' – what would all existence be like if 'time' were to physically exist?

There is an observation that expresses the following sentiment: "When removing the impossible from an equation, all that remains, no matter how improbable, must be the truth." An improved version of this statement would be: "When removing all that is physically impossible from an equation, what remains is the physical reality." The second version is more apt, because it discards 'improbable', because improbable is a conjecture, an opinion, rather than a fact of physical reality. Physical reality is never 'improbable' for it is a physical reality that represents all physical existence.

Physical reality exists without any need of the impossible. That is why the impossible can be removed without affecting the physical reality. Whereas if the physical reality were to be removed from the equation, that equation would represent an unreality, which could not physically exist, because unreality would have no limitations that apply to physical reality. The concept of 'time' is a perfect example of the impossible.

Were 'time' to exist then it would be everywhere as a physical presence: constantly variable but distinctly felt and witnessed.

It would have to be physical because it would need to physically grab, touch, hold, pull or push some components of objects and bodies, or the whole of the objects and bodies, so as to manipulate them with its actions, in order to bring about change.

In case of slowing or increasing the speed of clocks in space capsules, satellites, and spaceships, 'time' would have to physically get hold of the mechanical, digital, and atomic systems of these devises, in order to physically alter their indications.

As to the slowing of the spacemen's aging, 'time' would have to physically manipulate the atoms, cells, the genetics and all the internal organs and functions of their bodies by altering them, so as to slow down all the functions of their metabolisms.

With 'time' having such direct physical impact and control of any body, this would mean that evolutionary change on Earth would not be evolutionary change at all, but a change devised according to 'time'. After all, how could there be any evolutionary change to any internal and external functions and structures of a body, when 'time' can intervene and alter anything, depending on location and speed? Why would evolutionary change be even necessary when 'time' can control all change? So with 'time' existing, evolutionary change would be unnecessary.

All this, of course, pales into insignificance when considering that were time' to exist, it ('time') would directly manipulate every single 'muon' particle and atom and molecule all over the whole Universe, so that everything would behave according to the dictates of 'time'.

If 'time' were to be responsible for change that had a significance and order, then it would need to possess intelligence. If 'time' comprised of intelligence, then this would signify a highest form of intelligence in existence, the source of which would be located lounging next to the all-knowing, all-creating, all-controlling almighty god in heaven.

And if 'time' had no intelligence – as it never could – and simply produced mindless change for the sake of change, then there would be no physical reality as it is now known and experienced by humans on Earth. All existence would be in disarray according to whatever state of 'time' that was present.

All physical existence would be that of a distortion, because 'time' would not be a rigid, constant entity. Instead, it would ebb and wane as tides and currents, as whirlpools and vortexes throughout the Universe, for it would not be denied by anything from doing so. This means that this planet, the Sun, and our Milky Way galaxy would not be excluded from its sweeping control, being caused to compress or to expand, together with the surrounding 'space-time continuum'.

With its movements accelerating or slowing down, there would be no constant of 'time', causing same-period produced or born entities to age and expire differently.

All existence would comprise of inconsistencies and fluctuations. For example, one side of a street would cease to exist, replaced by a void of another existence, which would wobble and sway as it too would be in transit of 'time' change. A human hand or part of a whole body inadvertently move into the void of different 'time' zone and then disappear, only to make an appearance somewhere else in 'time'. These lost parts of bodies would not bleed at the cut-off ends, but simply be gradients of regions, where one region and speed of 'time' finished and another began. In this manner various body parts, as well as parts of flora, parcels of air or water, or dirt, or boulders, would suddenly materialize anywhere, at any moment, as flotsam and jetsam of 'time', floating about, for even gravity would be controlled by 'time'.

Humans, animals and objects would vanish into 'time', only to be transported to an exactly same location in the past or future, and by that find themselves wedged into walls, or mountainsides, or at the bottom of seas, because the locations they left had different landscapes to those in which they emerged.

There would be sudden appearances from the past of humans wielding swords or spears, and those from the future, firing their deadly weapons. There would also be dead bodies that were caught for centuries in streams of 'time', with 'time' not being considerate enough to supply such prisoners of 'time' with nourishment and air.

In such mad, irrational existence, humans would cease live with any secure constancy of existence, because 'time' could alter anything at any 'time'. This would include entities outside Earth, where Suns would suddenly appear in the middle of the Solar System or next to a planet, while whole galaxies, or parts of them would be shifted by 'time' into different 'time' periods.

Were 'time' to be real and physical there would be no reality to anything because all existence would be surreal, in which all life on Earth would come to an end in the midst of the mad reign of 'time'.

Currently, those in sciences, especially those in physics and mathematics, perceive all that they imagine as taking place far away in the Universe, never at our own Earth's doorstep. Out there in far distance are wondrous events are occurring: black orifices in vacuum space of nothingness with event horizons are everywhere. And so are multitudes of various dimensions, together with the expanding space-time continuum, where 'time' and space can expand, shrink and stretch and bend and perform various distortions. It is also out there in far, far distance of a distant past that an entity the size of a pinhead had erupted in the so-called Big Bang, and was capable of releasing all the matter and energy that equated to this Universe; allowing all the particles that form all the atoms and molecules to increase their mass in size, as if by magic.

Indeed, everything is possible out there, where there are no restrictions to any human imagination, and where there are no physical limitations that seem to apply to all physical existence on Earth.

For that is what existence on Earth represents: physical limitation. It begins with the inunen limitations of size, limitations of phases of change, and limitations of behaviour. Then these limitations go on to govern the limitations of all that physically exists on Earth be those that make up this Universe. So the physical facts are: whatever occurs on Earth and our Solar System is exactly that which takes place all over the Universe, no more and no less, despite the current scientists' presumption to the contrary.

While this planet – as all else in existence – experience inevitable, gradual or sudden physical change, this change is comprehensible, because it remains within parameters of expected normality of reality. There had never been any astonishingly bizarre and abnormal occurrences taking place on Earth, or anywhere else for that matter.

There had never been any physical events outside of our accepted normality of reality. If there had been such deviations from normality of reality it would have been noticed and would have set new standards to the currently accepted understanding of physical reality. It would have shown that impossibility was not a possibility but a certainty. Such state of existence would prevail were 'time' and all other claims believed by science to be a reality.

Fortunately for everyone and everything in physical existence, no such 'time', and 'time' related activity had ever happened, nor could it, because 'time', as a mechanism of change, never had any physical existence.

While 'time' does not exist, the measurements relating to Earth's physical process of change – comprising of years, days, hours, minutes, and seconds – are an essential tool for human management and navigation of their lives and the Earth's surface. Without these measurements of the physical process of change, as it applies to Earth's relationship with the Sun – maintained by clocks, watches, and calendars – humans would be at a loss from failing to know when to perform their tasks and duties.

Similarly, it would be very difficult for science to function without using these measurements in reference to physical actions and movements of objects and bodies taking place on Earth and beyond.

Therefore, there would be nothing wrong with using these forms of measure and calling them 'time'. But just as long as this address of 'time' does not, in any way, refer to an entity that can, in some mysterious way, influence any physical object or body, and any of their actions and reactions by its existence as 'time' or 'space-time'.

THEORIES OF NONSENSE
AND
DECEPTION OF 'BLACK HOLES'

There are many fantasies that current scientists, academics, and researchers uphold, apart from the 'space-time continuum', 'time', and "time travel". And yet despite their pursuit of fiction, presented as knowledge that may provide them with financial wealth and glory – but having little grasp of physical reality (physicalistics) – there have been many advances achieved by those in science for the benefit of mankind: or at least for those who can afford it.

Actually there is no mystery to how humans (conscios) have achieved so much without letting go of their fantasies and ancient beliefs. All the successes in science had not come from human (conscio) knowledge that can directly predict exact outcomes. Instead, achievements had come from billions upon billions of experiments. As this practice of experimentation continues to increase exponentially, there often remains a lack of knowing the reasons why outcomes are what they are. This causes those involved in sciences to explain what they do not understand by use of invented notions. Indeed, this has always been a human practice: when in doubt, fill that doubt with fantasy.

It is not surprising then, that those in sciences have extended their beliefs of 'time' into other contrivances, such as: "time travel", existence of multi-dimensions, and cavities in vacuum space in form of 'black holes'. This is blatantly done as if physical reality that surrounds everyone and everything on Earth does not exist.

Without knowing that there are no particles that are all-positive, or all-negative, or all-neutral, or that any combined mass (made of individual units of energy: inunens) can change from 'uniting' and 'united energy' into 'dispersing energy' only once per cycle, and as 'dispersing energy' can never revert to being 'uniting' or 'united energy' in that cycle, current participants in science continue to labor under a false presumption that energy equals mass, and as such the two can be converted from one to another. This is done in the face of physical reality where, for instance, an exploded nuclear device cannot be converted back into the nuclear atoms once they were converted to 'dispersing energy' in the explosion. No one and nothing in physical existence can perform such miracles.

But such reality does not compel those in science into questioning not just how they think, but how to interpret what they do.

For example, by believing in 'time', those in science presume that actions taking place far away are those of the past. Astronomers viewing images of light with their telescopes claim to peer into the 'past'. Whereas they are actually observing in the present-moment of 'now' the 'dispersing energy' of light – in the same instant of the present-moment of 'now' – belonging to the object or body the light had left long ago. This means that everything they witness 'now' is what is 'now' of what may have existed in the past but exists no longer, as 'past' has no physical existence. Or if that which was in the past still exists, then that object of body can still be seen as it appears 'now' but not how it looked in the past, because the 'past' has no physical existence.

With the 'past' not existing, and everything in existence existing only in the present-moment of 'now', all the light streaming from vast distance is doing so in the present-moment of 'now', with this having nothing to do with any observation of the 'past'.

Then there are the propositions by which those in science have attempted to convert the vacuum space of nothingness into various extra-dimensional structures of existence, with the fourth-dimension being allocated to 'time'.

Unlike the present-moment of 'now' that applies to all change occurring to all physical objects and bodies everywhere, there are no such applications to the vacuum space of nothingness. As an eternal vacuum space of nothingness, it cannot be physically affected by anything physical, just as it cannot physically affect anything physical. This means that while all physical objects and bodies are three-dimensional, the vacuum space of nothingness has no dimensionality, from being a nothingness.

PART 3. DISCLOSURE OF EVERYTHING

Were fourth or fifth or sixth, or whatever dimension to exist, as claimed by those in sciences, then these dimensions would need to be physical, and have an existence like that of three-dimensional objects and bodies of all existence. And if all the extra-dimensions were physical then it would be inevitable for them to make a physical impact on behaviour of all that is physical.

As the vacuum space of nothingness is the same everywhere, and as all that is physical applies to all physical existence near and on Earth as everywhere else, then all that takes place on the other side of the Universe would also occur on this side of the Universe. Therefore, all the extra-dimensional existence, that supposedly happens somewhere else, must also take place here.

In which case, together with the fourth-dimension of 'time', multitudes of other dimensions should be making a direct physical impact on Earth as all else. All existence should be weird, unpredictable, and full of unreality. With their twisting, distorting and contorting forms, the multi-dimensions should have humans experiencing their bodies being rolled inside out, or being squashed flat while remaining alive, or stretched out into infinity. They would have a psychedelic experience of living in distorting and fracturing space without need of oxygen and sustenance, requiring no growth or development, as objects and bodies would exist without anything or anyone being responsible for that existence.

Fortunately for all that physically exists, nothing like that had ever taken place anywhere, at any period. This is because apart from the so-called third-dimension, there are no other dimensions in physical existence, but those in the imaginations of scientists.

There is not even a first or second dimension, as there is no physical entity without thickness. Instead, everything on Earth – as anywhere in the vacuum space of Eternity – is formed from physical three-dimensional inunens into three-dimensional physical objects and bodies. All these physical three-dimensional objects and bodies produce physical behaviour that follows the inunen phases of uniting before separating, be that as atoms, galaxies, this planet, and all that is on this planet, including human bodies. On, and on, the physical process of change continues with the same assured physical reality, without the madness that a fourth or fifth dimension, or whatever dimension and 'time' would 'create'.

If the nothingness of vacuum space has no 'time' and no fourth or fifth dimension, then – as a nothingness – it is unlikely to have an ability to shrink or expand, bend or curve. But that is exactly what current science insists vacuum space does. This is done without providing a shred of evidence that the vacuum space of nothingness comprises of some physical attributes of somethingness – such as structure made up of physical entities, whatever they may be – for only physical somethingness with physical substance can undergo physical distortion.

Should the nothingness of vacuum space possess some kind of physical structure then it would no longer be a vacuum space. By having physical attributes it would impede with its presence not just on all physical activities but the objects and bodies engaged in physical activities. It could influence both the acceleration and deceleration of objects and bodies in vacuum space. It would provide grip and friction for objects and bodies in vacuum space, which would cause them to slow down and eventually stop altogether, all of which would be clearly observed. And if space provides waves, or curvature of itself, it would cause all constellations to bob up and down, while floating about in all directions. This, of course, would mean that light trajectories no longed held straight directions, but swayed about indecisively.

Once again it must be said that no such occurrences had ever been observed nor are they likely to be, because there is no physical substance to the nothingness of vacuum space. As with all that is physical, the physical reality prevails with vacuum space of nothingness, despite any desire by those in science to award vacuum space of nothingness with impossible abilities.

There is one characteristic that applies to nearly all humans (conscios) including those who call themselves scientists: violence. It is in the light of this aggressive trait that humans (conscios) view themselves and all that surrounds them. And while scientists may not practice violence on each other directly – apart from devising all the weapons of mass destruction with which to threaten each other – they certainly perceive all physical formation to be derived of it.

So whenever there is any kind of problems needing a solution, their first reaction is usually that of violent intent of a collision. Should an asteroid threat to Earth be considered, the immediate advice is to smash it to bits with a nuclear missile (and have millions of small bits of asteroids hit the planet everywhere). Need to obtain a sample from a passing comet? Crash a space probe onto it.

Not surprisingly, with this kind of thinking, they have concluded long ago that formation of all matter had been achieved by particles smashing violently into each other. Also not surprisingly, their perceptions do not include possibility that formation of all objects and bodies, beginning with solid bodies (protons) and 'energy bodies' (electrons), were and are the results of inunens coming together gently, rather than violently. Soft and gentle unions are necessary to prevent unions from becoming instant dispersals from physical contacts between inunens. While unions may be concluded rapidly, the basic intention of unions is not to produce vast amounts of 'dispersing energy' conversions, whether this is possible or not.

When the desire to achieve a heavier particle to that of a neutron ('combined body') was proposed, there was yet another pretext for them to apply their formula of violent collision to achieve a desired result. This concept required the building and operating of a Large Hadron Collider (LHC), built inside a mountain at vast financial cost.

The intention of LHC was to obtain added mass to other subatomic particles, producing a form of heavier, slow particle, by conducting a near-speed-of-light collision of protons from opposite sides so as to find the so-called Higgs boson (a subatomic particle).

After years of experiments, in 2013 the operators of LHC announced, to a great scientific acclaim, that a Higgs boson had been confirmed. However, this achievement remains a conjecture, as the experiment's success was tentative, if not doubtful, as it is still unclear which model the supposed particle supports and if multiple Higgs bosons exist.

Still, considering that physicists desired the discovery of such boson, they got what they wanted, if not by fact then by fiction. It is certainly not a fact, because ongoing replications of the same experiment fail achieve what was hailed as one of the greatest achievements in science. For if it were a fact, then the violent collisions of inunens would be a common method by which they united.

But that is what science, and especially physics, is all about: unsubstantiated claims, without any detailed information, of subatomic particles' appearance and behaviour. Instead, scientific jargon is used to present unrealistic, badly thought through notions, even when there are some real aspects to those concepts.

Take for example the perception of 'black holes'. When the notion of 'black holes' was fist raised, it was not as 'black holes' but rather as perfectly spherical remnants of a star that had collapsed inwards, onto itself. When years later Stephen W. Hawking presented his version of 'black holes', according to his views their structure was meant to be literally holes in nothingness of vacuum space. He even claimed that such holes could comprise of two holes, with their separate 'event horizons, merging into one large hole. Or they could be a 'wormhole', supposedly traversing two regions of the Universe together.

This is so typical of those in science to provide a change of perception from sublime to ridiculous, with the initially correct hypothesis of a lightless, spherical entity with a powerful gravitational attraction (classified in this book as: Energy and Matter Absorbing and Compressing System – EAMAACS) altered into that of physical holes existing in nothingness of vacuum space. And that is where this perception remains to this day.

(Incidentally, while EAMAACS can become formed from an inward collapse of a very large star – its 'united energy' inunen contents being instantly converted into fully spherical 'dispersed energy' inunens, without allowing for any 'dispersing energy' inunens to disperse outwards – EAMAACS are able to be formed by 'dispersed energy' inunens in nothingness of vacuum space between universes, before producing new universes with the dispersal of their contents, comprising of newly compressed 'uniting energy' inunens).

Such miscomprehensions by those in science are directly attributed to their current acceptance of 'time', multiple dimensions, and flexibility of vacuum space – all of which are physical impossibilities.

And yet, a simple examination based on common sense alone is sufficient to establish that a physical shape of a 'black hole' cannot exist in a vacuum space of nothingness. In order to examine a 'black hole' let us consider such a structure, remaining on its own, surrounded only by vacuum space of nothingness.

To begin with, what would constitute a perfect shape of a hole? What about that of a glass test tube? An elongated cylindrical tube, with an opening at one end and a closed base at the other. (After all, this is exactly the kind of structures that were envisioned when the notion of 'black holes' was first presented.) Now imagine such a test tube shape in vacuum space, but which is not made of glass. If such structure were to exist, here are a few examining questions regarding its shape and function:

• The term 'black hole' insinuates a vacant space within a vacuum space of nothingness. So if a so-called 'black hole' is filled with energy and matter then the term of 'black hole' is already incorrect because that structure would represent a 'container' of some kind.

• If a 'black hole' is capable of attracting matter and energy, where exactly does that attraction come from when a 'black hole' begins its existence as an empty hole in vacuum space of nothingness? After all, gravitation is derived from some kind of mass, be it the Sun or Earth, but an empty hole in vacuum space would have no mass. So having no mass, by what magic do 'black holes' begin to have gravity with which to attract energy and matter? Or do 'black holes' pop into existence already filled to the brim with energy and matter?

PART 3. DISCLOSURE OF EVERYTHING

- For EAMAARS (Energy and Matter Attracting and Releasing Systems) to have gravity requires two directions of heat attraction: Firstly that of inward direction of heat attraction, which leads an attraction into the core of the structure; and secondly, this inward direction of heat attraction develops polarities that provide the external direction of heat attraction, which is the 'boundary of influence', currently known as gravitation fields.

It is these physical directions of heat attraction that influence each and every object and body in the Universe to develop a spherical external shape, like that of a ball.

Without either the inward direction of heat attraction, and the internal polarity heat attraction there can be no gravitational attraction at all.

If this physical fact were to apply to a 'black hole' that resembles a tube, then what kind of system of attraction must apply to a 'black hole' to allow it to retain such a shape?

If it has only the inward direction of heat attraction then it would still develop a spherical ball shape but have no eternal gravity attraction (exactly as EAMAACS – Energy and Matter Absorbing and Compressing Systems – which do have as a spherical shape, but do not have the external gravitational fields).

Without at least an inward direction of heat attraction, would mean that a 'black hole' is operating on principles that do not apply to all else in the Universe.

The concluding sentiment of this physical fact must be that either 'black holes' are not holes at all, but are spheres of EAMAACS, or they are a physical impossibility, in the manner of 'time' and 'space-time'.

- Of what do the sides of a 'black hole' comprise? After all, vacuum space of nothingness is not like solid ground on Earth, where a hole represents an absence of substance in the ground. So of what would the sides of a 'black hole' be constructed to allow them to remain rigid, separating the inside of a 'black hole' from the vacuum space of nothingness?

Furthermore, if holes in vacuum space are supposed to consume all matter and energy, then what prevents them from widening, until they became as wide as the galaxies, considering that vacuum space of nothingness cannot intervene to halt spreading of their sides.

- What are the principles that allow the front end of the 'black holes' in vacuum nothingness to remain open? On Earth friction between two opposing currents of air or water can produce tornadoes and whirlpools. A vortex in the air or water will close with speed reduction of the currents. So then, what sort of external activity, involving friction, is required to produce the opening in any 'black hole', considering there are no opposing currents of any kind in the vacuum space of nothingness?

And if it is presumed that the surrounding suns and stars are causing friction to make a 'black hole' stay open, then a more accurate description would be that the attracted stars and suns are actually washing over an EAMAACS by any other name, be that 'black hole', without the EAMAACS been influenced by any friction whatsoever.

- It is presumed that a 'black hole' draws energy and matter at its opening of the hole. But what would prevent this function to be conducted by the sides of the hole, or even the bottom. After all, by having no precise external membrane made of any specific material, a 'black hole' shape should be able to absorb anything from all sides and ends.
- 'Black holes' are always depicted with round black facings. If 'black holes' circular facings represent the beginning of a hole, from which that hole continues on in a cylindrical shape, then where do these cylindrical shapes hide? Or is vacuum space of nothingness has panels or curtains that hide the sides of black holes, as they are never mentioned or represented? Is it because the vacuum space of nothingness is actually opaque and not transparent as a nothingness? Or, perhaps, the vacuum space of nothingness has some beyond-the-vacuum-space-vacuum-space, into which 'black holes' slot in, so as to position the sides of a 'black hole' out of sight?
- What constitutes the depth of the hole, or the tube? Are all 'black holes' of same depth or are they all different in depth? How can this be recognised if the sides of 'black holes' are somehow hidden? And how come they are never depicted as endless 'worm holes' which they are supposed to be?
- How come all the indications of 'black holes' show them head-on, and never a quarter or three-quarter view? Surely some 'black holes' would be facing in all various directions, appearing as ellipses or even facing away? In which case, can a 'black hole' be recognised as such when it faces away from the viewer, and what does the bottom of a black hole look like?
- If 'black holes' originate with collapsing stars, which are spherical in shape, what causes these collapsed stars to alter their shape from a sphere to that of a tube?

Needless to say, there are many other similar questions that could be posed to all those who accept this concept of holes in vacuum space, to which there can be no serious answers as the notion of 'black holes' itself is full of holes.

But not to physicists, astronomers, and mathematicians. These scholars need no serious answers to anything, as they much rather devise their own perception of physical reality. For them, existence beyond Earth comprises of fantasies devised by their imaginations.

PART 3. DISCLOSURE OF EVERYTHING

They have what they call a "fabric of 'space-time' continuum", where space is directly connected to a fourth-dimension, which is 'time' that supposedly advances at one second increments, which is quite amazing, considering that a second is a human devised segment of one hour, with an hour being a segment of Earth's daily rotation before the Sun.

(How this adhesion between the nothingness of vacuum space and fourth-dimension 'time' is achieved is unclear to all, as scientists do not like such questions. To them it is a matter of their suppositions being accepted as 'gospel', without any presentation of concrete evidence).

According to scientists, objects, like our Sun, supposedly make a dent or a distortion in the fabric (whatever that may be) of this 'space-time continuum'. Therefore, if there were no such dents, or distortions in the fabric of 'space-time' then a passing comet would simply pass by. But because of the dent in the 'space-time', that comet would inevitably alter its direction. Obviously the comet would also cause dents in the 'space-time', as would the Earth and other planets.

This perception of how the so-called fabric of 'space-time' is supposed to function also leaves a few holes in logic, so to speak.

Firstly, if the Sun is to leave a dent in the fabric of 'space-time', then that dent would need to encapsulate the Sun from all sides, because this Universe is not laid out on a flat plane, but happens to be three-dimensional, where there is no conventional 'up' or 'down'. Therefore, if a dent – causing an inward slope – were to surround the Sun from all sides, then nothing could roll out or into it, or past it: it would be totally encased in the fabric of 'space-time'. That would mean the Sun's light and particles would be unable to disperse away from the Sun.

Secondly, if there were to be a local dent in the fabric of the 'space-time' surrounding the Sun, then anything surrounding the Sun, including all planets, would simply roll into the Sun. That is because this theory takes no account of the fact that gravity needs to have a component of 'grappled attraction', by which the planets remain at a distance from the Sun. In fact, if the notion of objects causing dents in 'space-time' were to be factual, then this Universe would comprise of one enormous object. This would happen as smaller objects rolled into the dents of larger objects to collide with them, and those into larger objects still, until finally there was just one solitary huge object.

Considering that such events are not taking place anywhere in the Universe, where, on the whole, objects and bodies in vacuum space remain apart despite their gravitational attractions, the only conclusion that can be gained is that 'grappled attraction' exists, while the fabric of 'space-time' does not.

Continuing with the current scientific beliefs, gravity is supposedly a local moving distortion in the fabric of 'space-time'. This supposedly occurs because any accelerating mass would give off gravitational waves, which would then ripple through the background fabric of 'space-time' at the speed of light. This infers that while electromagnetic waves travel across the 'space-time' of the Universe, gravitational waves are a travelling distortion of the actual 'space-time' of the Universe.

This would indicate in reality that if gravitational waves are distortions of 'space-time', with space being a vacuum nothingness, then gravitational waves would be travelling distortions of nothingness of vacuum space in the nothingness of vacuum space: a nothingness moving in a nothingness.

Such current scientific beliefs have no perception that external 'boundaries of influence' (gravitation fields, or rather gravitational shells) of large objects in space are but a large version of the external 'boundaries of influence' of inunens – individual units of energy. And the reason that the external 'boundaries of influence' exist is because they are external extension of the internal direction of heat attraction movements. They are all one and the same. And just as 'uniting energy' inunen will activate its external direction of heat attraction – which is the eternal 'boundary of influence' – once it senses another 'uniting' or 'united energy' inunen nearby, so do the large suns and planets.

In each instance their respective 'boundary of influence' has a limited range, but which can be extended by use of 'boundary of influence' of other nearby objects. An easy example of this would be an ordinary bar magnet and a number of paper clips.

A magnet is just a piece of metal that allows its internal structure of atoms to have a 'transfer of influence' passing of ebitransits – a movement of 'uniting energy' inunens, currently classified as an electric current, when allowed to do so in a closed circuit – moving from direction of high heat, which is where ebitransits are, towards the low heat, which is where they are not. When the passing of ebitransits in such metal leaves a permanently rigid 'transfer of influence' on all the surrounding atoms, this metal becomes a permanent magnet (until it is heated by 'dispersing energy', when it ceases to remain a magnet). When the ebitransit 'transfer of influence' is temporary, the function of this metal is called electromagnet. It is the permanent or temporary 'transfer of influence' that causes the attraction from some foreign metal objects with compatible atomic structures, resulting in the so-called electromagnetic attraction, where the 'boundary of influence' of foreign metal objects are activated to be attracted to the 'transfer of influence' within a magnet.

When a paperclip is placed pointing with one end towards a magnet, there will be no reaction until the paperclip is moved sufficiently close to that magnet, so that the paperclip's 'boundary of influence' becomes activated to feel its attraction to the magnet's internal 'transfer of influence'. If the length between the magnet and where the paperclip became attracted to magnet is marked, it would show just a small distance. However, if extra paperclips are attached to the first paperclip joined to the magnet, by pulling the magnet to one side, all the attached paperclips would be dragged along, like carriages following a train engine. When the distance is measured between the magnet and the last paperclip, that distance would be of much longer length to that of the original distance between the first paperclip's attraction to the magnet.

What this little experiment illustrates is that by use of the 'boundary of influence' of other objects, the 'boundary of influence' of the magnet had been expanded far beyond its actual ability. The same principle applies to the 'boundary of influence' gravity function of the Sun, where the 'boundary of influence' of the planets increase the reach of the Sun's 'boundary of influence'.

And the other important factor to note from this experiment is that when the overall extended 'boundary of influence' ended – where no more paperclips would join the others because their 'boundaries of influence' would not be activated from lack of the magnet's ability to attract with its internal 'transfer of influence' – that was the limit of the whole 'boundary of influence.' This means that there is a limit to the reach of every gravity, because when the range of any object's and body's 'boundary of influence' reaches its limit then no further attractions can be achieved.

Therefore, a 'boundary's of influence' shells (gravity's field) stop at a specific distance from the body of an object, like that of the Sun, without any 'gravity waves' being able to be sent into the vacuum space.

Besides this 'boundary of influence' limitation, there is yet another important relevance to the 'boundary of influence' which should have been understood by now: the external 'boundary of influence' shells (gravity fields) are only **indicators** of the external directions of heat attraction between the two external polarities of an object or a body in vacuum space: **they are not particles of any kind.**

The 'boundary of influence' shells (gravity fields) are like an invisible but physically felt pattern of signs to other 'boundary of influence' patterns of signs. They only exist as indicators without being physical particles in their own right. Therefore, these 'boundary of influence' indicator signs of external direction of heat attraction cannot be cast into the vacuum space as, for instance, like radio waves, because radio waves comprise of 'dispersing energy' inunens physical bodies with their own external 'boundary of influence' shells.

This then should make it clear that irrespective of the speed of any object of body in vacuum space, no part of their 'boundary of influence' shells (gravitational field) can be "given off" and sent as ripples into the vacuum space or the fictitious fabric of 'space-time', as there is nothing there to be "given off".

As such physical facts still elude current science, those in science are still able to declare discoveries of exactly these kinds of impossibilities. For example, in 2015, astronomers had announced the discovery of gravitational waves.

These gravitational waves were supposedly the result of two 'black holes' colliding billions of years ago, and 1.3 to 1.4 billion light years away. Supposedly, these two gigantic 'black holes' had originally orbited each other at a vast distance for 10 billions of years, while constantly coming nearer at an accelerated rate. While doing so, they both had supposedly emitted energy, in the form of gravitational waves. It is claimed that because of this constant loss of energy – gravitational waves – their orbits were becoming smaller.

It is claimed that just prior to their collision they were rotating around each other at 75 orbits per second, or at half the speed of light. The gravitational waves from the last eight orbits had travelled to this planet, and had been recorded by instruments on Earth, duration of which lasted for less than one fifth of a second.

As these last eight gravitational waves had rippled through Earth, they had supposedly caused the mass of the whole planet to experience eight compressions with side expansions, by a very tiny amount – one hundredth of a millionth of a millionth of a metre.

While there are more details to this "discovery", the above outline should suffice a rational examination:

For instance, if the 'black holes' had been emitting gravitational waves for billions of years, and the last eight of these gravitational waves had been recorded, this would mean that all the preceding gravitational waves from those two 'black holes' had been reaching this planet for billions of years, but were not recorded.

In which case, if the last eight gravitational waves were recorded by apparatus set up for many years, why were earlier gravitational waves reaching Earth not recorded previously, considering that a tenth and ninth last gravitation waves would not have been much different to that of eighth last gravitational wave?

Or is the inference here is that the last eight gravitational waves just made it to Earth, before having no more ability to travel any further in the fabric of space-time'?

PART 3. DISCLOSURE OF EVERYTHING

But is the fabric of 'space-time' not its own entity that functions with no limitations, so that any gravitational waves it carries would go on forever?

As a proof that this is but a fabrication, if the last eight gravitational waves had caused this planet to concertina its shape minutely, then all the previous gravitational waves – that had to have been arriving to Earth, and travelling past Earth for billions of years, just like the last eight gravitational waves – would have been doing the same: causing a constant distorting vibration to this planet's body. Over billions of years these ongoing gravitational waves would have caused this planet to disintegrate.

Considering that this planet is still in existence would validate, beyond any doubt, that no such gravitational waves had ever existed, had ever reached Earth, or had ever distorted its shape.

Furthermore, if the two 'black holes' – and now supposedly one combined 'black hole' – are 1.3 to 1.4 billion light years away, and the gravitational waves travel simultaneously outwards, like a wave on a pond's surface, that would mean those gravitational waves ripple trough all the planets, moons and the Sun of the Solar System, for billions of years, affecting their structures in the same manner to that of Earth.

Yet, again, as no harm had been done to the Solar System, this would indicate that no gravitational wave event from some distant 'black holes' had ever taken place.

Besides all these speculations, how do those in physics and mathematics can provide physical proof that the two former gigantic 'black holes' had orbited each other at 75 revolutions per second? They cannot, for they do not actually know. This claim is just another guess plucked from their imagination.

And then again, what about the supposition of "special relativity" that a speeding body encounters the slowing down of 'time' and enlargement of its mass? So if that were to be true then the two 'black holes' would increase in size while drastically slowing down their orbits of each other.

As with many other instances, those in science feel free to announce anything they want to because they know full well that they will have no need to produce physical proof to confirm their claims, as it is all beyond human reach or observation. Announce that the gravitational waves caused the planet to concertina? Why not? Provide proof of this and where exactly did the planet concertina? Not likely. Nor is it likely that any proof will be supplied as to how the gravitational waves look like, their content, their thickness and their size. And from exactly where and what direction did those gravitational waves were supposed to come from? Oh, just say: 1.3 to 1.4 billion light years away, that is enough explanation, even if no one has any idea what that means exactly. Make proper research to learn exactly what caused the apparatus to record a short segment of sound? Why bother? Just claim it is the gravitational waves and be famous.

What is most peculiar about those in science is their approach to research. For example, consider events such as global warming. Global warming is not a recent phenomenon. It had been studied for decades by thousands of scientists in many various scientific fields: everything from gathering and assessment of atmospheric data, to water temperature research. While most scientists agree now that global warming is caused by human activity, that is: uncontrolled emission of carbons – caused by needs of overpopulation –there still remain those who are not convinced, who argued that global warming is nothing but a conspiracy and a natural occurrence.

The points of such contention in science represent a fact that there had always been, and remain, debates and disputes between scientists (with many defending their altering points-of-view with allocated research grants provided by interest groups) in regards to events taking place on Earth.

And yet, when it comes to out-of-this-world events, such as 'time', 'space-time' and claims of gravitational waves and 'black holes', there is an absolute absence of any scientific contradiction and debate. No one publishes papers with alternative views and theories, and no one presents arguments based on logic and physical reality. It is as if any theory presenting illogical, silly, and physically impossible notions are given full blessing by all the scientific community: appearing incapable of rational thought.

With this kind of distinct difference of attitude towards on-Earth and off-Earth scientific study, questions instantly arise: why do scientists have constant disputes about what is under their scientific noses on Earth, but maintain an accord on all that which is somewhere out there, far, far away from this world and beyond any physical proof? Why do intelligent and highly educated people devise and apply nonsense to science; the very science that they vocally profess to respect, uphold and expand?

Well, the answer lies within another, similar question: why do many intelligent and highly educated people believe in gods? From that there is but one answer: belief. It is 'belief' that applies similarly to religious theists as to scientists. The religious faithful believe in what they had been taught to believe by representatives of their religions, while all current scientists believe in what they had been taught to believe by their science tutors.

Whether religious or scientific beliefs, the results of such indoctrinations were devised (with assistance of their brains) by earlier generations of conscios, who had little understanding of physical reality. Nonetheless, this

PART 3. DISCLOSURE OF EVERYTHING

ignorance did not prevent either the religious or scientific participants from accepting and embracing their forms of fiction.

There is, however, a major difference between the faith of theists and of those in science, which concerns hypocrisy. The religions do not presented themselves as being at the forefront of human advancement, while science presumes to do just that. Science claims to inspire its participants to search for the uncompromising truth of all existence, while actually basing its knowledge on physical unrealities, including that of 'time': an ancient notion devised by figment of ancient human imagination. And now, by continuing to accept these unrealities, those in science seriously compete in inventing even more ludicrous theories then before, with an absolutely dismissal of the reality of all physical limitations that constitute all physical existence.

DECEPTION OF SPACE VOYAGING

In allowing all the resources that physically exists on this planet to be used up as quickly as possible, the business enterprises, politicians, and scientists assure those whom they control that this planet is but a stepping-stone to the stars. They all proclaim that humans have the abilities to voyage into space, so as to colonize neighbouring planets and moons. Once there, it is expected that humans can exploit the mineral resources found on those planets and moons. In response to such presumptions there are already those on Earth who demand the right to trade in real estate of other planets and moons.

But these who propagate the human dreams of ruling the stars, do so not because they care whether humans achieve this goal or not. Their only intent is to obscure the reality of human wasteful and needless over-consumption of this planet's natural resources, by presenting a reassuring fiction that promises infinite human life-support on other planets and moons.

And yet it is inevitable that sooner or later humans shall realise a very important fact about their existence: despite all their very expensive short-term space voyaging and the permanent maintenance of space stations, and dispatching of satellites and probes to other planets, they, as humans, are all completely bound to a spaceship lifepod, called planet Earth.
Earth is in a group of other planets, which (like Earth) orbit the Sun, but which (unlike Earth) are inhospitable to life forms. Whether any of these planets, like the Earth's neighbor planet Mars, have remnants of any bacterial life forms is immaterial. The fact is that Mars is not suitable for human life. This is not just due to an absence of air and water, and the presence of inhospitable weather

conditions of fine dust, extreme cold, and high levels of 'dispersing energy' radiation from the Sun. For there is yet another vital ingredient imperative to all Earth life forms. This 'must have' entity is the Earth's 'boundary of influence': its gravity. The lack of Earth's gravitation would be the most damaging factor to any attempts of living on places like Mars.

On Earth, all life forms have incorporated an awareness of Earth's 'boundary of influence' (gravitation) into their physical body structures; the designs of which intend to provide both a resistance to, and an acceptance of Earth's gravitation, so as to produce unique gravity-influenced actions with which to maintain their survival and existence. This equally applies to human bodies.

Human bodies are structured to overcome influences of Earth's gravitation by use of muscular and skeletal structures that absorb stress and loss of 'dispersing energy'. Simultaneously, the positioning of internal organs takes advantage of Earth's gravitational attraction, so as to achieve an economy of effort in processing the intake of 'uniting' and 'united energy' (nourishment) and converting that into 'dispersing energy' – an essential process for any living organism – continuing this all the way to the release of any remaining body waste.

Even the Earth's daily rotations before the Sun had been incorporated into human brain and body structures, so as to assist their brains and bodies to rest and recover during the night from daylight hour's influences of physical stress applied to their bodies by the Earth's gravity.

Bodies evolved to be exclusively attuned to Earth gravity place a limitation on their ability to live where gravity is different or is absent altogether. Considering there is no other planet or moon in the Solar System with a gravitational attraction to match that of Earth signifies that this is an insurmountable problem to human expansion beyond Earth. Mars' gravity, for instance, is only one-third of Earth's, while the Moon's gravity is one-sixth of Earth's. These differences means that even a few years spent away from Earth's gravity on the Moon, or Mars, or in vacuum space with zero gravity, would result in that human body experiencing a substantial, and possibly irreparable damage. This is because such environments will not provide assistance to normal body functions, like the circulation of blood in a body that produce bone and mussel tissue regeneration, and the transfer of Dioxygen to a brain.

Bodies deprived of such support result in rapid body and brain deterioration, due to their inability to function and to recuperate properly in a foreign gravity. And any physical devices proposed by humans to provide life-supporting aids to humans living in foreign gravity shall never, ever prevent these humans from experiencing accelerated body malfunctions and premature mental and physical deteriorations, leading rapidly to an early end of life.

As humans have a total physical dependency on Earth's levels of gravitational attraction, then (together with other logistical problems arising from human requirements essential for life) any proposals of them colonizing other planets and undertaking long-distance space travel are a pure fantasy.

Our Solar System is very far from its neighbors. Considering that the nearest star to Earth is some four and a half light years away (that distance being equivalent to travel conducted for four and a half Earth years at the speed of light), and considering that there are no space vehicles that can instantly provide such speed of light motion, then it would seem that a very long period of Earth years would be required to travel that distance. Even if presuming that it would require human travellers some 500 Earth years to reach their destination by a space vehicle capable of maintaining incredible speeds (and ignoring the human need for Earth gravity) there would be other concerns needing a consideration.

For instance: there would be the human problems with personal relationships, including: love, sexuality, lust, envy, selfishness and hatred. There would be problems with materialism and capitalism (indoctrinated by most societies into most humans from early age), which teach intolerance towards socialism and possession-sharing, while glorifying selfish individualism of capitalism, endorsing greed for possessions and encouraging quests for power. Then there would be problems with reproductions and diseases. And, of course, there would be the demands of basic human needs for life: air, food, water, followed by fuel and other energy supply storage, requiring to last at least 500 Earth years.

Even if a superficial comparison is made between the presumed requirements of long distance space travel to that of actual human existence on Earth since the last ice age (of only about 14,500 Earth years), whose diets were based upon a vast diversity of fresh foods derived from fruit, vegetables and cereals, domestic and game animals, fish, birds and the honey of the bees – all of them depending on the influences of the Sun, Earth gravity, and the Earth environment – then the only conclusion to be reached is that to provide even but a fraction of all this for a long space voyage is nothing short of human delusion, equivalent to humans being able to fly by flapping their arms.

But there are yet other physical obstacles to any presumed long-distance space travels by humans. The first of these is the presence of solid body objects and bodies in the vacuum space, and the second is the non-imposing nothingness of the vacuum space.

While the vacuum space of Eternity – be that even the region of the Solar System between planets – may appear as empty, it is actually full of various physical objects, ranging in size from large asteroids, to miniscule rocks and pebbles, as well as particles of atoms, grununs, ununens, and inunens. All these objects, bodies, and 'dispersing energy' radiations move freely everywhere, due to having no restrictions of movement in the vacuum space of nothingness.

While the space objects move freely in vacuum space of Eternity without any preordained plans – colliding with each other if that is unavoidable – the intention of human spaceship would be to avoid any such physical contacts. But because the vacuum space of nothingness presents no physical impediments, this meant that unlike the spaceship acrobatics depicted in make-believe motion films, actual spacecraft cannot stop or manoeuver quickly in vacuum space: the faster the speed of the craft, the less is its ability of manoeuvring.

At a very high speed, the slightest correction to a spaceship's forward direction would cause the craft to spin out of control – mashing everything inside, like a kitchen shredding machine – while continuing in the same, virtually unchanged direction, at the same rate of speed, with continued rate of spin. To make any directional changes in vacuum space a spaceship would need to come almost to a halt, slowly make the required adjustment of direction, before accelerating once again. Such a procedural complexity would be unlikely to enable a speeding spacecraft to avoid a contact with some space debris; the consequences of such collision being that an impact with even a tiny object would result in a destruction of the whole spaceship.

Then again, there are also the possibilities of being absorbed by one of countless small EAMAACS, which cannot be detected in vacuum space as they absorb everything, including any exploratory 'dispersing energy' radars employed for detecting objects and bodies in vacuum space.

Finally, and most decisively, here is the fundamental reason why humans can never ever become vacuum space voyagers.

It is already know that human bodies deteriorate in zero gravity. The longer they are exposed to zero gravity the more their bodies become weaker, despite all precautions taken to counter this problem. Current astronauts maintain a daily regime of physical exercises and mineral supplements to assist the muscles in the body to retain their strength and body mass.

Unfortunately, the problem with lack of gravity for human bodies is more profound than is currently known or understood.

Gravity, temperature, and air pressure have fundamental influences on inunen behaviour. While increase in temperature and pressure cause inunens to 'relax' their 'boundary of influence', and by that become more likely to leave their unions, and with that more likely to be more liable for dispersals, a colder environment with less pressure, or a stronger gravity, influences inunens to retain their unions with more vigor, and with that being less involved with dispersals.

This means that while on Earth, all the atoms that form all the human bodies are affected by Earth's gravity, temperature and air pressure to function as they do. This also means that the reason why all the atoms function as they do is because the same Earth gravity is influencing all the inunens, that form all the atoms, to behave in a constant manner. Because of the particular Earth gravity – but also Earth's temperature and air pressure – all the inunens retain a constant level of behaviour, represented by the volume of their dispersals and the level of strength of their dispersals.

In order to survive on board any spacecraft, human brains and bodies require an average temperature of Earth's heat and air pressure, similar to what they experience on Earth, but without the Earth's gravity.

Once the influence of Earth gravity is removed from a human body, all the inunens alter their behaviour. They no longer retain the same levels and volumes of dispersals, because they become 'relaxed', or less inclined to retain unions, due to an absence of Earth gravity.

While this may appear to be inconsequential, in physical reality this signifies a breakdown of atoms, because inunens cease to maintain their unions. This takes place slowly but definitely. At first it is the muscles that suffer the lack of gravity. Gradually organs begin to fail. As the body decomposition becomes more pronounced, the body liquids begin to congregate within the body into a single mass.

Because liquids have the ability to maintain their own gravity, which is slightly stronger to that of Earth's gravity, they can gather into a three-dimensional mass in zero gravity environment. This can be witnessed with some water taking shape of a blob in the zero gravity of a space station, or a spacecraft. And that is exactly what eventually occurs within a human body when the bodily fluids begin to congregate inside a body, leaving much of the body dehydrated.

With an extended duration of a human body remaining with an absolute zero gravity, away from all gravities of the Sun and other planets, a human perishes, while the body itself continues to take on a spherical shape, as the body decomposes and all the liquids – that used to be a body, come together into a single rounded form.

After human flesh becomes the first casualty of exposure to an extended zero gravity, eventually even the structure of the spaceship, and all its contents, will undergo similar deterioration, and change of its overall shape.

These factors may not be appealing to the human (conscio) dream of colonizing the stars and being masters of the Universe, but unfortunately for humans, physical existence is what it is, and cannot fulfill human imagination.

There is no doubt that there are those in science who shall claim that an artificial gravity is possible to achieve. Also, no doubt, that there will be those who shall believe them. But according to physical reality nothing that humans may do, thanks to their brains, can ever replicate Earth gravity. Certainly not a centrifuge, to produce centrifugal force.

So where does all this leave humans (conscios)? Absolutely and completely bound to Earth, which is absolutely and completely isolated in vacuum space. We, humans, may have projections and dreams of conquering the planets, moons and even the Universe, but neither our unsuitable-for-space-travel bodies, nor our deep isolation in vacuum space will ever give us too many options on where we can leave our permanent footprints, other than Earth.

This physical reality must convince us – hopefully sooner than later – that apart from this planet our lives have no other solution for a lengthy survival.

For the moment, Earth – our only spaceship lifepod – is still sound and steady, but only just. The resources on their spaceship lifepod are finite. These need to be treasured and carefully managed – not needlessly squandered – for they are irreplaceable. These resources, including air, water, and land, need to be integrated and shared between all life forms on Earth, and not just between us, humans; for each life-form has a relevant and vital function for the benefit of the whole.

They have to be cared for, if they are to continue providing their life-support to us, humans. Once a link in nature is broken it will impact on the ability of our survival. Isolated in vacuum space on our spaceship lifepod Earth, we have a choice: act in discord or act in harmony, where discord leads to a speedy termination, while harmony leads to longevity of existence.

But as it is not possible for us, humans (conscios) to leave Earth, why should we even worry about longevity of our existence? Why not just use up all there is and be done with it?

We can do that, of course, as there are no gods to stop us. Many humans are doing just that, right now, in the present-moment. These highly selfish humans always feel justified in proclaiming and practicing rampant consumption, avarice and greed, with no respect given to anyone or anything, apart from their own existence. They even entice the less selfish to join them in an orgy of dividing, developing, buying, selling, consuming, wasting, and destroying all that they can get their hands on.

Alternatively, as any human condemned to die – as all humans are – we can learn to savor every moment, not in greed and avarice but in wondrous celebration of our lives, of this whole planet, and of this Universe. We can begin to accept having less while enjoying it more. We can conduct our life's journeys on this spaceship lifepod Earth, by sharing; and by doing so, dispensing with enemies to confront and fear. For how can one have enemies when one is all and all are one?

26

Deceptions in the name of gods

It is currently presumed that science cannot disprove the existence of god – any god. This is perfectly true. It is true not because gods actually exist – which they do not – but simply because current science is flawed. These flaws stem not only from those in sciences accepting ancient religious notions devised thousands of years ago, but also from their inability to differentiate between physical reality and fictitious nonsense.

There is a shared belief that exists between those in business of science and those in business of religions. Both the theists and the scientists believe in an invisible entity, which, while never being seen or physically examined, is presumed to control everything. Apart from theist belief in gods, this shared belief is that of 'time'. Those in science believe in 'time' that affects everything, including past and future, while the religious believe in 'time' that supposedly connects the beginning and the end of the world's existence, while delivering (at certain periods and at certain locations) the preordained godheads and messiahs.

As science and religions are linked to one another in their common beliefs in something that does not physically exists, it is very unfair to accuse the religious of being deluded when those in sciences are not only deluded but are actually oblivious to the fact that they are deluded.

However, it is possible that those in science may recognise just how much their current understandings are flawed and correct these mistakes. The same cannot be expected of those involved with religions. They much rather uphold their beliefs of nonsense with pride at being blindly obedient to the ancient texts of fiction. But just because obstinacy is prevalent in their acceptance of beliefs, this does not make their beliefs a representation of any actual reality.

A human (conscio) strong and habitual liking of presenting lies to themselves – so as to avoid facing the physical reality – inevitably end up causing a great deal of mischief and waste for humanity. In which case it should be vital to provide a choice to those who may want to review their current opinions. And in order to do this it is necessary to present logic and reason why all that humans (conscios) believe to exist does not. This information, therefore, is not

PART 3. DISCLOSURE OF EVERYTHING

presented to convince zealots to alter their beliefs, but to those who may actually want to know why gods cannot have a physical existence.

The physical facts, as they apply to this planet, is that Earth is in a constant physical process of change. The same can be said of the Sun and the whole Solar System. Any physical changes occurring on Earth are immediately noticed, because we, humans (conscios), reside on this very planet and become instant witnesses. Should an earthquake or a volcanic eruption take place somewhere on Earth, these seismic changes are noted but not feared as some form of supernatural act imposed by some supernatural being.

Similarly, while any physical process of change affecting the Moon may not be as quickly noticed as those on Earth, be it a comet strike or a ground tremor, whatever those physical changes may be, they will still be accepted as a natural physical events taking place in the ever changing environment of physical existence.

As all such physical events are understood and accepted by human (conscio) comprehension to be representations of physical reality, this signifies that physical reality is the normal physical state of all existence on Earth.

Therefore, reinforced with the knowledge that physical reality governs all events on Earth, can any human (conscio) point to any period of human existence and declare a witness to any kind of unnatural events taking place on Earth, which were contrary to the accepted and normal physical reality? Events, such as: a god floating in the air before human crowds, and appointing other individuals to be god's representatives on Earth; circumstances of humans, animals and buildings being sucked into a fifth-dimensional void; or an appearance out of nowhere of a futuristic city with its futuristic occupants and futuristic technology, being planted onto some area on Earth by a 'time' vortex?

No.

The reason that there had never been any unnatural occurrences is because everything in physical existence is comprised of the very same inunens (individual units of energy) that physically form everything physical. By being responsible for all physical existence and for all the physical activity of that physical existence, their physical limitations prevent anything unnatural, or supernatural, or impossible – such as 'time' – from occurring. So even if inunens somehow wanted to supersede their physical abilities in order to produce some unusual predicament or existence, the limitations of their physical abilities at changing their body phases would prevent this from happening.

By observing that there are no examples of unreality occurring anywhere on Earth – as anywhere else – even if the reason for that is not currently known (because current science knows not of inunens) this could still be thought of as a sufficient cause to reject any notions that proclaim any existence of unreality. Not for human conscios, it seems.

Despite all the human conscio claims and assurances that they practice reason and logic, not only do they fail to do so, they fail to register what is plainly in their sights: the constant view of all the physical reality that surrounds them day-in and day-out, without there ever being even a miniscule sample of unreality.

This failure to adhere to all the surrounding physical reality is caused by one basic factor: a selfish conscio desire to stipulate that physical reality is not what it is, but is that which a conscio wants it to be. Such selfish attitudes have moulded human conscios to accept the dictum that it is human (conscio) opinions – in the form of a body or action obtained from imagination – that represents reality and not the surrounding physical reality. The surrounding reality may remain as it is: the same old boring reality, whereas the imagined super-abled entity has an exciting presence, with powers – granted merely by imagination – that supposedly supersedes anything that physical reality has.

The problem of attitudes that disregard physical reality is in that they succeed in influencing humanity to ignore, if not dismiss, the causes of physical change that begin to take place in physical reality, such as the climate change. By believing in super-beings and super-powers that supposedly can protect humanity from events occurring in physical reality, the faithful fail to take necessary actions for self-preservation, preferring, instead, to ignore the obvious change, pretending that it is of no consequence.

Unfortunately, when such accepting approach to unreality provides a false sense of security, as if life can go on as before without any remedial actions being taken, this begins to directly threaten human existence. It is one thing to be entertained by fiction of fables, fairy tales and sagas of, say, Greek gods and the like, and quite another to be subjugated into mental slavery of unreality by the Judaic, Christian, and Islamic religious lies: lies that should be understood exactly how and why they are lies, because they are not, and never were of any benefit to humanity.

One prevalent human (conscio) characteristic is a weakness for lies. They do this because they like lies. Lies provide many advantages. Lies hide their intentions and liberate them from all their physical, social, and financial shortcomings. This makes lies comforting, and equalizing. Lies are means of pacifying others. Because physical reality is often boring in its repetitious sameness, lies reassure liars that there is hope of escaping the reality of life, even if just at the end of life. Lies are means of reassurance in diplomacy and in giving of compliments. Lies are means of subversion, where lies of liars can reduce the high achievements, abilities, and prowess of others, while enhancing their own meager efforts and abilities. Lies are weapons of domination. Lies are tools of converting a disadvantage into an unfair advantage. Lies are weapons of destruction. Lies are the source of distrust, suspicion, and disharmony. Lies are means of impressing, selling, promoting, hiding, distorting, and avoiding anything that is unpleasant but true. Lies are thrilling, challenging, and liberating to liars, as they free them from conventions, from morals, and the burdens of honesty and responsibility. Humans (conscios), as they presently exist, cannot do without the use of distortions of lies; big lies; little lies; white lies – lies invented for themselves or for others.

When confronted with lying, liars lie again and again in their efforts to evade their actions of lying, by trying to re-affirm their lies to be the truth. Even when provided with irrefutable proof of lying, liars try to lie their way out, by placing the blame onto others.

Fools are not fools because they behave foolishly, but because they allow liars to fool them over and over again, just like the electorates who allow political liars to keep fooling them.

Because lies give hope, liars know that in seeking hope the gullible will accept their lies. That is how cultural traditions became established: lies fabricated by past liars being embraced as truths, so that even the current liars believe the same concoctions of lies, and by lying in the present, base their lies upon the earlier lies. Therefore, currently, ranging from political claims and promises, religious claims and promises, and advertising claims and promises, lies continue to rule by liars.

Most humans had, and continue to remain smug in their self-assurance of being able to lie with impunity, because proof has to be physically provided to show that an event had physically taken place when they claim it did not, or did not physically occur when they claim it did. It is these kinds of imputations that were used in books of religious fantasy.

But while many humans have become so used to lying that they willingly accept religious lies (just as others accept scientific lies) this should not stand in the way of those who may want to distinguish lies from physical truth, so as to be aware of lies.

The basic technique to do this is to examine how claims treat physical reality. Lies will either avoid physical reality, or distort it, or belittle or over-exaggerate it. Any such avoidance and disregard of physical reality confirms those claims to be lies.

In applying this method to theology, the Hebrew god, Yahweh – a precursor to Christian and Muslim gods – can be used as an example.

While human verbal statements can be – and often are – claimed at a later date to have been misunderstood or misrepresented, a written or printed word is as unchangeable as that carved in stone. This means that in such non-retractable form, any ancient writings – especially those relating to gods – can be scrutinized in comparison to physical truth, or physical reality: the very physical reality that every human on Earth experiences daily, and where nothing unnatural or supernatural ever happens. That is: there are no human masses and their property being sucked into a 'time' warp of a fourth or fifth dimensions; no miraculous regrowth of amputated limbs; no booming voices coming out of nowhere, giving instructions from god or from space aliens.

By applying strict meaning to the text, all ancient written claims that happen to be lies become easily exposed as lies, from having been composed by ignorant individuals lacking any knowledge of physical reality as it applies to structure of human body; this knowledge obtained from fields of study of anatomy, chemistry, and biology.

This method of uncovering lies applies especially to the devious authors of the *Hebrew Bible*, who (while stealing the intellectual property of the many myths and spiritual beliefs from many surrounding national and religious cultures of that period) wrote their lies with intentions of mentally – and by that physically – enslaving their own people, by presenting their lies as being an absolute truth. But just because those early faceless scribes had succeeded in convincing most humans – for thousands of years – that their claims (lies) represented the 'Godly' truth (a lie in itself), this does not mean that all these lies cannot be exposed for the lies that they are.

The proof to the god Yahweh lie is present in the very precise claims, made by the Hebrew authors, that their writings are depicting a real, one-and-only god, who was responsible for real, physical events. One of the fundamental claims attributed to Yahweh was that as a singular god, he, Yahweh, 'created' everything out of nothingness. This includes the 'creation' of the first human, Adam, in his (Yahweh's) own image. The authors do this not as a turn of phrase or a symbolic gesture, but as a definite, unambiguous claim that their god, Yahweh, had 'created' man in a form of Yahweh.

[Genesis 1:26] *Then God said, "Let us make man in our image, after our likeness;"*

PART 3. DISCLOSURE OF EVERYTHING

And:

[Genesis 1:27] *So God created man in his own image, in the image of God he created him; male and female he created them.*

In writing this, the inventors of Yahweh had made an irreparable error: they directly linked their spiritual god to that of a physical Homo Sapient human body; a human body whose structure has organs, and whose functions rely upon the influence of Earth gravity ('boundary of influence') and the Earth's environment, including air for breathing and food for energy, obtained by passing the food though the digestive organs of the body.

There is also the fact that god is quoted to say, "Let us make man in our image, after our likeness;" before Adam, the first human, was even supposedly 'created'. This shows just how dismissive of physical reality the Hebrew writers were – and the faithful still are – for who would have recorded any of god's words before there was any 'man' before Adam to record those words? If this were a physical reality, there would be no one at all to record the words of god. Furthermore, how could the words of god be recoded before early humans developed speech and languages, and before writing was invented?

Such disdain for physical reality is present throughout the *Hebrew Bible* and *The Old Testament*. This, of course, illustrates that it is the authors – the inventors of Yahweh's adventures with the Hebrews – who took upon themselves to devise the lie of some god taking credit for hundreds of thousands of years of human evolution.

But just as the Hebrew scribes knew nothing of logic that demands a witness to have existed **before an event** of god giving speeches, so as to be possibly present at that event, they also knew nothing of human body.

Every part of a human body has a purpose for being and looking as it does; where every limb and organ provides a physical function necessary for the whole of the body to live and function specifically on Earth. A male human body requires a mouth and teeth to consume food; an anus and a penis to release body waste from within the body; with the penis also being a necessary device for a reproduction function. The heart circulates the blood which has Dioxygen obtained by lungs and passed in blood to the brain, without which the brain dies. Additionally, all human bodies have been adapted to Earth's gravity ('boundary of influence'), without which the bones, muscles and organs deteriorate rapidly, causing a difficulty for life control. Then, and most importantly, there are the human brains, without whose functions their bodies would be lifeless.

Now, in order to establish how much physical truth there is in the old Hebrew claims that their god, Yahweh, had 'created' man in his own image, here are a few simple but probing questions.

If god 'created' man in his own image then his body has all the organs and structures of a male human body. In which case, how could an almighty god, himself – who, supposedly, 'created' the Universe as well – have a body requiring Earth's gravity and surface pressure? Would that mean god is an earthling, stuck on Earth just like humans, unable to exist anywhere else but Earth?

Considering that every human has a need of a nose to smell with and to breathe through, a mouth full of teeth to chew with, stomach, intestines and other organs to process food, and a rectum to expel bodily waste, then why would an all-powerful god in space (or in his heaven) need the same organs that are imperative for human life? Having teeth in a mouth, which humans use to chew food, beginning a process of food conversion that ends up as body waste, leaving the body via the rectum. Does that mean eternal god needs to eat daily and to pass body waste? If god excretes body waste, what is that excreta; where does it get buried, or does god uses a toilet as many humans do? In which case who has built a toilet for god? And how and on what does Yahweh clean his rectum? Do angels come and wash his backside, as was the custom with ancient kings? And despite being an immortal entity, how often would god need to expel excreta, something that humans do at least daily?

As to having a penis – as human males do – the function of which is not only to urinate but to be used as a tool of impregnation, a question inevitably arises: with whom would god have sexual intercourse with, considering that he is supposed to be the one and only god? Would he use the angels as his sexual partners? And would they become pregnant, considering that the intent of ejaculation of sperm is to impregnate. And if god had angels to impregnate, (whom god also would have needed to 'create') then the only ones that could become pregnant would be of female persuasion, as males cannot become pregnant no matter how much god may have tried. If having a penis himself – in the image of Adam – and knowing that it is used for procreation, why would god need to 'create' Eve out of Adam's rib?

Then again, if angels were also in god's image, why would they need human bodies, the kind of bodies that humans need to sustain life on Earth? And if angels had human bodies but with wings (considering that no animal on Earth has wings, arms and legs as part of the same body structure), then what muscles would they use to flap the wings, considering that it would require huge wingspan with enormous muscles to propel a heavy human-type body through the air. That leads to a supposition that Heaven is full of air, as there would be no need for wings if there were no air to fly in.

Returning to god, why would an immortal god have the need of lungs – so necessary for human bodies – or does that mean he really does require Dioxygen, without which he could not live, just like humans? And if that is so, then how could an immortal god be immortal when that immortality had a limitation in needing air? After all: no air, no god?

If Yahweh needs internal organs with which to digest food – as human bodies do – then what nourishment does his body receive to be eternal?

But then again, being in "god's own image" – and having a similar organ structure – why are humans not immortal?

And to what use would spleen, liver and kidneys – so necessary to every human body – be to an eternal god? How could he have living hair follicles in order to have a beard and hair on his head if he has no blood (or can he bleed?), and how, and where, and by whom, do they get trimmed, or is he overgrow by hair? And how and where does god wash and comb his hair and beard?

(In mentioned 'blood'– something that Adam would have needed to live on Earth, while simultaneously having blood 'in the god's own image' – a question can be asked as to by what means would a god have blood? Furthermore, considering that blood freezes at low temperatures and disperses in vacuum space of Eternity, how could a god needing blood live anywhere but on Earth, where there is no 'heaven' to be found?

There is also the question of god's brain. To have created humans in his own image means that god would also had to have a brain. A human brain maintains its conscio (commanding override negotiator, selectively controlling input-output) consciousness system. Would that also apply to god? If so, then god-the-brain would be co-dependent on its conscio and, therefore, would not be omnipotent, because human brains hide a great deal of information from their conscios. Furthermore, having a human brain means education for a young brain, as no human brain acquires instant knowledge. In having a human-type brain, who exactly educated god to be omnipotent? Did this one-and-only god have tutors?

And if god with a brain could supposedly produce 'creations', and humans are "in his own image", then why are human brains incapable of producing 'creations' out of nothingness but can only produce what they need out of that which already physically exists?

Then again: who created god in the first place? When a newborn emerges from its mother's womb it is still attached to her by the umbilical chord, which, when severed, heals into a belly button. "In his own image" would signify that god also has a belly button. To whom then was god connected by his umbilical chord? If Yahweh was created by some other entity, would not this make that other entity supreme over Yahweh, and in the process making Yahweh neither singular nor almighty, as claimed? And if 'Adam' was created in god's own image, this would mean that he too had a belly button. So, therefore, to whom had Adam been attached by his umbilical chord if he was 'created' by a singular god?

Every human male shares a female physical characteristic of having two nipples on their chest. Nipples on human males are simply there because in the process of establishing a body gender they are either used as teats for breast milk-secreting mammary glands, as in case of females, or they do not provide this function, as in case of males. If god created a human male "in his own image", that would mean the singular god has a body that has nipples as all human males do, which would be active if the body developed into a female – this variable occurring only in the female embryo. In which case, god with nipples could have developed only in a female embryo, the male-female outcome determined by chromosomes containing DNA of the male sperm and the female egg. If that is so, then god had to have been the result of a sexual intercourse, resulting in sperm and the egg producing an embryo, then a fetus. If that is the case, then the singular god – who is always depicted as an elderly male, never a teenager or a young man – must have been a child, an infant, and a teenager in his past. If so, then who were the two beings that produced god? Who took care of him when he was a baby and an infant? Who nursed god with breast milk, and changed his diapers?

If god ages, does that mean he will reach a stage of being docile and decrepit? If he ages then eventually the eternal god has to reach a point of death. When the eternal god dies, who then shall take care of his burial rites? Will it be the same angels in heaven? After a singular god, who would take his singular place? Another singular god, perhaps? Which would mean that gods are not eternal.

In 'creating' a human male "in his own image" would also mean that god has growing fingernails and toenails, as do all humans. But as Yahweh is indestructible, then, by such inference, so would be his fingernails and toenails. Were an indestructible god to have indestructible fingernails and toenails, then, by what method does he trims them? Perhaps cutting them off with scissors or a knife? But where would these come from? Perhaps god had also 'created' them? Or could there be ore mines to produce metal in heaven, for angels to invent and make knives and scissors? Perhaps they have a heavenly forge where they also had made Gabriel's horn? And if the eternal god does not trim his nails, does this mean that, by now, he has fingernails and toenails that are thousands of millions of miles long, considering that Yahweh is immortal?

In any address to Yahweh's overall appearance, there had never been any indication that he had presented himself naked to his chosen priests. Why does a god require clothes in space or heaven, if preservation of body heat or protection of indestructible body is not required by an eternal body? If it is a question of modesty, then why does a god, who could 'create' all, including a perfect body for himself, be fearful of being seen publicly in the nude? Why would an almighty god experience insecurity of own nakedness by embracing modesty? Would such a fearless god be afraid of being scoffed at, or laughed at by mere mortals?

As Yahweh was presumed to wear clothing reflecting the clothing of the period, would that mean he wore animal skins when the early humans wore them, and now wears suits, shoes and a ties? Or has he nothing new to wear, and consequently is hiding in old rags, embarrassed to show himself, and by that avoids all public appearances, which is why he is never seen? But if he has new garments, where did the raw materials came from, then as now? Who produces the fabric, designs his outfits, tailors the fittings, and sews the garments? Or is there just one sheet, as indestructible as god – but how could that be? A god equal to a sheet? And where did that sheet come from? Another of god's 'creations', with him declaring, "Let there be a sheet for me to wear!" And when god supposedly appears and disappears – as he was supposedly had done in the ancient days – this would make his garments equal to god in ability to appear and disappear: so equal to god in ability.

And why would a supreme god, who flies about, and supposedly made appearances to Moses, have the need of limbs and digits? After all, there is no description of Yahweh walking or doing anything with his hands – not even for Moses (according to the conversation between god and Moses, when they were supposedly all by themselves, but still recorded by yet another unknown third party), with Moses having to do all the writing of the commandments to god's dictation, lasting a month, with no food or water. So what is the reason a god would have hands with fingers and feet with toes if they are never used in a way that humans use them? Just to watch the nails grow?

If Yahweh was responsible for having 'created' Hebrews-Israelites-Jews in his own image, then in whose image were all other races 'created', and by whom, including all those earlier species that existed well before Homo Sapiens? And who 'created' all the life forms for hundreds of millions of years before the earliest humans evolved? An ape god, perhaps?

If Yahweh, as an almighty god, could 'create' all that he supposedly did, why then did he not give Hebrews all the lands in America, and North Europe, and India, and Australia? Better still, as an almighty god, why did he not 'create' a whole planet just for Hebrews-Israelites-Jews alone?

With Yahweh being presented as all-able, living somewhere in heaven, why did he not inform "his people" that he 'created' a planet, considering he claimed 'creating' everything else? Is it because he was as ignorant of the Solar System as his faithful?

With Yahweh being presented as all-knowing god and having 'created' everything, why then did he not know of all the regions unknown to ancient Hebrews, such as those of Asia, Africa, and the Americas, with people who had their own racial characteristics, and their own gods? If Yahweh was so able, why then did he not introduce himself to people of all other nations in person, as he did to Moses?

It is possible to go on and on examining the ignorant and unrealistic claims the ancient Hebrews made in regards to their god, and by comparing these ludicrous claims to physical reality of human bodies and the location of human existence prove these claims to be nothing but fiction and intentional lies.

Yet, is it only the ancient Hebrew fraudsters who are to blame for all the human (conscio) conflicts and misery that their religion – and those that were later based on it – have caused? Or is it all the following generations of humans (conscios) who either accepted the lies they were preached, or those who ran the business of religion for personal profit of wealth and influence?

There is no doubt that the fabricators of Yahweh did not envisage having their inventions make an impact on human existence in some distant future. Their aim was always intended to impress the culture of their day. That is why all gods reflect the views of the societies of the age when they were invented. The inventors of Yahweh were no different. In the age when gods were thought to be everywhere, their god, Yahweh, was theirs, and theirs alone to do with as they pleased, in order to achieve the power and wealth they sought for themselves. They could make the most outrageous claims and have these accepted as facts by other Hebrews, with no scrutiny conducted by them from not knowing any better.

The only element that had to be included, so as to make the whole lie appear authentically genuine, were the male attitudes of that age. This is why they approached the subject of 'creation' of Adam in the way that they did: to Hebrews a god would have to be superior to that of a man, so that by making Adam in his own image gave them an esteem of being associated with god but beneath god, and supposedly owing him gratitude for being 'created' by him.

But had the inventors of Yahweh intended to foil any attempts by future generations to disprove the physicality of their god, they would have needed Yahweh to take on the same body as that of Adam: a form of covenant between himself and man. By doing that, it could have been said that god had no need of human body structure or organ operations, keeping instead just the exterior resemblance of a human body out of respect for his 'creation' of man.

This of course the inventors of Yahweh did not do, because had they done so in their day, it would had been thought that a god taking on a human body would be indicative of a god inferior to man. Besides, they composed their inventive lies not to fool the future generations but those of their day.

Not that this prevented their lies from fooling all the future generations of humans (conscios) nonetheless. Not because their lies were that hard to dispel but because humans (conscios) had always preferred lies to that of truth: fiction to that of physical reality.

The fact that the old Hebrew fantasies and lies – and those of more recent religious fabrications – remain relevant to current generations of humans, confirms that real knowledge is of little importance to humanity. They much rather wallow in self-delusion of religious lies without giving the slightest thought to what there would be were gods to exist.

Whether they believe in gods as theists, or not as atheists, humanity views the concepts of gods in context of life on Earth being the same whether gods were to be real or not. Just like with the tale of god Yahweh, when he supposedly presented himself to Moses, He made an appearance out of nowhere, gave his instructions, and then disappeared into nothingness of his heaven. An unusual event, no doubt, but of no great significance, right? Not to physical reality it is not. For physical reality does not allow for unreality to occur anywhere. It is due to the limitations of physical reality that the physically impossible remains just that: physically impossible. That is why physical reality cannot support the existence of gods performing impossible feats: be that appearing out of nothingness, or 'creating' man, or walking on water, or awakening the dead.

So could gods exist?

Only if physical reality to cease its existence, which is in itself a physical impossibility, considering that inunens can only behave in the way that they do: with no deviations or alterations.

But for the sake of those who choose to believe, what would happen if god, or gods were to exist?

No doubt that those who believe in gods have an expectation that were god to make his presence be seen and felt on Earth, life would be wonderful, like that in paradise (whatever that is like). He would take care of the poor, the frail, the sick, and the unjustly treated. Life would be harmonious and peaceful, with all humanity being kind and loving to each other.

DECEPTIONS IN THE NAME OF GODS

Unfortunately for such day dreams, the reality of no physical reality would be somewhat different.

With no physical reality, unreality would be the normal state of events.

All gravitation would cease to exist, as would physical reality of that which can and cannot occur. This would allow god and his angels to make appearances out of nowhere, and to make 'creations' out of nothing. Such existence would become normal, with different-dimensions making a contributions of distorting everything physical, including that of 'time', allowing god to move back and forth in 'time'. Life on Earth would be very interesting when all the saints and the Devil's monsters from Hell begin visitations, while the angel choirs provide the background music. Then there is the wrath of god to content with by naughty children, especially those who write letters to Santa Clause, who lives with his reindeer at what used to be the North Pole.

If all such impossibilities were not sufficient for theists, all their agonies and ecstasies would also come trues, because god would fulfil all their prayers.

But there is another element to god's presence on Earth, which directly concerns human behaviour.

What would human (conscio) reactions be, were they actually to be dictated by their god as to what they could and could not have, and what they should and should not do? Without doubt, the reaction would be malice, contempt, and rebellion!

You disagree? You think that humans (conscios) would abide the will of god?

Well, just imagine a god making a physical appearance – not unlike an extra-terrestrial being – claiming an omnipotent control and presence over all humans; threatening their lives with his unlimited power to oppress, suppress, and subjugate them, something humans had always practiced on each other. Would this situation go well with humans (conscios)? Would they timidly submit, and praise the lord – their dictator?

Unlikely.

If humans (conscios) submit but cannot abide authority, always rebelling against it, what makes anyone presume, even for an instant, that they would tolerate the meddling of any god into their affairs? The physical reality is that were gods to exist, humans would be driven to kill them all, if not send them back to kingdom come!

27

The ultimate truth of all physical existence

There is no doubt that the human brain imagination is an extraordinary ability, the likes of which no other animal brain possesses. This ability allows human brains to project to their conscios possible outcomes of real future events. Imagination-ability can also project to a conscio's awareness fictitious events, those that had been invented by the brain to fulfill its conscio's want. Despite such definite distinctions between the various forms of imagination, conscios often fail or refuse to differentiate between them, preferring instead to blend all imagination into an accepted-by-them reality. This composite of probable, possible, and the impossible had become the normal way that conscios view themselves, their lives, and their future existence.

It may appear that such avoidance of differentiation to imagination is of little importance to development of human species. After all, they had developed all their cultures and customs precisely on such blending of imagination, where the reality of their daily lives is fully incorporated with fantasy. Whether it is their religious or scientific beliefs, or both, conscios continue rely on them to imagine that they can achieve anything they desire: including life after death, or an unlimited life, or a life amongst the stars. So if humanity had, and has no desire to differentiate variances in their imagination, nothing in physical existence can possibly intervene to prevent this.

There may be, however, some humans (conscios) who would care to know the scope of all physical existence, which needs but little imagination. This is not to say that imagination is not required to perceive this physical existence, as it is outside the bounds of human abilities to view all that is too small, or too large, or too far away. But this imagination is that which is used in conjunction with calculation and reason, whose first realisation is that everything in physical existence does so due to one vital attribute: limitation.

A limitation is the difference between that which exists and its usage, and that when it no longer exists as it did. The best example of this, are all the inunens and the phases of change they undergo. In each phase they are limited to what they can physically do because of the shapes of their bodies and the direction of their change. Compressed bodies allow them to form everything in physical existence. But once they are caused to expand their compressed bodies, in the process performing a function of 'work', they convert to the 'dispersing energy' phase of their existence, dispersing as heat and light; from which they cannot physically revert to their previous phases, as they, themselves, cannot recompress their expanded bodies.

So this physical inability to recompress their own bodies limits 'dispersing energy' inunens to remain as 'dispersing energy' and only 'dispersing energy', until the next phase of becoming 'dispersed energy' inunens. It is precisely this limitation of being unable to reverse the shape of their bodies to what they had previously been that provides the physical reality of all physical existence, where nothing physical can revert to that which it had once been.

There are other physical limitations to inunens: their inability to alter their size and their inability to be anything but what they are supposed to do.

Primarily, the structure of all inunens is that of their internal direction of heat attraction. This represents a movement of internal heat, where the size of the whole structure is fundamental unit of a 'single constant'. Such 'single constant' is eternal, and while it is able to alter its shape from a compressed inunen to an expanded inunen when that is necessary, it is physically incapable of being penetrated. This prevents any inunen to be enlarged or reduced in size, because nothing can be added to it or subtracted from it.

But even their 'single constant' size is so small that this has caused humans (conscios) – especially those in science – to misunderstand all that the inunens form and disperse in all the physical existence.

Because 'uniting' and 'united energy' inunens are minute they can never be witnessed by human sight, so as to be seen as physical bodies that form everything in existence. Their physical existence can only be accepted from an understanding that everything they form resembles their compressed and expanded shapes, and replicates their phases of behaviour.

Only when the structures, that inunen form, unite into sufficiently large formations so as to be visible to human sight, that humans (conscios) acknowledge existence, but not the inunens that form them. Then, when 'uniting' and 'united energy' inunens are caused to alter their shape to that of 'dispersing energy' inunens, and depart from the objects and bodies they are forming as heat and light, this can be witnessed in totality but again not as individual units of energy (inunens).

This means that the limitation of their physical size prevents them from being recognised as being physical, and the cause of all physical existence.

This size limitation has allowed humans – including those in science – to presume that physical objects and bodies exist and behave according to some unseen, metaphysical forces, such as gods and 'time', which are supposedly responsible for providing physical change to physical objects and bodies, while ignoring all the contrary evidence provided by all the physical reality that surrounds them.

The results of such erroneous thinking, and disregard of physical reality, can be witnessed by more deluded and unreality-accepting scientific perceptions. By failing to work out the existence of inunens and their physical limitations – which apply to everything, for it is all made of inunens – those currently in science consider that the very opposite is true: where everything is possible with no limitations and no restrictions. That is why current science accepts the physically impossible to be possible, in the form of 'time', space-time continuum, various multi-dimensionalities, and gravitational waves, as if physical reality is either of no consequence or of no existence.

As proof of this, simply consider the following scientific beliefs.
To begin, there is the notion that all pre-existing matter had been compressed into the size of a pinhead before 'Big Bang' had occurred. This indicates that as far as current science is concerned there are no fundamental particles that are indivisible, impenetrable and from that are incapable of being reduced in size with no limitation. If this were to be true then there would be no fundamental particles, as inunens are, because all particles could be of different mass size, and that would result in every Hydrogen atom being of differing size from being made up of different sized particles. As a consequence of this size variation of the most basic of atoms, all atomic structures would be of random size, as would be all structures they came to form. Fortunately, such is not the case for all physical existence, where all atoms retain a uniformity of size, because, obviously, that which they are formed of – which is inunens – also have a uniformity of size.

What current science also fails to answer, is how exactly all matter was compressed into the size of a pinhead, even if there were reducible fundamental particles? What was the power that caused all the compression into one pinhead? And what exactly caused the impossibly reduced particles to expand after the 'Big Bang'? Was there some kind of spring-release mechanism? One minute everything is somehow is being compressed and the next it is all released – amazing! And what prevented these impossibly reduced particles from expanding to impossible size, considering that if they could reduce to impossible small size then they could do the same in an expansion?

Of course, the real question is from where were the reducing and expanding particles gathered before the 'Big Bang', and who actually organized it, and how? The reason this question is important is because current science considers that space and 'time' were also compressed into the same pinhead as the matter. So if prior to the moment of the 'Big Bang' the vacuum space was compressed (shrunk) into the pinhead as all else, then that would mean the pinhead (and all its contents, including the vacuum space) existed not in vacuum space – as that would have been inside the pinhead – but in a nothingness of nothingness, which equates to a non-existence. Therefore, if it was all a nothingness of nothing prior to the 'Big Bang', from where did the matter and vacuum space originate in an non-existence, and who or what was able to compress them into a pinhead while being in an non-existence?

As to the component of 'time': were 'time' to have been part of the compressed matter and space located in the pinhead, then this would indicate that prior to the moment of 'Big Bang' 'time' did not exist. If 'time' is supposedly responsible for all progression of physical bodies and objects, and 'time' did not exist before 'Big Bang' moment, then what caused the moment of 'Big Bang'? If 'time' did not activate the 'Big Bang' (as it was not there) then that would indicate that 'time' was of no essence prior to and at the moment of the 'Big Bang'. And if the 'Big Bang' moment took place with no need of 'time', then why would 'time' be needed by anything physical after the moment of 'Big Bang'?

A final question is: did the scientists who came up with all the 'Big Bang' nonsense, and those who believe in it to this day, rationally consider the logic that if there was a nothingness of nothingness then that would constitute an absolute nothingness of anything, even that of nothingness of vacuum space? Therefore, if the nothingness of nothingness were to be, then there would be nothing in that secondary nothingness: no pinhead containing the compressed matter, no vacuum space and no 'Big Bang' moment.

So what is the explanation as to what was there before the Great Dispersal (or if you prefer, Big Bang) took place?

In order to provide the explanation to this ultimate question – relating to all physical existence – requires a consideration of what current science accepts as a fact: "Energy can neither be created nor destroyed, but only changed from one state to another". While this is perfectly true, it is not quite the fact current science perceives it to be.

The current understanding of energy is that mass can be converted to energy, which is incorrect. Mass **cannot** be converted to energy, because mass is energy, or rather: **energy has mass**, because all inunens – individual units of energy – are units of energy, which possess bodies that constitute mass. The only conversion that can be made is for energy-mass to convert its body to a different phase: from 'dispersing' or 'dispersed energy' inunen to that of a 'dispersing energy' inunen.

But then an additional assertion had been made to infer that after mass was converted to energy, energy can be converted back into mass, as if – according to the current "theory of relativity" – energy can change between the two states, back and forth. This is another physical impossibility.

Once energy-mass (comprising of 'uniting' and 'united inunens' that form everything in physical existence) have been converted into 'dispersing energy' inunens –producing work prior to dispersing as heat and light – nothing in physical existence can revert the process. Only an EAMAACS compressing (flattening) of 'dispersed energy' inunens into 'uniting energy' inunens can provide a new inunen cycle of forming all the bodies and objects, before again dispersing as 'dispersing energy' inunens and ending up as part of another EAMAACS.

So the rejection of "special relativity" has to be made on another physical principle that applies to all physical existence, of which current science has no idea: active, inactive, and dormant states of existence.

Because those in science have failed to appreciate the structure of all physical existence, comprising of inunens – results of which are present everywhere in plain view – it is not surprising that when expressing their opinions of physical existence, such as "laws of thermodynamics", many omissions were made. For instance, the first of these so-called laws is but partially correct, because it has to be part of all three Universal Constants: that of vacuum space of Eternity, physical process of change, and simultaneity. If a correct statement is to be made of depicting how ALL the transformations in physical existence actually take place, then the expression should be as follows:

"*Energy can neither be created nor destroyed but only changed cyclically from one phase to another, whilst being in an active/inactive phase; this **active/inactive phase** being convertible to an **active/dormant states** that can interchange; with all this various and varying physical activity, inactivity and dormancy taking place simultaneously, but individually, everywhere in the vacuum space of Eternity.*"

This means that an inunen is an active individual unit of energy, but between its periods of activity the very same inunen goes though periods of either physical inactivity or that of dormancy. During its period of inactivity, an inunen converts from being an active 'uniting energy', 'united energy' or 'dispersing energy' inunen to that of a 'dispersed energy' inunen. Whereas in its dormant **state** an inunen converts to the dormant state of a **dor**mant **en**ergy: **doren**.

The shape of dorens resembles that of round coins. These flattened bodies maintain no internal or external attractions to heat or space, remaining non-active in their dormancy. Apart from physically existence, they do nothing, absolutely nothing. They react to no physical contacts as if they did no exist, nor do they attract or are attracted to anything. For that reason they are not absorbed by EAMAACS. They simply fill the vacuum space of nothingness with their presence: an almost nothingness in an absolute nothingness.

This doren (dormant energy) state of dormancy is indefinite, and may last longer than the existence of many Universes.

Dorens exist in vast numbers throughout the vacuum space of nothingness of eternity. They are absolutely everywhere, so that apart of filling the vacuum space their presence is within all human bodies as within every physical object or body. But because they are dormant they do nothing, and possess no ability of action or reaction. As such, they can be pushed aside by active inunens without causing active inunens any need to activate a phase change of their bodies. That means any physical body or object simply pass dormant dorens through their structures with no knowledge of this.

Unlike active inunens, which cannot be seen individually but can be perceived in all the physical bodies and objects that undergo physical change, dorens cannot be witnessed or perceived in anything, even thought they exist.

Having explained that active inunens can also exist in a dormant state, it is now possible to refer to inunens – active or dormant – as being a 'solid space', in contrast to the vacuum space of nothingness.

This means that: **all 'solid space' remains neither created nor destroyed, but only changes from one STATE to another**, where it can change from dormant state to active state, then again to dormant state before returning, eventually, again to an active state, and so on.

This also means that all 'solid space', be it active or dormant, is eternal, so that all physical existence is temporary in phases but eternal in its states. This applies to all that is on this Earth, and even Earth itself, our Solar System, the Milky Way galaxy, and all that constitutes this Universe. This also applies to humans: we humans are temporary but the inunen particles we are composed of are eternal.

While there is no real purpose to physical existence – which comprises of dormant dorens and active inunens 'solid space' – but to physically exist, its function is to constantly circulate the eternal 'solid space' within the eternal vacuum space. In this way all physical existence can be viewed as one endless and eternal physical system of cyclic events within other cyclic events.

It all functions in the following manner: dormant dorens convert to that of active inunens. Active inunens produce systems of distribution: 'energy and matter attracting and releasing systems' (EAMAARS) attract inunens, in formations of atoms, in vast numbers and then release them, primarily as 'dispersing energy' of heat and light. Then the inactive 'dispersed energy' inunens produce another system of distribution: 'energy and matter absorbing and compressing systems' (EAMAACS). EAMAARS unite inunens as 'dispersed energy' before dispersing inunens over large distances as 'dispersing energy'. This process continues in perpetuity, until some active inunens revert to dormant dorens, while elsewhere in vacuum space dormant dorens convert to active inunens, whose cycle of inunen movement repeats within the overall cycle of 'solid space' activity.

An important point to all this physical activity is in that there was never a moment that every 'solid space' was either dorens or inunens. This means that at no period of all physical existence was there a beginning, the kind that we, humans (conscios), understand or expect. Instead, it always was, is, and always shall be physical change in all its complexity, where 'solid space' fluctuates between the states of being active and dormant.

The other factor to all physical existence is its sheer scale. Despite our extraordinary imagination ability, this is beyond the scope of any human imagination.

While current science still views all existence as being limited to our Universe, no doubt presuming: "if we, on Earth, could not be the centre of the Universe then, as far as we are concerned, this Universe will be our one and only exclusive Universe in existence" – the physical reality is very much different.

Firstly, the vacuum space of nothingness of Eternity is three-dimensional: it is endless in all directions simultaneously. So it is not just horizontal but all directional, where there is no 'up' or 'down', as we view life on Earth.

Secondly, if it were possible to view the complete existence of our Universe, it would resemble an expanding, glowing, sphere-like shape, made up of all the glowing and twirling galaxies; the whole structure resembling an exploding firework, flaring-outwardly in all directions and dispersing into the blackness of the vacuum space of Eternity.

If it were possible to imagine all physical existence then it would require the projection of countless numbers of such exploding universes filling the vacuum space of Eternity. Some of these exploding spheres are larger and others smaller, all separated by vast distances.

If it were possible to step back from this image, it would become obvious that the grouping of these universes resembles a cluster of universes. And around this multiverse cluster there are countless other similar multiverse clusters, and all these multiverse clusters forming even larger groupings of multiverse clusters, and these forming even larger multiverse clusters, and so on forever into infinity.

All these multiverse clusters constantly shimmer as multitudes of universes dim with dispersals while multitudes of new universes light up.

And throughout all these flashing and shimmering endless multiverse clusters are endless numbers of unseen dorens – the always-present alternative to active inunens responsible for all the universes.

Therefore, in response to the question of what was there before our Universe came into existence, the answer is: there were countless of multiverse clusters everywhere. And after our Universe disperses and in no more a universe, there shall remain multiverse clusters everywhere.

This then is the panorama of all physical existence: 'solid space' in its constant physical process of change: constant in its simultaneity, constant in its physical existence within the vacuum space of nothingness of Eternity. Not just a single universe but part of countless multiverse clusters in their temporary existence by which the eternal individual units of 'solid space' of somethingness spread themselves in the eternal vacuum space of nothingness.

They all do this with no involvement of 'time', for were 'time' to actually exist it would still be absolutely inconsequential to any physical object or body, and to any physical action. All physical existence behaves according to physical abilities and limitations of their physical relationships – dormant or active 'solid space' in vacuum space. These physical relationships need no 'time' as they are all timeless – eternal – needing no 'time' to exist or to function. And with 'time' not being needed, 'time' has no physical existence to be had.

There is no 'space-time continuum', as there never ever had been a 'space-time continuum'. There is only the 'solid space' – vacuum space continuum, where the 'solid space' is capable of eternal conversion between active and inactively dormant states of being.

And what of the change, or the physical process of change? While being never-ending, change – all physical change brought about by active inunen phases – is constantly ongoing. It may have variances in speed, duration and quantity, but the basic process of change remains eternal. It remains cyclical, and it remains restricted to the physical behaviour that retains the parameters of physical reality, irrespective of where it takes place in the eternal vacuum space of nothingness of Eternity. All physical change – which is the physical process of change – shall remain like that no matter how much we, human conscios, want to substitute it with our own perceptions of physical behavior.

INDEX

Aaron, 148, 163, 197
Abraham, 142, 150, 189-190, 197
Adam, 142, 145, 151, 159, 163, 181, 190-191, 391, 395, 397
addiction/s, 32, 43, 68, 73-76,
adrenaline, 31-32, 72-73
Aegeans, 136
Africa/n, 26, 87, 125, 133, 135, 397
after death, 66, 83, 171, 176, 184, 186, 195, 198, 204, 400
Agrippa, 164
Ahab, 153
Ahriman (Satan), 159
Ahura-Mazda, 159
air, 39, 65, 77, 274, 319, 325, 327, 361, 364, 373, 381-382, 384-385, 388, 392-394
aircrft, 334, 351, 353
Akkadian/s, 135-139, 148
Alaska, 129
Alexander the Great, 160-161
algae, 329
Amorite/s, 138, 144
anger, 20, 33, 42-44, 49, 99, 114
Antioch, 161
Antiochus III, 161
Antiochus IV, 161
Antony, Mark, 162
apocalyptic (literature), 160
Arab/s, 136, 141, 161, 188-190, 192, 195-195, 198-200, 219
Archelaus, 162
Armenian, 161
Asarhaddon, 157
Asharbanipal, 167
Asia/n, 129, 158, 202-203, 218-219, 220, 397
Assyrian/s, 139, 157
atom/s, atomic, 39, 84, 172, 208, 240, 252, 257-259, 261, 264-268, 275-277, 280, 291-293, 296 305, 313-314, 316-319, 320-329, 330, 340-341, 349, 350-354, 361, 363, 365, 367-368, 370, 375-376, 383- 384, 402, 405
Australia, 129, 396

Baal, 137, 152
Babylon - Babylonian, 136, 138-139, 152, 157-159, 161, 200, 202, 208,
bacteria - bacterial, 3, 83, 282, 313, 328-329, 381
Bel, 139, 152
Benjamin the Egyptian, 164
black hole/s, 278, 366, 367, 370-373, 377-379

'boundary of influence' (gravity), 21, 245-249, 257-258, 278-279, 281, 283 285, 287-293, 295-303, 317-318, 320-325, 327, 332, 338, 340, 350-352, 372, 375-377, 381, 384, 392
(see: magnetic field)
brain/s, 2-34, 42-49, 51-87, 90-91, 93-96, 105-121, 124-126, 128, 130, 130, 134, 151, 158, 172, 181-182, 188, 207-208, 222-224, 228, 233, 259, 329-330, 336-337, 340-341, 343-344, 355, 380-381, 384-385, 392, 394, 400
brain-body/ies, 4, 47, 281
brain-conversing, 66-67
Buddha, 202, 205, 207-209, 215, 218, 220
Buddhism, 116-117, 209, 214, 218-219
business/es, 8, 26, 36, 38, 39, 40, 45, 52, 64, 90, 96-99, 101-103, 109, 117, 131-132, 138, 163, 186, 220, 227-231, 235-236, 380, 387
Byzantine/Byzantium, 187-188

Canaan, Canaanite/s, 135-139, 141-142, 144-145, 151-152, 154, 157, 200
Catholic.Catholicism, 50, 54, 185, 228, 337
Chaldean, 157
China/Chinese, 82, 116, 125, 133, 198-202, 209-210, 214 219, 220
Christian/s/Christianity, 50, 84, 117, 125, 160, 166, 172-173, 177-178, 181-191, 193, 195-198, 200-201, 219, 221, 222, 227-228, 232, 337, 341, 389, 391
chromosome/s, 45-46, 329, 395
circuit/s, 244-245, 247, 250, 252, 254, 274, 286, 295, 298, 299, 300, 303, 319, 327, 375
'combined body/bodies' (neutron), 258-259, 261, 295, 314-317, 370, 300, 320, 327
combined mass, 252, 254, 268, 279, 283, 285-287, 291-293, 314, 318, 321, 367
commanding override negotiator, selectively controlling input-output, 11, 13, 15, 17, 26-27, 78, 106, 113, 121, 394, 340 (see: conscio)
communism, communist, 163, 173, 175-176
Confucius, 202, 209-110, 215-218, 220
conscio/s, 11-87, 90-99, 101,103-105, 109-121, 123-129, 133-134, 137, 141, 143, 149, 151, 154-156, 158, 169, 120, 172, 181-182, 188, 201, 205-208, 219,

INDEX

222-228, 231-237, 259, 325, 330-332, 335-338, 340, 342, 344-345, 350-351, 354-355, 366, 369, 380, 385, 387,390, 394, 397, 398-401
Constantine, 171, 185-188, 221
Cyrus II, 158-159

Darius, 158-159
David, 143, 145, 152-153, 163, 169, 196
Dead Sea Scrolls, 214, 163
Deuteronomy, 144, 150
devil, 20, 159, 182, 207, 227, 399
direction of space attraction, 262
direction/s of space travel, 332-334
'dispersal code', 260-263, , 267, 269-270, 273, 324, 343, 349, 350 (*see:* responsive 'dispersal code')
'dispersed energy' (inunens), 242, 255, 260, 275-281, 343, 354, 361,371, 401, 403-406
'dispersing energy' (inunens), 242-243, 252, 255-256, 259-280, 286-287, 291-292, 295-301, 317-318, 321-322, 324, 326-330,332, 343, 349-350, 353-353, 361-362, 367, 370-371, 377, 381, 383, 401, 403-405
DNA, 2, 7, 22, 45, 55, 328, 395 (*see:* 'transfer of life' code)
'Dominant Male/s', 2, 7, 22, 45, 55, 129, 131-135, 140, 145, 147, 157, 159-161, 185, 187, 198, 199, 202, 210, 214, 216, 218-221, 225, 228, 232, 234, 328, 395
drug/s, 3, 27, 56, 60, 69, 72-75, 78, 80, 110, 115, 117

EAMAACS, 277-282, 293, 331-333, 371-373, 383, 404-406
EAMAARS, 227, 278-280, 282-283, 287, 293, 331-332, 372, 405-406
ebitransit/s, 289-301, 303, 327, 329, 375
Egypt, 134-136, 145, 147-152,
Egyptian/s, 125, 133-134, 136, 147-149, 152, 154, 157, 164, 186, 337, 343
electric/electrical, 250, 271, 299, 300-301, 303, 319, 329, 353, 375
electricity, 295, 298, 303, 327
elecromagnet/s, 288, 275
electromagnetic field/s, 288, 292, 319-320, 332, 350, 353
electromagnetic spectrum, 270-274, 349
electromagnetic waves, 27, 30

'energy body' (electron [shell/s]), 258, 280, 294-303, 305, 313-320, 322, 326-327
Eohippus, 6
Essenes, 163
Euphrates, 133, 135, 138
Europe, 202, 396
European/s, 82, 138, 190, 199, 219
evolution / evolutionary, 2, 4-6, 23-24, 89, 106, 330, 341, 363, 392
exodus, 148, 151
external direction of heat attraction, 245-251, 255, 257-258, 274-275, 281, 297, 303, 321, 372, 375, 377
external space, 254-255, 289

fear/s, 15-18, 20-21, 26-42, 45, 49, 62, 69, 73-75, 77, 92, 95-96, 98-99, 103, 105, 110, 115-116, 120, 125-127, 133, 208, 222-224, 386
'fear code', 17-18, 28-31, 60, 62
'fear code' index, 17-18, 28, 30
dimension/s, 84, 86, 91, 263, 365, 367-368, 371, 391, 399

Gandhi, the Mahatma, 208
Gautama, 208-209, 205-208 (*see:* The Buddha)
genes, 5, 7, 22
Genghis Khan, 219genius, 309, 316
glass prism, 266
god/gods, 7-8, 25, 38, 65, 77, 79, 82, 92-94, 96, 117-119, 125, 128-129, 132-152, 155-161, 164-168, 170-171, 174-178, 182, 186, 188-191, 195-198, 200-205, 207, 209, 214-215, 217, 220, 221-223, 224-231, 233, 237, 335, 337-339, 341-342, 364, 379, 386-389, 391-399, 401
Goliath, 153
'grappled repulsion', 287-288, 290, 293
gravity, gravitation/al, (*see:* 'boundary of influence'), 239, 249-250, 257-258, 277-278, 283-285, 287-293, 303, 319-325, 327, 332, 338-339, 348-354, 359, 361, 364, 371-372, 375-379, 381-385, 392-393, 399, 402
Great Dispersal, 280-281, 293, 332-333, 403
Greece, 161
Greek/s, 125, 143, 154, 157-161, 165, 172-173, 178, 187, 337, 389
grunun/s (grouped units of ununens), 259, 264, 276, 353

409

INDEX

Hadrian, 182
Hammurapi, 138
Hapiru, 136, 139 (*see:* Hebrew)
Hasmonean, 160-161
Hebrew/s, 136-148, 150-151, 154-155, 158-160, 162-165, 167-169, 172, 189-191, 193, 195, 200, 391-392, 396-398
Hebrew Bible, 143-145, 147, 150, 154-155, 159-160, 162-163, 167, 172, 190-191, 193, 195, 200, 391-392
Hellenes, 160 (*see:* Greek/s)
Henry VIII, 228
Herod, 161-164, 168-169, 173
Hezekiah, 157
Hindus, 203-205, 208
Hindus River, 133, 158, 202-203
Hittites, 136, 139, 144, 151
homosexual, 47, 179 (*see:* sabogeat)
Holy Roman Empire, 186-187
Hurrians, 139
hypnosis, 65-66
Hyrachius, 6

India / Indian, 82, 133, 202-205, 208, 214, 218-219, 396,
individual units of energy (heat [*see:* inunen/s]), 240-241, 270, 280, 326, 330, 342, 349, 354, 367, 375, 388, 401, 403
'infant fear code', 16-18, 28, 30-31, 60, 62, 68
'infant fear code' index, 16-17, 28
inunen/s (individual unit of energy), 227, 240-249, 251-287, 291-301, 303, 324, 326-332, 342-345, 350-355, 360-362, 367-368, 370-371, 375, 377, 383-384, 388-389, 398, 400-407
intelligence, 2, 3, 5, 8, 56, 90-91, 96, 121, 259-263, 269, 282, 326, 328-330, 350, 355, 363-364
internal direction of heat attraction, 245-247, 250, 254, 273, 275, 281-282, 284-285, 288, 292, 297, 375, 401
internal space, 244, 251-252, 254-255, 303, 313-315, 317
Iran / Iranian, 138-139, 143, 158-159, 198, 203
Iranian, 227, 263
Iraq, 135, 198
Islam / Islamic, 125, 135, 189, 191, 194-201, 219, 222, 232, 389
Israel / Israeli, 136-145, 147-154, 157-158, 168-169, 173, 190, 193, 195, 196, 203, 396

Israelite/s, 138-145, 147-154, 157-158, 168-169, 190, 193, 195-196, 203, 396

Jerusalem, 153, 157, 161-163, 182, 194, 197
Jesus, 159, 165-190, 193, 196-200, 205-207, 337
Jew/s/Jewish, 125, 138, 149, 154, 157-167, 169-182, 188-190, 195-197, 200, 202, 208, 221, 227, 229, 396
John, 165-168, 178, 180, 197
Josephus, 172
Joshua, 145, 152
Judah, 152, 157, 172
Judaism, 160, 162, 172, 200-201, 219
Julius Caesar, 162, 337

Kassites, 138-139
'kinetic' (energy), 165

Lao-tzu, 209-210, 214-215, 218
Lao-tzu Pien-hua Ching, 214
Large Hadron Collider (LHC), 370
larger-space-beyond (attraction), 251, 254, 257, 274-275, 277-278, 282
Levi/s, 140-142, 147, 166, 203
lie/ lies, 36-38, 50, 57, 60, 69, 120, 125, 133, 144-145, 154-156, 159, 162, 172, 181, 187, 224, 227, 387, 389-392, 397-398, 146
light, 241, 256, 260, 263-278, 280, 291-292, 300-302, 321, 323, 327, 329, 343, 346, 349, 354, 359-362, 367, 369, 370, 374-375 377, 382, 401, 404, 405
Luke, 178, 180

Maccabean, 160-162
magnet/s, 250, 282, 290, 300-301, 353, 375-376
magnetic field/s, 288, 292, 299, 319-300, 332
Mark, 165-169, 172-179, 181, 188, 221
Marduk, 137, 139, 152, 157-158
Mars, 381
Mary (virgin), 167, 173, 178, 190, 196-197
Mattathias, 160-161
Matthew, 166-179,
Mecca, 188, 193-194
memory code/s, 53-54, 56

INDEX

memory recall, 54-56, 58
memory syntax, 56, 81
memory system, 54-56
Merenptah (Pharaoh), 152
Menes (Pharaoh), 134
mental code phrases, 60
mental code statements, 60
Mesopotamia/n, 133, 135-136, 138-139, 142, 151, 157, 162, 202-203,
Milky Way galaxy, 278, 331-333, 364, 405
Moon/moon, 143, 325-326, 340, 351-352, 378, 380-381
Moses, 142, 145, 147-148, 150-152, 165-168, 189-190, 197, 209-210, 396-398
messiah, 163-166, 169, 181, 185, 196, 341, 387
messianic, 163-164, 169
Muslim/s, 189, 192-195, 197-200, 219, 391
Mount Sinai, 150
Muhammad, 188-195, 197-198, 200, 206

Nabopolassar, 157
Nebuchadnezzar II, 157-158
neutron, 258-259, 261, 295, 300, 341-317, 320, 327, 370 (see: 'combined body')
New Testament, The, 160, 164, 173, 176, 178, 181, 191, 195, 209
Newton, Isaac, 87
Noah, 142, 145, 189-190, 196-197
nuclei/nucleus, 3, 280, 295-298, 300, 302-304, 313-317, 319-320, 326-329, 349

Old Testament, The, 143-145, 153-155, 159, 181, 195, 197, 392

Pandeira, Yeishu ben, 163
Parthinian, 161
Passover, 148-149
Paul, 173 (see: Saul)
Persia/Persian/s, 135, 157-159, 162
Pharisees, 162
Philistines, 152, 154, 157
Phoenicians, 136
physical process of change, 50, 53, 83, 92, 96, 116, 155, 205, 227, 241, 227, 282, 284, 331, 339-341, 343, 357-359, 366, 368, 388, 404, 407

physical truth, 113, 155-156, 172, 227, 390, 391, 392,
Pilate, Pontius, 162
planet/s, 5, 7-8, 17, 25-26, 40, 42, 50, 52, 83, 89-91, 96, 100, 102, 110, 121, 124, 129-130, 133, 154-155, 170, 172, 183, 185, 201, 219, 220, 222, 227-228, 235-236, 249-250, 254, 263, 277-278, 282-284, 287, 290, 292-293, 313, 325, 331, 333, 336, 338, 340-347, 351, 356, 358, 364-365, 368-369, 374-378, 380-383, 385-386, 388, 396-397
Pompey, 161-162
present-moment, 53, 56-58, 62, 86, 94, 120-121, 241, 335, 343-344, 346-348, 355, 359-360, 367, 386
profit-from-fear, 38-40, 45, 49

Queen of Sheba, 153
Qur'an, 189-197

radiation, 184, 271, 274, 381, 383
Ramses III (Pharaoh), 152
rate-of-speed-of-dispersal, 262
Red Sea, 147, 149, 153
relativity, 346, 352, 354, 359-360, 362, 378, 404
repulsion/s, 246, 249-250, 256, 279-280, 283-285, 287-288, 290-293, 295-296, 303, 314, 333, 352
responsive 'dispersal code', 262, 266-267, 349
Roman/s, 125, 161-165, 170-174, 177, 181-185, 187, 228, 337-338
Rome, 161-162, 171, 182, 185-187

sabogeat, 47-49, 179
Sadducees, 163
Sargon, 138, 148, 157
Saul, 152, 173 (see: Paul)
science, 8, 10, 82, 93, 97-101, 109, 115, 117, 209, 235-238, 241, 271-272, 283, 292, 301-302, 319, 125, 330, 335-336, 338, 341-342, 344, 346-349, 353-354, 362, 365-371, 377-380, 385, 387, 389, 401-404, 406,
scientific lie, 390
'selfish fear/s', 29, 35-41, 44-45, 49, 51, 60, 62, 68, 83, 96, 125
semi-conscious (conscio), 62-63, 65-67, 80, 82
Semite/s, 136, 138-141, 189-200

411

INDEX

sexual, 21, 23, 29, 40, 45-52, 62, 126, 199, 232, 393, 395
sexuality, 44-51, 382
Sheshonk (Pharaoh), 157
Shia (Islam), 198-199
Shroud of Turin, 173, 184-185
Siberia, 129
simultaneity, 241, 282, 331, 355, 404, 407
sleep, 24, 28, 55, 60-62, 64-66, 197
socialist, 163, 177
Solar System, 249, 278, 283-284, 287, 333, 351, 365, 378, 381-383, 388, 397, 405
'solid body/s' - (proton/s), 257-258, 261, 293-296, 298, 300, 314-317, 320-321
solid space, 239-240, 244, 251-254, 266, 278, 405-407
Solomon, 145, 152-154, 157
soul/s, 8, 25, 77-79, 81, 83, 96, 119, 196, 208
space attraction, 262
specie/s, 2-4, 6-7, 10, 22, 26-27, 51-52, 56, 84, 90, 95, 98, 100, 106, 108-109, 112-113, 124-126, 336-337, 400
speed of light, 272-273, 354, 359-362, 375, 377, 382
'Spiritual Male/s', 128-129, 132, 134, 142, 144, 150, 185, 187, 194, 198-200, 208, 223-228, 337-338
'strings of attraction', 256
Sumer / Sumerian/s, 135-136, 138, 236
Sun, 249-250, 268, 277-280, 282-285, 287, 299, 300, 314, 321, 326-328, 331-334, 338-341, 350-353, 357-358, 364, 366, 372-374, 376, 378, 380-382, 385, 388
Sunni (Islam), 199, 198
supernatural, 7, 25, 65, 77, 81, 128, 189, 263, 334, 388, 391

TaNaKh, 143, 145, 155, 160 (*see:* Hebrew Bible)
Tao, 202, 209-215, 220
Taoism, 210-211, 214-215, 218-219
Tao-te Ching, 209-210
temperature, 4965, 99-100, 109, 235-236, 300, 317, 327, 358, 379, 284, 394
Tetrapods, 5
Theudas, 164
three-dimensional thinking, 84-91
Tigris (river), 133, 135
'time', 93, 164, 263, 335-342, 344-360, 362-369, 371-372, 374, 378-380, 387-388, 391, 399, 401-403, 407

Titus, 182
'transfer of influence', 266, 296-301, 303, 324, 375-376
'transfer of intent', 261-263, 267, 269, 271-273, 300, 317, 321, 324, 327, 329
'transfer of life' code, 328-329 (*see:* DNA)
'tropic of civilization', 133-134
Turkey, 151, 187
Turkish, 139

'united body' (atom), 261, 326, 330
'united' energy' (inunens), 242, 249, 252, 255-256, 258-270, 273, 275-278, 280-282, 284-285, 287, 292, 295, 297-298, 328-329, 343, 349, 350-351, 353, 360-361, 367, 371, 375, 381, 401, 404
'uniting energy' (inunens), 242, 246, 248-249, 255-260, 264, 266, 274, 278-281, 287, 293-301, 303, 327, 327, 332, 354, 371, 375, 404
universe, 17, 19, 203, 212, 227, 232, 239, 277, 279-281, 283, 331-332, 342, 345, 347, 363-365, 368, 371-372, 374-375, 385-386, 393, 405-407
ununen/s (united units of energy), 256-257, 259, 261, 264, 276, 280, 291, 293, 353, 383
utopian, 163, 165, 173

vacuum space, 84, 239-241, 244, 246-247, 252-254, 255-257, 263, 268-269, 273-274, 277-280, 282-284, 290, 292, 296, 314, 322, 331-332, 334-335, 341, 344-352, 354, 356, 359-362, 365, 367-377, 381, 383, 385, 394, 402-407
'vegetative state', 72, 79
velocity, 265, 332, 335, 384, 350-351, 353-355

'War Male/s', 144, 188, 194, 198, 201, 222-223,
wavelength/s, 270-273
'work', 260-261, 265, 268, 275, 298-299, 327, 329, 342-343, 349, 361, 401

Xerxes, 159

Yahweh, 137, 139, 141-144, 148, 152,
    158-159, 161, 165-168, 177, 182,
    189-190, 195, 200, 228, 391-398
Yahuda of Galilee, 164, 169
Yellow River, 202, 133

Zealots, 162-163, 178, 200, 388
Zen Buddhism, 216, 218
Zeus Olympus, 161,
Zoroaster, 159, 177

www.ingramcontent.com/pod-product-compliance
Lightning Source LLC
Chambersburg PA
CBHW021827220426
43663CB00005B/160